Medical Nutrition Therapy

A Case Study Approach

Medical Nutrition Therapy

A Case Study Approach

Third Edition

Marcia Nahikian Nelms, PhD, RD, LD
Southeast Missouri State University

Sara Long Roth, PhD, RD, LD
Southern Illinois University

Karen Lacey, MS, RD, CD
University of Wisconsin–Green Bay

Australia • Brazil • Japan • Korea • Mexico • Singapore • Spain • United Kingdom • United States

Medical Nutrition Therapy:
A Case Study Approach,
Third Edition
Marcia Nahikian Nelms, Sara Long Roth, Karen Lacey

Publisher: Yolanda Cossio

Acquisitions Editor: Peggy Williams

Development Editor: Elesha Feldman

Editorial Assistant: Sarah Farrant

Technology Project Manager: Lauren Tarson

Marketing Manager: Tom Ziolkowski

Marketing Communications Manager: Belinda Krohmer

Project Manager, Editorial Production: Trudy Brown

Creative Director: Rob Hugel

Art Director: John Walker

Print Buyer: Linda Hsu

Permissions Editor: Bob Kauser

Production Service: Newgen–Austin

Copy Editor: Cherie Wilkerson

Cover Designer: John Walker

Compositor: Newgen

Library of Congress Control Number: 2008925818
ISBN-13: 978-0-495-55476-9
ISBN-10: 0-495-55476-6

Wadsworth
10 Davis Drive
Belmont, CA 94002-3098
USA

Printed in the United States of America
4 5 6 7 12 11 10

To our students—past and present—who continue to challenge us, teach us, and guide us as we strive to become better educators.

CONTENTS

Unit Four

NUTRITION THERAPY FOR PANCREATIC AND HEPATOBILIARY DISORDERS 169

Unit Five

NUTRITION THERAPY FOR NEUROLOGICAL AND PSYCHIATRIC DISORDERS 209

Unit Six

NUTRITION THERAPY FOR PULMONARY DISORDERS 237

Unit Seven

NUTRITION THERAPY FOR ENDOCRINE DISORDERS 267

Unit Eight

NUTRITION THERAPY FOR RENAL DISORDERS 319

Unit Nine

NUTRITION THERAPY FOR HYPERMETABOLISM, INFECTION, AND TRAUMA 347

Unit Ten

NUTRITION THERAPY FOR HEMATOLOGY–ONCOLOGY 387

PREFACE

In teaching, we seek to promote the fundamental values of humanism, democracy, and the sciences—that is, a curiosity about new ideas and enthusiasm for learning, a tolerance for the unfamiliar, and the ability to critically evaluate new ideas.

We believe it is our mission to provide the environment that will support students in their quest for integration of knowledge and support the development of critical thinking skills. Thus, we strive to develop these "laboratories" and "real-world" situations that mimic the professional community to build that bridge to clinical practice.

The idea for this book actually began more than fifteen years ago as we began teaching medical nutrition therapy for dietetic students, and now as this third edition publishes, we hope that these cases reflect the most recent changes in nutrition therapy practice. Entering the classroom after being clinicians for many years, we knew we wanted our students to experience nutritional care as realistically as possible. We wanted the classroom to actually be the bridge between the textbook and the clinical setting. In fashioning one of the tools used to build that bridge, we relied heavily on our clinical experience to develop what we hoped to be realistic clinical applications. Use of a clinical application or case study is not a new concept. It is common to see the use of case studies in nutrition, medicine, nursing, and many other allied health fields. The case study places the student in a situation that forces integration of knowledge from many sources, supports use of previously learned information, puts the student in a decision-making role, and nurtures critical thinking.

What, then, makes this text different from a simple collection of case studies? It is our hope that the pedagogy we have developed with each case takes the student one step closer as he or she moves from the classroom to the real world. The cases represent the most common diagnoses that rely on nutrition therapy as an essential component of the medical care. Therefore, we believe these cases represent the type of patient with whom the student will most likely be involved. The concepts presented in these cases can apply to many medical conditions other than those presented here. Furthermore, the instructor can choose to use a variety of questions from each case even if he or she chooses not to have the student complete the entire case. The cases represent both introductory- and advanced-level practice, allowing the instructor to choose among many cases and questions that fit the students' level of expertise.

The cases cross the life span, allowing the student to see the practice of nutrition therapy during childhood, adolescence, pregnancy, and adulthood through the elder years. Because placing nutrition therapy and nutrition education within the appropriate cultural context is crucial, we have tried to represent the diversity of individual patients we encounter today.

The instructor will find it helpful to begin by orienting students to the components of a case. We have provided an outline of this introduction below (see "Introducing Case Studies"). The benefit of teaching students how to use this book is that they become more autonomous learners.

The medical record provides the structure for each case. This is different from most case studies that have been previously developed for publication. The student will seek information to solve the case by using the exact tools he or she will need to use in the clinical setting. As the student moves from the admission form to the physician's history and physical, to laboratory data, and to documentation of daily care, he or she will need to discern the relevant information in the medical record.

In this third edition, questions for each case are organized using the nutrition care process. Introduction to the pathophysiology and principles of nutrition therapy begin each case and are followed by each component of the nutrition care process. Integrated throughout are questions prompting the student to identify nutrition problems and synthesize PES statements.

The appendices for this text do not provide an extensive amount of reference material. This is purposeful. To be consistent with the philosophy of the text, each case requires that the student seek information from multiple resources to complete the case. Each case includes an extensive reference section, which should provide an adequate framework that the student may use to begin his or her research. These selected articles provide excellent background information pertinent to the case. Many of the articles provide essential data about diagnosis and treatment in each case. We have found that when students learn how to research the case, their expertise grows exponentially. In this third edition, we have also added online resources that will guide the student to appropriate reference sources on the Internet.

The cases lend themselves to several important uses. They fit easily into a problem-based learning curriculum. They also can be used as a summary for classroom teaching of the pathophysiology and nutrition therapy for each diagnosis. In addition, the cases can be integrated into the appropriate rotation for a dietetic internship, medical school, or nursing school. Objectives for student learning within each case are built around the nutrition care process and competencies for dietetic education. This allows an additional path for nutrition and dietetic faculty to document student performance as part of program assessment.

TEACHING STRATEGIES

Instructors can find cases to emphasize topics that are part of the curriculum for pathophysiology and medical nutrition therapy (a list of cases by topic is provided below). We have found that by selecting specific questions for each case, instructors can modify the cases to assist in the pedagogy for numerous classes.

Nutrition Assessment: Case 2 Rheumatoid Arthritis; Case 13 Diverticulosis with Incidence of Diverticulitis; Case 29 AIDS

Fluid Balance/Acid–Base Balance: Case 5 Polypharmacy of the Elderly; Case 11 Infectious Diarrhea with Resulting Dehydration; Case 21 COPD with Respiratory Failure

Genetics/Immunology/Infectious Process: Case 2 Rheumatoid Arthritis; Case 12 Celiac Disease; Case 16 Acute Hepatitis; Case 27 Renal Transplant; Case 29 AIDS

Hypermetabolism/Metabolic Stress: Case 28 Traumatic Brain Injury; Case 30 Metabolic Stress and Trauma; Case 31 Lymphoma Treated with Chemotherapy

Dysphagia: Case 28 Traumatic Brain Injury; Case 32 Esophageal Cancer Treated with Surgery and Radiation

Nutritional Needs of the Elderly: Case 5 Polypharmacy of the Elderly; Case 13 Diverticulosis with Incidence of Diverticulitis; Case 19 Alzheimer's Disease

Pediatrics: Case 1 Childhood Overweight; Case 3 Cystic Fibrosis; Case 28 Traumatic Brain Injury

Supplement Use/Complementary and Alternative Medicine: Case 2 Rheumatoid Arthritis; Case 17 Cirrhosis of the Liver with Resulting Hepatic Encephalopathy; Case 29 AIDS; Case 31 Lymphoma Treated with Chemotherapy

Enteral Nutrition Support: Case 8 Congestive Heart Failure with Resulting Cardiac Cachexia; Case 21 COPD with Respiratory Failure; Case 28 Traumatic Brain Injury; Case 30 Metabolic Stress and Trauma; Case 32 Esophageal Cancer Treated with Surgery and Radiation

Parenteral Nutrition Support: Case 14 Crohn's Disease; Case 21 COPD with Respiratory Failure; Case 30 Metabolic Stress and Trauma

ACKNOWLEDGMENTS

We first need to thank our developmental editor, Elesha Feldman, who provided expert guidance in all steps of this revision. We would also like to thank the following reviewers who provided invaluable suggestions for updating the existing cases and refining the new ones.

Jayne Byrne, College of St. Benedict/St. John's University
Christina Campbell, Montana State University
Catherine Christie, University of North Florida
Brenda M. Davy, Virginia Tech
Joan Fischer, University of Georgia
Susan Fredstrom, Minnesota State University–Mankato
Susan Fullmer, Brigham Young University
Melissa Hansen-Petrik, University of Tennessee–Knoxville
Brenda Malinauskas, East Carolina University
Sandra Mayol-Kreiser, Arizona State University
Deborah McCafferty, California State University–Chico
Tania Rivera, Florida International University
Charlene Schmidt, University of Wisconsin–Stout
Vicki Schwartz, Drexel University
Julie Shertzer, Indiana University
Maria T. Spicer, Florida State University
Kelly Tappenden, University of Illinois at Urbana-Champaign

ABOUT THE AUTHORS

Marcia Nahikian Nelms, PhD, RD, LD

Professor
Director, Didactic Program in Dietetics
Southeast Missouri State University

Marcia Nahikian Nelms is currently a professor, registered dietitian, and director of the dietetics program at Southeast Missouri State University. She comes to her academic career after practicing as a clinical dietitian and public health nutritionist for over 25 years. Her clinical practice expertise centers on gastrointestinal diseases and hematology–oncology. Dr. Nahikian Nelms continues to consult in these practice areas as she guides new practitioners into the profession of dietetics. Dr. Nahikian Nelms is the lead author of *Nutrition Therapy and Pathophysiology,* published by Wadsworth/Cengage Learning (2007), and author of "Nutritional Care of Lymphoma" in *Nutritional Issues of Cancer Care,* published by the Oncology Nursing Society (2004), as well as numerous peer-reviewed journal articles and chapters for other texts.

Dr. Nahikian Nelms has most recently been honored as the recipient of the PRIDE award at Southeast Missouri State University. This award is given to a faculty member who has demonstrated excellence as a teacher and an extraordinary level of scholarship and service. She additionally was named Outstanding Dietetics Educator in Missouri and received the Governor's Award for Outstanding Teaching.

Sara Long Roth, PhD, RD, LD

Professor/Director, Didactic Program in Dietetics
Animal Science, Food, and Nutrition
Southern Illinois University–Carbondale

Dr. Long is a professor and director of the Didactic Program in Dietetics in the Department of Animal Science, Food, and Nutrition at Southern Illinois University–Carbondale. Prior to obtaining her PhD in health education, she practiced as a clinical dietitian for 11 years. Dr. Long also served as the nutrition education/counseling consultant for Carbondale Family Medicine for 18 years. She has been an active leader in national, state, and district dietetic associations, where she has served in numerous elected and appointed positions, including president of the Illinois Dietetic Association, Council on Professional Issues Delegate (Education) in the American Dietetic Association House of Delegates, member of the Commission on Accreditation for Dietetics Education, and most recently, member of the Commission on Dietetic Registration.

Dr. Long is coauthor of *Nutrition Therapy and Pathophysiology* and three other nutrition texts. She has received various awards and honors for teaching, including Outstanding Dietetic Educator (American Dietetic Association) and Outstanding Educator for the College of Agricultural Sciences.

Karen Lacey, MS, RD, CD

Senior Lecturer and Director of Dietetic Programs
University of Wisconsin–Green Bay

Karen Lacey received her Bachelor of Science degree in Foods and Nutrition from Valparaiso University–Valparaiso, Indiana, and her Master of Science in Dietetics with an emphasis in Clinical Dietetics and Management from Mt. Mary College–Milwaukee, Wisconsin. She completed a dietetic internship at the University of Michigan Hospitals in Ann Arbor, Michigan.

Ms. Lacey has been and remains very active with the national, state, and local dietetic associations. She has served as a dietetic program reviewer and is currently a board member of the Commission on Accreditation for Dietetic Education (CADE). She is a former delegate from Wisconsin and has also chaired the ADA's Quality Management Committee. She was a member of the ADA's Standardized Language Task Force and chaired the work group that developed the Nutrition Care Process and Model.

She has authored several articles for the *Journal of the American Dietetic Association* and has written a chapter in the textbook *Promoting Wellness: A Nurse's Handbook.* Her most recent publication is a chapter on the nutrition care process in *Nutrition Therapy and Pathophysiology.*

Ms. Lacey is a frequent presenter to the American Dietetic Association, the Wisconsin Dietetic Association, and several district and state dietetic associations, as well as other organizations. She is the recipient of the ADA's Outstanding Dietetic Educator Award for both Didactic Programs and Dietetic Internships from Area 2, as well as Wisconsin Dietetic Association's State Medallion Award.

ABOUT THE CONTRIBUTORS

Deborah Cohen, MMSc, RD

Deborah Cohen is a clinical faculty associate at Southeast Missouri State University where she teaches both undergraduate- and graduate-level dietetics courses. She obtained a Master of Medical Science from Emory University and is a doctoral candidate in the Doctoral of Clinical Nutrition program at the University of Medicine and Dentistry of New Jersey. She has 15 years experience as a clinical dietitian, primarily in the area of nutrition support of critically ill patients. Recently she worked with trauma ICU physicians and developed a research protocol that will assess the energy expenditure of trauma patients with an open abdomen. Ms. Cohen has written several textbook chapters, most recently on the topic of neoplastic disease in *Nutrition Therapy and Pathophysiology.* In addition, she has published a review article about zinc and the common cold in *Topics in Clinical Nutrition.* She recently received Outstanding Teacher of the Year for the College of Health and Human Services at Southeast Missouri State University.

INTRODUCING CASE STUDIES, OR FINDING YOUR WAY THROUGH A CASE STUDY

Have you ever put together a jigsaw puzzle or taught a young child how to complete a puzzle? Most everyone has at one time or another. Recall the steps that are necessary to build a puzzle. You gather together the straight edges, identify the corner pieces, and match the like colors. There is a method and a procedure to follow that, when used consistently, leads to the completion of the puzzle.

Finding your way through a case study is much like assembling a jigsaw puzzle. Each piece of the case study tells a portion of the story. As a student, your job is to put together the pieces of the puzzle to learn about a particular diagnosis, its pathophysiology, and the subsequent medical and nutritional treatment. Although each case in the text is different, the approach to working with the cases remains the same, and with practice, each case study and each medical record becomes easier to manage. The following steps provide guidance for working with each case study.

1. Identify the major parts of the case study.
   - Admission form
   - Medical record
   - Laboratory data
   - Bibliography

2. Read the case carefully.
   - Get a general sense of why the patient has been admitted to the hospital.
   - Use a medical dictionary to become acquainted with unfamiliar terms.
   - Use the list of medical abbreviations provided in Appendix A to define any that are unfamiliar to you.

3. Examine the admission form for clues.
   - Height and weight
   - Vital signs (compare to normal values for physical examination in Appendix B)
   - Chief complaint
   - Patient and family history
   - Lifestyle risk factors

4. Review the medical record.
   - Examine the patient's vital statistics and demographic information (e.g., age, education, marital status, religion, ethnicity).
   - Review the chief complaint (Is it the same as on the admission form?).
   - Read the patient history (remember, this is the patient's subjective information).

5. Use the information provided in the physical examination.
   - Familiarize yourself with the normal values found in Appendix B.
   - Make a list of the values that are abnormal.
   - Now compare abnormal values to the pathophysiology of the admitting diagnosis. Which are consistent? Which are inconsistent?

6. Evaluate the nutrition history.
   - Note appetite and general descriptions.
   - Evaluate the patient's dietary history: calculate an average kcal and protein intake, and compare them to MyPyramid.
   - Is there any information about physical activity?
   - Find anthropometric information.
   - Is the patient responsible for food preparation?
   - Is the patient taking a vitamin or mineral supplement?

7. Review the laboratory values.
   - Review the hematology reports.
   - Review the chemistry reports.
   - What other reports are present?

- Compare the values to the normal values listed. Which are abnormal? Highlight those and then compare them to the pathophysiology. Are they consistent with the diagnosis? Do they support the diagnosis? Why?

8. Use your resources.
- Use the bibliography provided for each case.
- Review your nutrition textbooks.
- Use any books on reserve.
- Access information on the Internet, but choose your sources wisely: stick to government organizations, not-for-profit organizations, and other legitimate sites. A list of reliable Internet resources is provided for each case.

Medical Nutrition Therapy

A Case Study Approach

Unit One

NUTRITION FOR LIFE CYCLE CONDITIONS

This section introduces medical and nutritional conditions commonly found in specific life cycle groups. Each life cycle group has its own unique nutritional concerns related to such factors as age, pregnancy, and growth. You will apply your knowledge of pathophysiology and nutrition therapy to the nutrition care process for each of these clients. The life cycle cases also require knowledge of the unique psychosocial concerns for each group. The admitting diagnoses do not necessarily have to be the primary health concern for these cases, but significant health problems are uncovered as a result of the hospital admission. These cases certainly could be encountered in outpatient clinics, private physician offices, or public health clinics.

The first case highlights an increasing public health problem in the United States—childhood obesity. The rate of obesity has increased steadily over the last several decades. It is now estimated that 58 million individuals are classified as being overweight, 40 million are classified as obese, and as many as 3 million are morbidly obese. Prevention and treatment of obesity are crucial components of medical nutrition therapy.

The second case uses the diagnosis of rheumatoid arthritis as the context for understanding the immune system, the inflammatory response, and the abnormalities of autoimmune disease. This case also highlights physical symptoms that interfere with adequate oral intake. These symptoms, common in arthritis, appear in numerous medical conditions. Finally, this case addresses difficulties with supplementation and drug–nutrient interactions.

Cystic fibrosis is the most common fatal autosomal recessive disease affecting Caucasian populations. One in 20 Americans is a carrier of the defective gene. The third case, in particular, addresses the needs of a growing adolescent and the manifestations of complex nutritional problems that present with cystic fibrosis.

The goals of nutrition therapy coincide with treatment goals for the disease: maximize nutritional status, minimize side effects of disease and treatment, and enhance quality of life. You can apply these same goals to almost any diagnosis. The fourth case focuses on eating disorders. The age of the 30-year-old patient in this case emphasizes the lifelong struggle with eating disorders that many individuals face.

The final case within this section focuses on the unique needs of an older individual. Nutrition plays a large role in the health status of this group. The older adult population may be at nutritional risk because of the physiological effects of aging, complex medical problems, and potential psychosocial and economic concerns. This final case lets you address these issues within the context of polypharmacy, a common problem for the elderly.

Case 1

Childhood Overweight

Objectives

After completing this case, the student will be able to:

1. Discuss the physiological effects of pediatric overweight.
2. Interpret laboratory parameters for nutritional implications and significance.
3. Analyze nutrition assessment data to evaluate nutritional status and identify specific nutrition problems.
4. Determine nutrition diagnoses and write appropriate PES statements.
5. Prescribe appropriate nutrition therapy for weight loss in the pediatric population.
6. Develop a nutrition care plan with appropriate measurable goals, interventions, and strategies for monitoring and evaluation that addresses the nutrition diagnoses of this case.

Missy Bloyd is taken to see her pediatrician by her parents, who have noticed that she appears to stop breathing while sleeping. She is diagnosed with sleep apnea related to her weight and referred to the registered dietitian for nutrition counseling.

Name: Missy Bloyd
DOB: 10/9 (age 10)
Physician: D. Null, MD

ADMISSION DATABASE

BED #	DATE: 11/20	TIME: 1000	TRIAGE STATUS (ER ONLY): ☐ Red ☐ Yellow ☐ Green ☐ White

Initial Vital Signs

TEMP: 98.5	RESP: 17	SAO$_2$:	
HT (in): 57	WT (lb): 115	B/P: 123/80	PULSE: 85
LAST TETANUS 8 years ago		LAST ATE breakfast	LAST DRANK 7:30 AM

PRIMARY PERSON TO CONTACT:
Name: Dominick Bloyd (father)
Home #: 985-555-2636
Work #: 453-555-7512

ORIENTATION TO UNIT: ☐ Call light ☐ Television/telephone ☐ Bathroom ☐ Visiting ☐ Smoking ☐ Meals ☐ Patient rights/responsibilities

CHIEF COMPLAINT/HX OF PRESENT ILLNESS

Parents are concerned after noticing episodes when child appears to stop breathing while sleeping

PERSONAL ARTICLES: (Check if retained/describe)
☐ Contacts ☐ R ☐ L ☐ Dentures ☐ Upper ☐ Lower
☐ Jewelry:
☐ Other:

ALLERGIES: Meds, Food, IVP Dye, Seafood: Type of Reaction

N/A

VALUABLES ENVELOPE:
☐ Valuables instructions

PREVIOUS HOSPITALIZATIONS/SURGERIES

N/A

INFORMATION OBTAINED FROM:
☐ Patient ☒ Previous record
☒ Family ☒ Responsible party

Signature *Dominick Bloyd*

Home Medications (including OTC) Codes: A = Sent home B = Sent to pharmacy C = Not brought in

Medication	Dose	Frequency	Time of Last Dose	Code	Patient Understanding of Drug

Do you take all medications as prescribed? ☐ Yes ☐ No If no, why? N/A

PATIENT/FAMILY HISTORY

☐ Cold in past two weeks
☐ Hay fever
☐ Emphysema/lung problems
☐ TB disease/positive TB skin test
☐ Cancer
☐ Stroke/past paralysis
☒ Heart attack Maternal grandmother
☐ Angina/chest pain
☐ Heart problems
☒ High blood pressure Maternal grandmother
☐ Arthritis
☐ Claustrophobia
☐ Circulation problems
☐ Easy bleeding/bruising/anemia
☐ Sickle cell disease
☐ Liver disease/jaundice
☐ Thyroid disease
☒ Diabetes Mother and grandmother
☐ Kidney/urinary problems
☐ Gastric/abdominal pain/heartburn
☐ Hearing problems
☐ Glaucoma/eye problems
☐ Back pain
☐ Seizures
☐ Other

RISK SCREENING

Have you had a blood transfusion? ☐ Yes ☒ No
Do you smoke? ☐ Yes ☒ No
If yes, how many pack(s)?
Does anyone in your household smoke? ☐ Yes ☒ No
Do you drink alcohol? ☐ Yes ☒ No
If yes, how often?______ How much?
When was your last drink? ______/______/______
Do you take any recreational drugs? ☐ Yes ☒ No
If yes, type:______ Route:
Frequency:______ Date last used:______/______/______

FOR WOMEN Ages 12–52

Is there any chance you could be pregnant? ☐ Yes ☒ No
If yes, expected date (EDC):
Gravida/Para:

ALL WOMEN

Date of last Pap smear:
Do you perform regular breast self-exams? ☐ Yes ☒ No

ALL MEN

Do you perform regular testicular exams? ☐ Yes ☐ No

Additional comments:

x *Rita Hays, LPN*
Signature/Title

Stature-for-Age and Weight-for-Age Percentiles: Girls, 2 to 20 Years

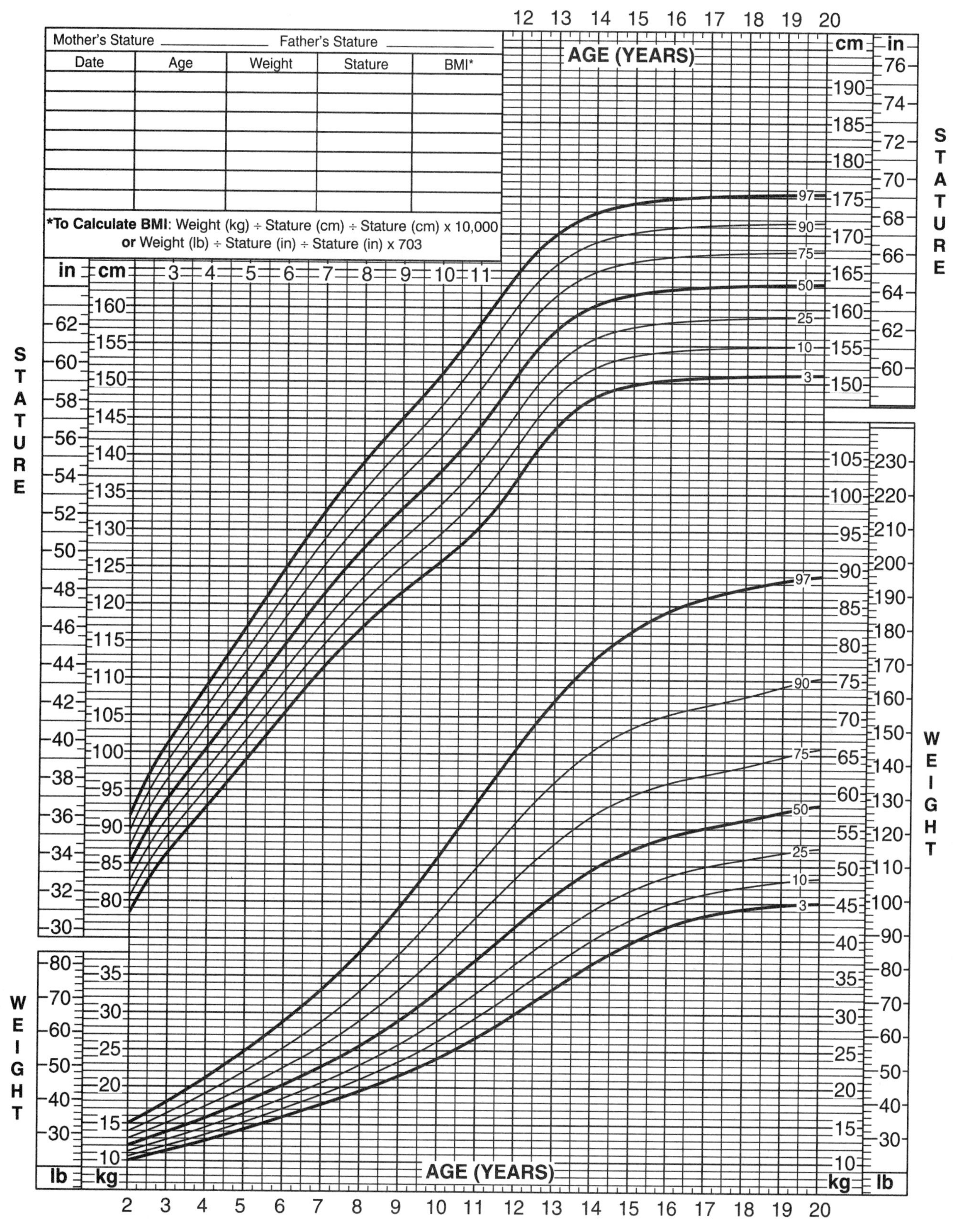

Source: Centers for Disease Control and Prevention. National Center for Health Statistics. 2000 CDC Growth Charts: United States. Available at http://www.cdc.gov/growthcharts. Accessed April 10, 2008.

Body Mass Index-for-Age Percentiles: Girls, 2 to 20 Years

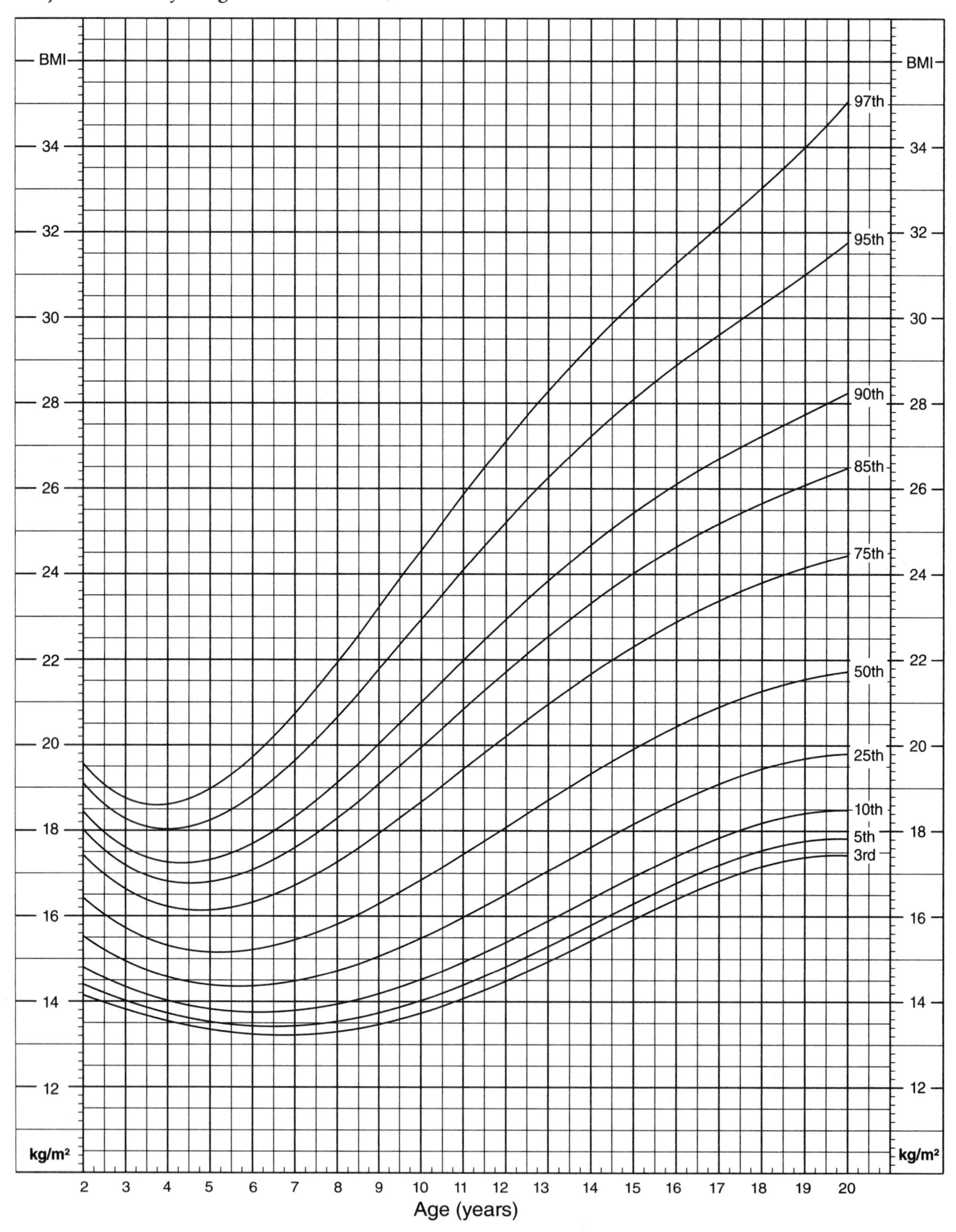

Source: Centers for Disease Control and Prevention. National Center for Health Statistics. 2000 CDC Growth Charts: United States. Available at http://www.cdc.gov/growthcharts. Accessed April 10, 2008.

Client name: Missy Bloyd
DOB: 10/9
Age: 10
Sex: Female
Education: Less than high school *What grade/level?* 5th grade
Occupation: Student
Hours of work: Regular school hours
Household members: Father age 36, mother age 35, sister age 5
Ethnic background: Biracial (African American and Caucasian)
Religious affiliation: Catholic
Referring physician: D. Null, MD

Chief complaint:
"We've noticed that Missy appears to stop breathing for several seconds several times a night. She is really cranky when she gets up for school. Her teacher says Missy gets very sleepy during school. . . . She fell asleep in class yesterday."

Patient history:
Parents describe sleep disturbance in their daughter for the past several years, including: sleeping with her mouth open, cessation of breathing for at least 10 seconds (per episode), snoring, restlessness during sleep, enuresis, and morning headaches. Parents discussed changes in Missy's grades and overall success in school. They state that Missy's teacher has described deficits in attention span at school. Additionally, she has been overweight since she was born.
Onset: Actual date of onset unclear; but parents first noticed onset of the above-mentioned symptoms about 1 year ago.
Type of Tx: None at present
Meds: None at present
Smoker: No
Family Hx:What? Possible gestational diabetes; type 2 DM *Who?* Mother and grandmother

Physical exam:
General appearance: Somewhat tired and irritable 10-year-old female
Vitals: Temp 98.5°F, BP 123/80 mm Hg, HR 85 bpm, RR 17 bpm
Heart: Regular rate and rhythm, heart sounds normal
HEENT:
  Eyes: Clear
  Ears: Clear
  Nose: Normal mucous membranes
  Throat: Dry mucous membranes, no inflammation, tonsillar hypertrophy
Genitalia: SMR (Tanner) pubic hair stage 3, genital stage 3.
Neurologic: Alert, oriented × 3
Extremities: No joint deformity or muscle tenderness, but patient complains of occasional knee pain. No edema.
Skin: Warm, dry; reduced capillary refill (approximately 2 seconds); slight rash in skin folds
Chest/lungs: Clear
Abdomen: Obese

Nutrition Hx:

General: Very good appetite with consumption of a wide variety of foods. Missy's physical activity level is generally low. Her elementary school discontinued physical education, art, and music classes due to budget cuts 5 years ago. She likes playing video games and reading.

24-hour recall:

AM:	2 breakfast burritos, 8 oz whole milk, 4 oz apple juice, 6 oz coffee with ¼ c cream and 2 tsp sugar
Lunch:	2 bologna and cheese sandwiches with 1 tbsp mayonnaise, 1-oz pkg Frito corn chips, 2 Twinkies, 8 oz whole milk
After-school snack:	Peanut butter and jelly sandwich (2 slices enriched bread with 2 tbsp crunchy peanut butter and 2 tbsp grape jelly), 12 oz whole milk
Dinner:	Fried chicken (2 legs and 1 thigh), 1 c mashed potatoes (made with whole milk and butter), 1 c fried okra, 20 oz sweet tea
Snack:	3 c microwave popcorn, 12 oz Coca-Cola

Food allergies/intolerances/aversions: NKA
Previous nutrition therapy? No
Food purchase/preparation: Parent(s)
Vit/min intake: Flintstones vitamin daily

Dx:

R/O Obstructive sleep apnea (OSA) secondary to obesity and physical inactivity

Tx plan:

Polysomnography to diagnose OSA, FBG, HbA1C, lipid panel (total cholesterol, HDL-C, LDL-C, triglycerides), psychological evaluation, nutrition assessment

UH UNIVERSITY HOSPITAL

NAME: Missy Bloyd
AGE: 10
PHYSICIAN: D. Null, MD

DOB: 10/9
SEX: F

CHEMISTRY

DAY:
DATE: 11/20
TIME:
LOCATION:

	NORMAL	11/20	UNITS
Albumin	3.5–5	4.8	g/dL
Total protein	6–8	6.2	g/dL
Prealbumin	16–35	33	mg/dL
Transferrin	250–380 (women) 215–365 (men)	254	mg/dL
Sodium	136–145	138	mEq/L
Potassium	3.5–5.5	4.2	mEq/L
Chloride	95–105	101	mEq/L
PO_4	2.3–4.7	4.6	mg/dL
Magnesium	1.8–3	2.1	mg/dL
Osmolality	285–295	288	mmol/kg/H_2O
Total CO_2	23–30	29	mEq/L
Glucose	70–110	108	mg/dL
BUN	8–18	8	mg/dL
Creatinine	0.6–1.2	0.6	mg/dL
Uric acid	2.8–8.8 (women) 4.0–9.0 (men)		mg/dL
Calcium	9–11	9.2	mg/dL
Bilirubin	$\leq$ 0.3	0.1	mg/dL
Ammonia (NH_3)	9–33	8	μmol/L
ALT	4–36	5	U/L
AST	0–35	6	U/L
Alk phos	30–120	99	U/L
CPK	30–135 (women) 55–170 (men)	72	U/L
LDH	208–378	220	U/L
CHOL	120–199	190	mg/dL
HDL-C	> 55 (women) > 45 (men)	50	mg/dL
VLDL	7–32	30	mg/dL
LDL	< 130	110	mg/dL
LDL/HDL ratio	< 3.22 (women) < 3.55 (men)	2.2	
Apo A	101–199 (women) 94–178 (men)		mg/dL
Apo B	60–126 (women) 63–133 (men)		mg/dL
TG	35–135 (women) 40–160 (men)	114	mg/dL
T_4	4–12	5	mcg/dL
T_3	75–98	78	mcg/dL
HbA_{1c}	3.9–5.2	5.5	%

Case Questions

I. Understanding the Disease and Pathophysiology

1. Current research indicates that the cause of childhood obesity is multifactorial. Briefly discuss how the following factors are thought to play a role in the development of childhood obesity: biological (genetics and pathophysiology); behavioral–environmental (sedentary lifestyle, socioeconomic status, modernization, culture, and dietary intake); and global (society, community, organizational, interpersonal, and individual).

2. Describe health consequences associated with an overweight condition. Describe how these health consequences differ for an overweight versus an obese condition.

3. Missy has been diagnosed with obstructive sleep apnea. Define sleep apnea. Explain the relationship between sleep apnea and obesity.

II. Understanding the Nutrition Therapy

4. What are the goals for weight loss in the pediatric population? Under what circumstances might weight loss in overweight children not be appropriate?

5. What would you recommend as the current focus for nutritional treatment of Missy's obesity?

III. Nutrition Assessment

A. Evaluation of Weight/Body Composition

6. Overweight or obesity in adults is defined by BMI. Children and adolescents are oftentimes classified as "overweight" or "at risk for overweight" based on their BMI percentiles, but this classification scheme is by no means universally accepted. Use three different professional resources and compare/contrast their definitions for overweight conditions among the pediatric population.

7. Evaluate Missy's weight using the CDC growth charts provided. What is Missy's BMI percentile? How would her weight status be classified by each of the standards you identified in question 6?

B. Calculation of Nutrient Requirements

8. If possible, RMR should be measured by indirect calorimetry. Identify two methods for determining Missy's energy requirements other than indirect calorimetry and then use them to calculate Missy's energy requirements.

C. Intake Domain

9. Dietary factors associated with increased risk of overweight are increased dietary fat intake and increased kilocalorie-dense beverages. Identify foods from Missy's diet recall that fit these criteria. Calculate the percentage of kilocalories from each macronutrient and the percentage of kilocalories provided by fluids for Missy's 24-hour recall.

10. Increased fruit and vegetable intake is associated with decreased risk of overweight. Using Missy's usual intake, is Missy's fruit and vegetable intake adequate?

11. Use the MyPyramid Plan online tool (available from http://www.mypyramid.gov/; click on "MyPyramid Plan") to generate a personalized MyPyramid for Missy. Using this eating pattern, plan a 1-day menu for Missy.

12. Now enter and assess the 1-day menu you planned for Missy using the MyPyramid Tracker online tool (http://www.mypyramidtracker.gov/). Does your menu meet macro- and micronutrient recommendations for Missy?

D. Clinical Domain

13. Why did Dr. Null order a lipid profile and a blood glucose test?

14. What lipid and glucose levels are considered to be abnormal for the pediatric population?

15. Evaluate Missy's lab results.

E. Behavioral–Environmental Domain

16. What behaviors associated with increased risk of overweight would you look for when assessing Missy's and her family's diets?

17. What aspects of Missy's lifestyle place her at increased risk for overweight?

18. You talk with Missy and her parents. They are all friendly and cooperative. Missy's mother asks if it would help for them to not let Missy snack between meals and to reward her with dessert when she exercises. What would you tell them?

19. Identify one specific physical activity recommendation for Missy.

IV. Nutrition Diagnosis

20. Select two high-priority nutrition problems and complete PES statements for each.

V. Nutrition Intervention

21. For each PES statement written, establish an ideal goal (based on signs and symptoms) and an appropriate intervention (based on etiology).

22. Mr. and Mrs. Bloyd ask about using over-the-counter diet aids, specifically Alli (orlistat). What would you tell them?

23. Mr. and Mrs. Bloyd ask about gastric bypass surgery for Missy. What are the recommendations regarding gastric bypass surgery for the pediatric population?

VI. Nutrition Monitoring and Evaluation

24. When should the next counseling session with Missy be scheduled?

25. Should her parents be included? Why or why not?

26. What would you assess during this follow-up counseling session?

Bibliography

American Dietetic Association. Treating Childhood Overweight. Evidence Analysis Library. Available at: http://www.adaevidencelibrary.com/topic.cfm?cat=2795. Accessed November 20, 2007.

Hoppin AG, Katz ES, Kaplan LM, Lauwers GY. Case records of the Massachusetts General Hospital. Case 31-2006: A 15-year-old girl with severe obesity. *N Engl J Med.* 2006;355(15):1593–1602.

Lee RD. Energy balance and body weight. In: Nelms M, Sucher K, Long S. *Nutrition Therapy and Pathophysiology.* Belmont, CA: Thomson/Brooks-Cole; 2007: 323–369.

Maahs D, de Serna DG, Kolotkin RL, Ralston S, Sandate J, Qualls C, Schade DS. Randomized, double-blind, placebo-controlled trial of orlistat for weight loss in adolescents. *Endocr Pract.* 2006;12(1):18–28.

Mullin MC, Shield J. *Childhood and Adolescent Overweight: The Health Professional's Guide to Identification, Treatment, and Prevention.* Chicago: American Dietetic Association; 2004;63–117,145–179.

Spear BA, Barlow SE, Ervin C, Ludwig DS, Saelens BE, Schetzina KE, Taveras EM. Recommendations for treatment of child and adolescent overweight and obesity. *Pediatrics.* 2007;120:S254–S288.

University of Texas at Austin, School of Nursing, Family Nurse Practitioner Program. Evaluation and treatment of childhood obesity. Austin, TX: University of Texas at Austin, School of Nursing; 2004. Available at: http://www.guideline.gov/summary/summary.aspx?ss=15&doc_id=5431&nbr=3725. Accessed November 21, 2007.

 Internet Resources

American Academy of Pediatrics: Overweight and Obesity. http://www.aap.org/obesity/

American Obesity Association: Childhood Obesity. http://obesity1.tempdomainname.com/subs/childhood/

American Sleep Apnea Association. http://www.sleepapnea.org/

Centers for Disease Control and Prevention: About BMI for Children and Teens. http://www.cdc.gov/nccdphp/dnpa/bmi/childrens_BMI/about_childrens_BMI.htm

Centers for Disease Control and Prevention: 2000 CDC Growth Charts: United States. http://www.cdc.gov/growthcharts/

eMedicine: Polysomnography: Overview and Clinical Application. http://www.emedicine.com/neuro/topic566.htm

Mayo Clinic: Childhood Obesity. http://www.mayoclinic.com/health/childhood-obesity/DS00698

MedlinePlus: Obesity in Children. http://www.nlm.nih.gov/medlineplus/obesityinchildren.html

MedlinePlus: Polysomnography. http://www.nlm.nih.gov/medlineplus/ency/article/003932.htm

The Sleep Well: Childhood Sleep Apnea. http://www.stanford.edu/~dement/childapnea.html

Case 2

Rheumatoid Arthritis

Objectives

After completing this case, the student will be able to:

1. Apply working knowledge of pathophysiology to the nutrition care process.
2. Analyze nutrition assessment data to establish baseline nutritional status.
3. Identify potential drug–nutrient interactions and appropriate nutrition interventions for preventing or treating drug–nutrient interactions.
4. Identify and explain the common nutritional risks for rheumatoid arthritis.
5. Establish the nutrition diagnosis and compose a PES statement.
6. Create strategies to maximize calorie and protein intake.
7. Analyze current recommendations for nutritional supplementation and determine appropriate nutrition interventions.

Mr. Richard Jacobs is admitted to University Hospital to evaluate the current status of his rheumatoid arthritis and to adjust his medical regimen. Mr. Jacobs was diagnosed with rheumatoid arthritis 5 years ago, and his current medications are not controlling his pain and symptoms. Mr. Jacobs is particularly interested in pursuing alternative medical treatments.

Name: Richard Jacobs
DOB: 5/4 (age 39)
Physician: K. Sanders, MD

ADMISSION DATABASE

BED #	DATE:	TIME:	TRIAGE STATUS (ER ONLY):
1	4/18	1300	☐ Red ☐ Yellow ☐ Green ☐ White

Initial Vital Signs

TEMP:	RESP:	SAO$_2$:	
98.8	18		
HT: 5′10″	WT (lb): 154 (highest wt 165 2 years ago)	B/P: 128/82	PULSE: 86
LAST TETANUS: unknown		LAST ATE: this AM	LAST DRANK

PRIMARY PERSON TO CONTACT:
Name: Peter and Myra Jacobs
Home #: 555-345-7890
Work #:

ORIENTATION TO UNIT: ☒ Call light ☒ Television/telephone ☒ Bathroom ☒ Visiting ☒ Smoking ☒ Meals ☒ Patient rights/responsibilities

CHIEF COMPLAINT/HX OF PRESENT ILLNESS

"My morning stiffness is worse and lasts almost till noon.
I have problems at night with pain, too."

PERSONAL ARTICLES: (Check if retained/describe)
☒ Contacts ☐ R ☐ L ☐ Dentures ☐ Upper ☐ Lower
☐ Jewelry:
☒ Other: glasses

ALLERGIES: Meds, Food, IVP Dye, Seafood: Type of Reaction

Eats no pork; eats kosher during religious holidays.

VALUABLES ENVELOPE:
☐ Valuables instructions

PREVIOUS HOSPITALIZATIONS/SURGERIES

none

INFORMATION OBTAINED FROM:
☒ Patient ☐ Previous record
☐ Family ☐ Responsible party

Signature *R. Jacobs*

Home Medications (including OTC) Codes: A=Sent home B=Sent to pharmacy C=Not brought in

Medication	Dose	Frequency	Time of Last Dose	Code	Patient Understanding of Drug
ibuprofen	500 mg	bid	this AM	C	yes
prednisone	10 mg	daily	this AM	C	yes

Do you take all medications as prescribed? ☒ Yes ☐ No If no, why?

PATIENT/FAMILY HISTORY

☐ Cold in past two weeks
☐ Hay fever
☐ Emphysema/lung problems
☐ TB disease/positive TB skin test
☐ Cancer
☐ Stroke/past paralysis
☐ Heart attack
☐ Angina/chest pain
☐ Heart problems
☐ High blood pressure
☐ Arthritis
☐ Claustrophobia
☐ Circulation problems
☐ Easy bleeding/bruising/anemia
☐ Sickle cell disease
☐ Liver disease/jaundice
☐ Thyroid disease
☐ Diabetes
☐ Kidney/urinary problems
☐ Gastric/abdominal pain/heartburn
☐ Hearing problems
☐ Glaucoma/eye problems
☐ Back pain
☐ Seizures
☒ Other Rheumatoid arthritis

RISK SCREENING

Have you had a blood transfusion? ☐ Yes ☒ No
Do you smoke? ☒ Yes ☐ No
If yes, how many pack(s)? $\frac{1}{2}$/day for 15 years
Does anyone in your household smoke? ☐ Yes ☒ No
Do you drink alcohol? ☒ Yes ☐ No
If yes, how often? weekly How much? 1-2 drinks
When was your last drink? 4/16
Do you take any recreational drugs? ☐ Yes ☒ No
If yes, type:______ Route:
Frequency:______ Date last used:______/______/______

FOR WOMEN Ages 12–52

Is there any chance you could be pregnant? ☐ Yes ☐ No
If yes, expected date (EDC):
Gravida/Para:

ALL WOMEN

Date of last Pap smear:
Do you perform regular breast self-exams? ☐ Yes ☐ No

ALL MEN

Do you perform regular testicular exams? ☐ Yes ☒ No

Additional comments:

x *Michelle Jenkins, RN*
Signature/Title

Client name: Richard Jacobs
DOB: 5/4
Age: 39
Sex: Male
Education: Bachelor's degree
Occupation: Employment counselor
Hours of work: M–F 9–6 PM
Household members: Lives alone
Ethnic background: U.S. born—Caucasian
Religious affiliation: Jewish
Referring physician: Kevin Sanders, MD (rheumatology)

Chief complaint:
"My morning stiffness is considerably worse and actually lasts almost to noon. I have noticed problems in the evening as well. I wake up during the night with pain. I have also wondered whether there is anything I can do regarding diet that might help me. I recently began taking an antioxidant combination as well as fish oil capsules. I really don't want to take more medicines."

Patient history:
Onset of disease: Approximately 5 years ago diagnosed with rheumatoid arthritis
Type of Tx: Motrin, 1,000 mg/day; prednisone, 10 mg/day
PMH: No significant illness prior to this diagnosis
Meds: See above.
Smoker: Yes
Family Hx: What? HTN *Who?* Father

Physical exam:
General appearance: Slim 39-year-old white male who moves with some difficulty and pain
Vitals: Temp 98.8°F, BP 128/82 mm Hg, HR 86 bpm, RR 22 bpm
Heart: Regular rate and rhythm, heart sounds normal
HEENT: Slight temporal wasting noted. All other WNL and noncontributory.
Genitalia: Normal
Neurologic: Alert and oriented, hesitant gait, normal reflexes
Extremities: Appearance of slight muscle wasting, joints tender to touch
Musculoskeletal: Mild limitation of motion, bony enlargement of the DIP (distal interphalangeal) joints of both hands consistent with Heberden's nodes. Shoulders, elbows, wrists, and small joints of the feet show evidence of swelling, warmth, and erythema of these joints.

Nutrition Hx:
General: Appetite fair—hungrier when pain is controlled

Usual dietary intake:

AM:	Coffee, juice with meds
Midmorning:	Doughnut, sweet roll, etc., with coffee
Lunch:	Occasionally skips, but if eats, it is at fast-food restaurant—burger, sub sandwich, pizza with coffee or soda

Dinner: Eats out at local restaurants 2–3 × per week (meat entree, pasta, salad), or when cooks, uses ready-to-eat boxed meals or frozen entrees. Likes to cook, but has not had time recently. States that he should follow a kosher diet but admits that he does not, except during religious holidays.

24-hour recall (in hospital):

AM: 1 slice whole-wheat toast with 2 pats margarine and 1 container of jam, 3 c black coffee

Lunch: Vegetable soup—1.5 c, 6 saltine crackers, 2 c black coffee

Dinner: Baked fish (2 tsp margarine), 6 oz rice—½ c, chocolate cake with icing—1″–2″ square, 2 c decaf coffee

Food allergies/intolerances/aversions: "I do not eat pork of any kind. I generally eat kosher during religious holidays."

Previous nutrition therapy? No

Food purchase/preparation: Self

Vit/min intake: Centrum multivitamin antioxidant combination; fish oil capsules

Dx:

R/O Exacerbation of Rheumatoid Arthritis

Tx plan:

Evaluate current status of rheumatoid arthritis. Adjust medical regimen as necessary. Begin treatment with methotrexate in addition to current medications.

UH UNIVERSITY HOSPITAL

NAME: Richard Jacobs　　　　DOB: 5/4
AGE: 39　　　　SEX: M
PHYSICIAN: K. Sanders, MD

CHEMISTRY

DAY:		1	
DATE:		4/18	
TIME:			
LOCATION:			
	NORMAL		UNITS
Albumin	3.5-5	3.8	g/dL
Total protein	6-8	6.0	g/dL
Prealbumin	16-35	32	mg/dL
Transferrin	250-380 (women) 215-365 (men)	220	mg/dL
Sodium	136-145	142	mEq/L
Potassium	3.5-5.5	4.2	mEq/L
Chloride	95-105	104	mEq/L
PO_4	2.3-4.7	4.0	mg/dL
Magnesium	1.8-3	1.8	mg/dL
Osmolality	285-295	295	mmol/kg/H_2O
Total CO_2	23-30	28	mEq/L
Glucose	70-110	119 H	mg/dL
BUN	8-18	18	mg/dL
Creatinine	0.6-1.2	0.8	mg/dL
Uric acid	2.8-8.8 (women) 4.0-9.0 (men)	9.2 H	mg/dL
Calcium	9-11	9.5	mg/dL
Bilirubin	≤ 0.3	0.5	mg/dL
Ammonia (NH_3)	10-80		mcg/dL
ALT	4-36		U/L
AST	0-35		U/L
Alk phos	30-120		U/L
CPK	30-135 (women) 55-170 (men)		U/L
LDH	208-378		U/L
CHOL	120-199	190	mg/dL
HDL-C	> 55 (women) > 45 (men)	50	mg/dL
VLDL	7-32		mg/dL
LDL	< 130	131 H	mg/dL
LDL/HDL ratio	< 3.22 (women) < 3.55 (men)		
Apo A	101-199 (women) 94-178 (men)		mg/dL
Apo B	60-126 (women) 63-133 (men)		mg/dL
TG	35-135 (women) 40-160 (men)	155	mg/dL
T_4	4-12		mcg/dL
T_3	75-98		mcg/dL
HbA_{1C}	3.9-5.2		%

UH UNIVERSITY HOSPITAL

NAME: Richard Jacobs DOB: 5/4
AGE: 39 SEX: M
PHYSICIAN: K. Sanders, MD

HEMATOLOGY

DAY:		1	
DATE:		4/18	
TIME:			
LOCATION:			
	NORMAL		UNITS
WBC	4.8–11.8	6.0	$\times 10^3/mm^3$
RBC	4.2–5.4 (women) 4.5–6.2 (men)	4.8	$\times 10^6/mm^3$
HGB	12–15 (women) 14–17 (men)	15	g/dL
HCT	37–47 (women) 40–54 (men)	41	%
MCV	80–96		μm^3
RETIC	0.8–2.8		%
MCH	26–32		pg
MCHC	31.5–36		g/dL
RDW	11.6–16.5		%
Plt Ct	140–440		$\times 10^3/mm^3$
Diff TYPE			
ESR	0–25 (women) 0–15 (men)	33 H	mm/hr
% GRANS	34.6–79.2		%
% LYM	19.6–52.7		%
SEGS	50–62		%
BANDS	3–6		%
LYMPHS	24–44		%
MONOS	4–8		%
EOS	0.5–4		%
Ferritin	20–120 (women) 20–300 (men)		mg/mL
ZPP	30–80		μmol/mol
Vitamin B_{12}	24.4–100		ng/dL
Folate	5–25		μg/dL
Total T cells	812–2,318		mm^3
T-helper cells	589–1,505		mm^3
T-suppressor cells	325–997		mm^3
PT	11–16		sec

Case Questions

I. Understanding the Disease and Pathophysiology

1. Describe the inflammatory response that plays a role in the pathophysiology of rheumatoid arthritis. How do corticosteroids and NSAIDs interrupt this inflammatory process?

2. What is an autoimmune disease? How is this immune response different from the normal response to a foreign antigen?

3. What is the proposed mechanism of methotrexate in the treatment? Relate it to the pathophysiology of an autoimmune response.

4. What is the proposed rationale for using antioxidant supplements and omega-3 fatty acids in treating rheumatoid arthritis? What does the current research recommend?

II. Nutrition Assessment

A. Evaluation of Weight/Body Composition

5. Calculate percent usual body weight (UBW), percent ideal body weight (IBW), and body mass index (BMI). Is Mr. Jacobs's weight of concern? Why or why not?

B. Calculation of Nutrient Requirements

6. Calculate energy and protein requirements for Mr. Jacobs. Identify the formula/calculation method you used and explain the rationale for using it.

C. Intake Domain

7. Evaluate the 24-hour recall using computerized dietary analysis.

8. Mr. Jacobs states his appetite is fair. What other questions might you ask to further assess his appetite? What are possible causes of his decreased appetite?

9. List possible intake-related nutrition problems that Mr. Jacobs might have by listing the terms that may fit into the nutrition diagnosis labels.

10. What is the history and rationale for the kosher diet? Does this diet have any nutritional consequences for the patient?

D. Clinical Domain

11. This patient will be started on methotrexate. What are the common drug–nutrient interactions with this medication? Are there any other drug–nutrient interactions with his other medications that are of concern? Explain.

12. What information in the physician's assessment may lead you to be concerned about muscle stores? What additional anthropometric indices might you evaluate to assess muscle mass or lean body mass?

13. What may be the possible reasons for any loss of lean body mass?

14. What laboratory measures correlate with wasting of lean body mass?

15. What laboratory values will be used to assess nutritional status? Are any significant? Are there others that might be important to assess for patients with rheumatoid arthritis? Explain.

E. Behavioral–Environmental Domain

16. List possible behavioral–environmental nutrition problems that Mr. Jacobs might have. (At this point list only the terms that are considered the diagnostic labels; do not attempt to write the entire PES statement.)

III. Nutrition Diagnosis

17. For each of the nutrition problems that you identified in this case, complete the entire PES statement. If there is insufficient data, briefly describe what additional data you would need to make an accurate nutrition diagnosis.

18. Prioritize the nutrition diagnoses by listing them in the order that you expect the interventions to be developed.

IV. Nutrition Intervention

19. For each of the PES statements that you have identified, establish an ideal goal (based on the signs and symptoms) and an appropriate intervention (based on the etiology).

Bibliography

American Dietetic Association. Kosher diet. In: *Manual of Clinical Dietetics.* 6th ed. Chicago, IL: American Dietetic Association; 2000:785–789.

American Dietetic Association. *Nutrition Diagnosis and Intervention: Standardized Language for the Nutrition Care Process.* Chicago, IL: American Dietetic Association; 2006.

Anderson LS, Hansen EL, Knudsen JB, Wester JU, Hansen GV, Hansen TM. Prospectively measured red cell folate levels in methotrexate treated patients with rheumatoid arthritis: Relation to withdrawal and side effects. *J Rheumatol.* 1997;25:830–837.

Ariza-Ariza R, Mestanza-Peralta M, Cardiel MH. Omega-3 fatty-acids in rheumatoid arthritis: An overview. *Semin Arthritis Rheum.* 1997;27(6):366–370.

Cleland LG, Caughey GE, James MJ, Proudman SM. Reduction of cardiovascular risk factors with long-term fish oil treatment in early rheumatoid arthritis. *J Rheumatol.* 2006;33(10):1973–1979.

Cleland LG, James MJ, Proudman SM. Fish oil: What the prescriber needs to know. *Arthritis Res Ther.* 2006;8(1):202. Erratum in: *Arthritis Res Ther.* 2006;8(4):402.

Escott-Stump S. *Nutrition and Diagnosis Related Care.* 5th ed. Baltimore, MD: Williams & Wilkins; 2007.

Frazier C. Immunology. In: Nelms M, Sucher K, Long S. *Nutrition Therapy and Pathophysiology.* Belmont, CA: Thomson/Brooks-Cole; 2007:290–293.

Future Trends in the Management of Rheumatoid Arthritis. Proceedings of the International Rheumatology Round Tables. March 27–28, 1998, and February 19–20, 1999. *Rheumatology.* 1999;38(Suppl 2):1–53.

Gómez-Vaquero C, Nolla JM, Fiter J, Ramon JM, Concustell R, Valverde J, Roig-Escofet D. Nutritional status in patients with rheumatoid arthritis. *Joint Bone Spine.* 2001;68(5):403–409.

Gould BE. *Pathophysiology for the Health-Related Professions.* Philadelphia, PA: Saunders; 1997:11–13.

Hansen GV, Nielsen L, Kluger E, Thysen M, Emmertsen H, Stengaard-Pedersen K, Hansen EL, Unger B, Andersen PW. Nutritional status of Danish rheumatoid arthritis patients and effects of a diet adjusted in energy intake, fish-meal, and antioxidants. *Scand J Rheumatol.* 1996;25:325–330.

Heliovaara M, Knekt P, Aho K, Aaran RK, Alfthan G, Aromaa A. Serum antioxidants and risk of rheumatoid arthritis. *Ann Rheum.* 1994;53(1):51–53.

Kremer JM, Bigaouette J. Nutrient intake of patients with rheumatoid arthritis is deficient in pyridoxine, zinc, copper, and magnesium. *J Rheumatol.* 1996;23:990–994.

Lamour A, Le Goff P, Jouquan J, Bendaoud B, Youinou P, Menez JF. Some nutritional parameters in patients with rheumatoid arthritis. *Ann Intern Med.* 1995;6:409–412.

Lee R. Diseases of the Musculoskeletal System. In: Nelms M, Sucher K, Long S. *Nutrition Therapy and Pathophysiology.* Belmont, CA: Thomson/Brooks-Cole; 2007:843–879.

Lipsky PE. Rheumatoid arthritis. In: Kasper DL, Braunwald E, Fauci AS. Hausner SL, Longo DL, Jameson JO, eds. *Harrison's Principles of Internal Medicine.* 16th ed. New York, NY: McGraw-Hill; 2005:1968–1977.

Martin RH. The role of nutrition and diet in rheumatoid arthritis. *Proc Nutr Soc.* 1998;57:231.

Nelms MN. Cellular and physiological response to injury. In: Nelms M, Sucher K, Long S. *Nutrition Therapy and Pathophysiology.* Belmont, CA: Thomson/Brooks-Cole; 2007:219–236.

Pronsky ZM. *Food Medication Interactions.* 13th ed. Birchrunville, PA: Food Medication Interactions; 2007.

Volker D, Fitzgerald P, Major G, Garg M. Efficacy of fish oil concentrate in the treatment of rheumatoid arthritis. *J Rheumatol.* 2000;27:2343–2346.

Yocum DE, Castro W.L., Cornett M. Exercise, education, and behavioral modification as alternative therapy for pain and stress in rheumatic disease. *Rheum Dis Clin North Am.* 2000;26:145–159.

Yokota S. Mizoribine: Mode of action and effects in clinical use. *Pediatr Int.* 2002;44:196–198.

Internet Resources

Arthritis Foundation: Rheumatoid Arthritis: What Is It? http://www.arthritis.org/disease-center.php?disease_id=31

Merck Manuals Online Medical Library: Rheumatoid Arthritis. http://www.merck.com/mmhe/sec05/ch067/ch067b.html

National Institute of Arthritis and Musculoskeletal and Skin Diseases: What Is Rheumatoid Arthritis? http://www.niams.nih.gov/Health_Info/Rheumatic_Disease/rheumatoid_arthritis_ff.asp

Case 3

Cystic Fibrosis

Objectives

After completing this case, the student will be able to:

1. Understand the genetic abnormalities found in cystic fibrosis.
2. Apply knowledge of the pathophysiology of cystic fibrosis to identify and explain common nutritional problems associated with the disease.
3. Apply knowledge of nutrition therapy for cystic fibrosis.
4. Describe the unique nutritional needs of adolescence.
5. Analyze nutrition assessment data to evaluate nutritional status and identify specific nutrition problems.
6. Determine nutrition diagnoses and write appropriate PES statements.
7. Develop a nutrition care plan with appropriate measurable goals, interventions, and strategies for monitoring and evaluation that addresses the nutrition diagnoses of this case.

Lily Johnson, a 14-year-old Caucasian female, is admitted to University Hospital with pneumonia. She was diagnosed with cystic fibrosis at the age of 6 months.

Name: Lily Johnson
DOB: 5/12 (age 14)
Physician: Nate Tyson, MD

ADMISSION DATABASE

BED #	DATE:	TIME:	TRIAGE STATUS (ER ONLY):
2	7/14	0100	☐ Red ☐ Yellow ☐ Green ☐ White

Initial Vital Signs

TEMP:	RESP:	SAO$_2$:	
99.6	18	99%	
HT (in):	**WT (lb):**	**B/P:**	**PULSE:**
5′5″	102	114/60	82
LAST TETANUS		**LAST ATE**	**LAST DRANK**
age 10		12 noon	30 minutes ago

PRIMARY PERSON TO CONTACT:
Name: Sylvia Johnson (mother)
Home #: 555-421-4435
Work #: 555-421-9234

ORIENTATION TO UNIT: ☒ Call light ☒ Television/telephone ☒ Bathroom ☐ Visiting ☒ Smoking ☒ Meals ☒ Patient rights/responsibilities

CHIEF COMPLAINT/HX OF PRESENT ILLNESS

"I must have caught a cold when I was at camp. I woke up this morning with trouble breathing. My inhaler didn't help. My doctor thinks I have pneumonia."

ALLERGIES: Meds, Food, IVP Dye, Seafood: Type of Reaction

NKA

PREVIOUS HOSPITALIZATIONS/SURGERIES

At least one time per year, usually for pneumonia. No surgeries.

PERSONAL ARTICLES: (Check if retained/describe)
☐ Contacts ☐ R ☐ L ☐ Dentures ☐ Upper ☐ Lower
☐ Jewelry:
☐ Other:

VALUABLES ENVELOPE:
☐ Valuables instructions

INFORMATION OBTAINED FROM:
☒ Patient ☐ Previous record
☒ Family ☐ Responsible party

Signature *Sylvia Johnson*

Home Medications (including OTC) **Codes: A=Sent home** **B=Sent to pharmacy** **C=Not brought in**

Medication	Dose	Frequency	Time of Last Dose	Code	Patient Understanding of Drug
Pancrease	1–3 caps	after every meal	12:30 PM	C	yes
Prilosec	20 mg	daily	0800	C	yes
Humabid	$\frac{1}{2}$ tablet q 12 hrs	daily	0800	C	yes
Proventil	PRN	100 mcg inhaled q 4 hrs	0800	C	yes
multivitamin	1 tablet	daily	last week		yes

Do you take all medications as prescribed? ☒ Yes ☐ No If no, why? N/A

PATIENT/FAMILY HISTORY

☒ Cold in past two weeks
☐ Hay fever
☐ Emphysema/lung problems
☐ TB disease/positive TB skin test
☐ Cancer
☐ Stroke/past paralysis
☐ Heart attack
☐ Angina/chest pain
☒ Heart problems
☐ High blood pressure
☐ Arthritis
☐ Claustrophobia
☐ Circulation problems
☐ Easy bleeding/bruising/anemia
☐ Sickle cell disease
☐ Liver disease/jaundice
☐ Thyroid disease
☐ Diabetes
☐ Kidney/urinary problems
☐ Gastric/abdominal pain/heartburn
☐ Hearing problems
☐ Glaucoma/eye problems
☐ Back pain
☐ Seizures
☐ Other

RISK SCREENING

Have you had a blood transfusion? ☐ Yes ☒ No
Do you smoke? ☐ Yes ☒ No
If yes, how many pack(s)?
Does anyone in your household smoke? ☐ Yes ☒ No
Do you drink alcohol? ☐ Yes ☒ No
If yes, how often?_______ How much?
When was your last drink? ______/______/______
Do you take any recreational drugs? ☐ Yes ☒ No
If yes, type:_______ Route:
Frequency:_______ Date last used:______/______/______

FOR WOMEN Ages 12–52

Is there any chance you could be pregnant? ☐ Yes ☐ No
If yes, expected date (EDC):
Gravida/Para:

ALL WOMEN

Date of last Pap smear:
Do you perform regular breast self-exams? ☐ Yes ☒ No

ALL MEN

Do you perform regular testicular exams? ☐ Yes ☐ No

Additional comments:

x *Angie Stagemeyer, RN, BSN, MA*
Signature/Title

Client name: Lily Johnson
DOB: 5/12
Age: 14
Sex: Female
Education: Just completed 9th grade
Occupation: Student
Hours of work: N/A
Household members: Mother age 41 (divorced), grandmother age 66 (widowed), half-brother age 5
Ethnic background: Caucasian
Religious affiliation: None
Referring physician: N. Tyson, MD

Chief complaint:
"I just got back from working at a camp for the past two weeks. I caught a cold, and it has just gotten worse. My regular treatments were not working, and my doctor says I probably have pneumonia."

Patient history:
Onset of disease: Patient is a 14-year-old Caucasian female who was diagnosed with cystic fibrosis at age 6 months. Since that time, she has had a rather uneventful disease course. She has been hospitalized several times for respiratory infections but otherwise has been maintained with outpatient therapy. She is seen in the CF clinic yearly at University Hospital but receives routine medical care from her local physician.
Type of Tx: Uses high-frequency chest compression vest for 1 hour twice daily.
PMH: Last hospitalization was over a year ago. Lily has had a successful first year of high school. She participates in ballet and jazz, and is a cross-country runner. She typically runs 3–5 miles 5–6 times a week. Three times a week she has a dance class.
Meds: Outpatient medications include Pancrease (1–3 caps after meals); Prevacid (20 mg daily); Humabid (½ tablet every 12 hours); multivitamin, Proventil PRN.
Smoker: No
Family Hx: What/Who? Type 2 DM/grandmother (maternal); CF/great aunt (paternal, deceased)

Physical exam:
General appearance: 14-year-old thin female, flushed, in no acute distress
Vitals: Temperature 99.1°F, BP 114/60 mm Hg, HR 82 bpm, RR 18 bpm
Heart: Regular rate and rhythm, heart sounds normal
HEENT:
Eyes: WNL; PERRLA, fundi without lesions
Ears: Clear
Nose: WNL
Throat: Pharynx reddened with postnasal drainage
Genitalia: Normal
Neurologic: Alert and oriented
Extremities: No edema
Skin: Skin pale without rash

Chest/lungs: Decreased breath sounds, percussion hyperresonant, rhonchi and rales present
Abdomen: Bowel sounds present, nontender

Nutrition Hx:
General: Patient states that appetite is not very good right now but had been fine until the last few days. Patient states that she really never knows how much Pancrease to take: "You know, I just guess if I have a lot of fat in my meal then I need to take more." States that she likes most foods, but when questioned, states that she never drinks milk. Really likes fruits and vegetables but doesn't have them "as much as I should." "My mom and grandmother really cook healthy, but it's just that I am not really home for meals that much. I am at school or hanging out with friends. I am babysitting a lot this summer as well as working at camp."

Usual dietary intake:
AM: Rarely eats
Lunch: 3 tbsp extra-crunchy peanut butter or 2 oz ham and 2 oz Swiss cheese sandwich, 2–3 oz chips, 1 orange or other piece of fruit, water
Dinner: 5–6 oz of chicken, pork or beef—usually grilled or baked, 1–2 c of raw vegetables on lettuce, ¼ c ranch dressing, 1 c pasta, potatoes, or rice, usually with 1–2 tbsp margarine, water—but really depends on whether she eats at home or over at a friend's house

24-hour recall (PTA):
AM: Nothing
Lunch: 2 oz hot dog on bun, 1½ c macaroni and cheese (Kraft boxed variety made with 2% milk)
Dinner: 5 oz Salisbury steak with ¼ c gravy, few bites of green beans, 1 roll with 2 tbsp margarine, grape juice about 2 c

Food allergies/intolerances/aversions: Patient says she will eat almost anything. Usually tries to avoid fried foods because they make her have diarrhea. States: "Sometimes it's worth it though because I like pizza and french fries."
Previous nutrition therapy? Yes *If yes, when:* States that she has had nutrition information given to her throughout the years when she has visited the CF clinic. She describes meeting with a dietitian who has asked her diet history. "It has just been recently that I have really started thinking about my diet. My family has really made most of those decisions. I know I need to know more about my diet, and I really want to make sure I stay healthy."
Food purchase/preparation: Self, mother, and grandmother
Vit/min intake: Tries to remember to take multivitamin, but doesn't take it every day.
Anthropometric data: Ht 5′5″, Wt 102 lbs, UBW: 110–115 lbs (3 months ago): States that she has recently increased her running (5–7 miles, 3 to 4 times per week) and feels that is why she has lost weight.

Tx plan:
Activity: Bed rest *Diet:* Regular as tolerated
Lab: CBC, RPR, Chem16: I & O every shift; routine vital signs.
IVF D_5 @ 50 mL/hr
Vancomycin 10 mg/kg IV q6h
CXR—EPA/LAT.
Sputum cultures and gram stain.

Hospital course:
Lily was diagnosed with acute pneumonia confirmed by CXR and sputum culture. Intravenous antibiotics were initiated. Nutrition consult was initiated to assess current nutritional status and to ensure adequacy of current nutritional intake.

UH UNIVERSITY HOSPITAL

NAME: Lily Johnson DOB: 5/12
AGE: 14 SEX: F
PHYSICIAN: N. Tyson, MD

CHEMISTRY

DAY: 1
DATE: 7/14
TIME:
LOCATION:

	NORMAL		UNITS
Albumin	3.5-5	3.9	g/dL
Total protein	6-8	6.2	g/dL
Prealbumin	16-35	25	mg/dL
Transferrin	250-380 (women) 215-365 (men)	219	mg/dL
Sodium	136-145	142	mEq/L
Potassium	3.5-5.5	3.8	mEq/L
Chloride	95-105	105	mEq/L
PO_4	2.3-4.7	3.2	mg/dL
Magnesium	1.8-3	1.6	mg/dL
Osmolality	285-295	292	mmol/kg/H_2O
Total CO_2	23-30	26	mEq/L
Glucose	70-110	105	mg/dL
BUN	8-18	8	mg/dL
Creatinine	0.6-1.2	0.7	mg/dL
Uric acid	2.8-8.8 (women) 4.0-9.0 (men)	2.8	mg/dL
Calcium	9-11	9.1	mg/dL
Bilirubin	≤0.3	0.3	mg/dL
Ammonia (NH_3)	9-33	9	µmol/L
ALT	4-36	12	U/L
AST	0-35	8	U/L
Alk phos	30-120	99	U/L
CPK	30-135 (women) 55-170 (men)		U/L
LDH	208-378		U/L
CHOL	120-199	165	mg/dL
HDL-C	>55 (women) >45 (men)	55	mg/dL
VLDL	7-32		mg/dL
LDL	<130	125	mg/dL
LDL/HDL ratio	<3.22 (women) <3.55 (men)	3.0	
Apo A	101-199 (women) 94-178 (men)		mg/dL
Apo B	60-126 (women) 63-133 (men)		mg/dL
TG	35-135 (women) 40-160 (men)	120	mg/dL
T_4	4-12		mcg/dL
T_3	75-98		mcg/dL
HbA_{1C}	3.9-5.2	6.3	%

UH UNIVERSITY HOSPITAL

NAME: Lily Johnson DOB: 5/12
AGE: 14 SEX: F
PHYSICIAN: N. Tyson, MD

*****HEMATOLOGY*****

	NORMAL		UNITS
DAY:		1	
DATE:		7/14	
TIME:			
LOCATION:			
WBC	4.8-11.8	13	$\times 10^3/mm^3$
RBC	4.2-5.4 (women) 4.5-6.2 (men)	4.8	$\times 10^6/mm^3$
HGB	12-15 (women) 14-17 (men)	11.5	g/dL
HCT	37-47 (women) 40-54 (men)	33	%
MCV	80-96	87	μm^3
RETIC	0.8-2.8	0.9	%
MCH	26-32	30	pg
MCHC	31.5-36	31.7	g/dL
RDW	11.6-16.5	15	%
Plt Ct	140-440	320	$\times 10^3/mm^3$
Diff TYPE			
ESR	0-20 (women) 0-15 (men)		mm/hr
% GRANS	34.6-79.2		%
% LYM	19.6-52.7		%
SEGS	50-62		%
BANDS	3-6		%
LYMPHS	24-44		%
MONOS	4-8		%
EOS	0.5-4		%
TIBC	65-165 (women) 75-175 (men)		μg/dL
Ferritin	20-120 (women) 20-300 (men)	19	mg/mL
ZPP	30-80	35	μmol/mol
Vitamin B_{12}	24.4-100	98	ng/dL
Folate	5-25	19	μg/dL
Total T cells	812-2,318		mm^3
T-helper cells	589-1,505		mm^3
T-suppressor cells	325-997		mm^3
PT	11-16		sec

Case Questions

I. Understanding the Disease and Pathophysiology

1. Define cystic fibrosis.

2. Describe the most common populations affected by this disease, including age, gender, and ethnicity.

3. This disease is an autosomal recessive disorder affecting the *CFTR* gene on Chromosome 7. What does this mean? Describe what is currently understood about the genetic characteristics of this disease.

4. How is this disease diagnosed? List at least three methods that are used.

5. For each of the following organs or organ systems, describe the most common physical changes that occur as a result of the abnormality of the *CFTR* gene. Explain how these changes may affect Lily's nutritional status.

a. Respiratory
b. Reproductive
c. Pancreatic
d. Gastrointestinal

6. Lily was admitted and diagnosed with bacterial pneumonia. Why is this the most common hospitalization for patients with CF? Explain.

II. Understanding the Nutrition Therapy

7. What are the most common nutritional consequences of cystic fibrosis?

8. Describe the major modifications for carbohydrate, protein, and fat intake that would be needed as components of nutrition therapy for CF.

9. Is Lily at risk for electrolyte imbalances? Specifically, address her sodium and chloride requirements. Is there additional information from Lily's history that puts her at risk for changes in her sodium and chloride levels?

10. What is Pancrease? Lily mentioned that she did not know how much to take. What are the recommendations?

III. Nutrition Assessment

A. Evaluation of Weight/Body Composition

11. Assess Lily's weight and height. Plot her height and weight on the appropriate growth chart. Calculate her BMI. Calculate her %UBW. Explain what each of these assessments provides and why one or more provides the most relevant information for Lily.

B. Calculation of Nutrient Requirements

12. Determine Lily's energy and protein requirements. You see that she typically runs 5–7 miles 3–4 times per week as well as taking a dance class 3 times per week for 1 hour. Make sure this is taken into consideration when calculating her energy requirements. Your recommendations for Lily should include the appropriate macro- and micronutrients based on the requirements for an adolescent with cystic fibrosis.

C. Intake Domain

13. Analyze Lily's nutritional intake according to the usual dietary intake. Attach your computerized analysis for this assessment.

14. Compare your analysis to her estimated nutritional needs.

15. Identify three specific vitamins and minerals that are needed in increased amounts during adolescence. Explain why they are of special importance for an adolescent. Will Lily's CF affect the metabolism of these nutrients? Do Lily's diet history and 24-hour recall indicate that she consumes adequate amounts of these nutrients?

16. From the information gathered within the intake domain, list possible nutrition problems using the diagnostic term.

D. Clinical Domain

17. After reading the physician's history and physical, identify the signs and symptoms that are consistent with Lily's admitting medical diagnosis.

18. Evaluate each of the medications that Lily takes as an outpatient. Determine the function of each medication and identify any nutritional implications.

Medication	Function of Medication	Nutritional Implications

19. *Biochemical:* Evaluate Lily's laboratory values. In the following table, list any laboratory values that are abnormal. What is the most probable cause of the abnormality?

Abnormal Lab	Normal Value	Reason for Abnormality	Nutritional Implication

20. List possible nutrition problems within the clinical domain using the diagnostic term.

E. Behavioral–Environmental Domain

21. After reading the history and physical as well as the nutrition history, identify factors that may impact the success of Lily's current medical and nutritional care for her cystic fibrosis.

IV. Nutrition Diagnosis

22. Select two high-priority nutrition problems and complete the PES statements.

V. Nutrition Intervention

23. For each of the PES statements that you have written, establish an ideal goal (based on the signs and symptoms) and an appropriate intervention (based on the etiology).

24. What might be different about your nutrition interventions if Lily were a young teenager of different ethnicity and/or religion? Give an example and explain.

Bibliography

Beers MH, Berkow R, eds. *Merck Manual of Diagnosis and Therapy.* 17th ed. Whitehouse Station, NJ: Merck & Co., 2005. Available at: http://www.merck.com/mrkshared/mmanual/home.asp. Accessed June 19, 2007.

Bergman EA, Buergel NS. Diseases of the respiratory system. In: Nelms M, Sucher K, Long S. *Nutrition Therapy and Pathophysiology.* Belmont, CA: Thomson/Brooks-Cole; 2007:731–737.

Borowitz D, Baker RD, Stalling V. Consensus report on nutrition for pediatric patients with cystic fibrosis. *J Pediatr Gastroenteral.* 2002;35:246–259.

Brown K. Pneumonia. In: Strange GR, Ahrens WR, Lelyveld S, Schafermeyer RW, eds. *Pediatric Emergency Medicine: A Comprehensive Study Guide.* 2nd ed. New York, NY: McGraw-Hill; 2002:219–225.

Dodge JA, Turck D. Cystic fibrosis: Nutritional consequences and management. *Best Pract Res Clin Gastroenterol.* 2006;20(3):531–546.

Gilbert DN, Moellering RC, Eliopoulos GM. *The Sanford Guide to Antimicrobial Therapy.* 34th ed. Hyde Park, VT: Antimicrobial Therapy; 2004:25–29.

Hayek KM. Medical nutrition therapy for cystic fibrosis: beyond pancreatic enzyme replacement therapy. *J Am Diet Assoc.* 2006;106(8):1186–1188.

Kalnins D, Durie PR, Pencharz P. Nutritional management of cystic fibrosis patients. *Curr Opin Clin Nutr Metab Care.* 2007;10(3):348–354.

Lai HJ. Classification of nutritional status in cystic fibrosis. *Curr Opin Pulm Med.* 2006;12(6):422–427.

Marcason W. What are the calorie requirements for patients with cystic fibrosis? *J Am Diet Assoc.* 2005;105(4):660,555–556.

Mattfeldt-Beman M. Diseases of the hepatobiliary: Liver, gallbladder, exocrine pancreas. In: Nelms M, Sucher K, Long S. *Nutrition Therapy and Pathophysiology.* Belmont, CA: Thomson/Brooks-Cole; 2007:533–535.

Milla CE. Nutrition and lung disease in cystic fibrosis. *Clin Chest Med.* 2007;28(2):319–330.

Murphy AJ, Buntain HM, Wainwright CE, Davies PS. The nutritional status of children with cystic fibrosis. *Br J Nutr.* 2006;95(2):321–324.

Nelms MN. Assessment of nutrition status and risk. In: Nelms M, Sucher K, Long S. *Nutrition Therapy and Pathophysiology.* Belmont, CA: Thomson/Brooks-Cole; 2007:101–135.

Schall JI, Bentley T, Stallings VA. Meal patterns, dietary fat intake and pancreatic enzyme use in preadolescent children with cystic fibrosis. *J Pediatr Gastroenterol Nutr.* 2006;43(5):651–659.

Trabulsi J, Ittenbach RF, Schall JI, Olsen IE, Yudkoff M, Daikhin Y, Zemel BS, Stallings VA. Evaluation of formulas for calculating total energy requirements of preadolescent children with cystic fibrosis. *Am J Clin Nutr.* 2007;85(1):144–151

Internet Resources

Cystic Fibrosis Foundation. http://www.cff.org

GeneTests: Genetic Reviews. http://www.genetests.org

Merck Manuals Online Medical Library. http://www.merck.com/mrkshared/mmanual/home.asp

National Institutes of Health: Office of Rare Diseases. http://rarediseases.info.nih.gov/

U.S. National Library of Medicine: Genetics Home Reference: Educational Resources—Information Pages for Cystic Fibrosis. http://ghr.nlm.nih.gov/condition=cysticfibrosis/show/Educational+resources

Case 4

Anorexia Nervosa and Bulimia Nervosa

Objectives

After completing this case, the student will be able to:

1. Discuss the etiology of anorexia nervosa (AN), bulimia nervosa (BN), and compulsive eating.
2. Interpret laboratory parameters for nutritional implications and significance.
3. Analyze nutrition assessment data to evaluate nutritional status and identify specific nutrition problems.
4. Determine nutrition diagnoses and write appropriate PES statements.
5. Prescribe appropriate nutrition therapy for anorexia nervosa and bulimia nervosa.
6. Develop a nutrition care plan with appropriate measurable goals, interventions, and strategies for monitoring and evaluation that addresses the nutrition diagnoses of this case.

Paris Marshall was taken to the emergency room after she collapsed during Pilates class at the local gym. As she is being triaged, the nurse notes Paris's emaciated appearance after her baggy clothes have been removed. She also observes dry skin, lanugo, discolored teeth, and enlarged parotid and submandibular glands.

Name: Paris Marshall
DOB: 8/7 (age 34)
Physician: James Roth, MD

ADMISSION DATABASE

BED #	DATE: 11/23	TIME:	TRIAGE STATUS (ER ONLY): ☐ Red ☒ Yellow ☐ Green ☐ White
Initial Vital Signs			
TEMP: 98.1	RESP: 15	SAO$_2$:	
HT: 5′8″	WT (lb): 115	B/P: 90/70	PULSE: 50
LAST TETANUS		LAST ATE 11/22	LAST DRANK 2400

PRIMARY PERSON TO CONTACT:
Name: Nichole Marshall
Home #: 555-555-7684
Work #: 555-201-2157

ORIENTATION TO UNIT: ☐ Call light ☐ Television/telephone ☐ Bathroom ☐ Visiting ☐ Smoking ☐ Meals ☐ Patient rights/responsibilities

CHIEF COMPLAINT/HX OF PRESENT ILLNESS

Brought in through ER

PERSONAL ARTICLES: (Check if retained/describe)
☐ Contacts ☐ R ☐ L ☐ Dentures ☐ Upper ☐ Lower
☐ Jewelry:
☐ Other:

ALLERGIES: Meds, Food, IVP Dye, Seafood: Type of Reaction

Multiple food allergies: all meats, dairy foods, most desserts

VALUABLES ENVELOPE:
☐ Valuables instructions

PREVIOUS HOSPITALIZATIONS/SURGERIES

N/A

INFORMATION OBTAINED FROM:
☒ Patient ☐ Previous record
☐ Family ☒ Responsible party

Signature *Nichole Marshall*

Home Medications (including OTC) **Codes: A=Sent home** **B=Sent to pharmacy** **C=Not brought in**

Medication	Dose	Frequency	Time of Last Dose	Code	Patient Understanding of Drug
laxatives	varies	every other day	yesterday	C	very good

Do you take all medications as prescribed? ☐ Yes ☐ No If no, why?

PATIENT/FAMILY HISTORY

☒ Cold in past two weeks
☐ Hay fever
☐ Emphysema/lung problems
☐ TB disease/positive TB skin test
☐ Cancer
☐ Stroke/past paralysis
☐ Heart attack
☐ Angina/chest pain
☐ Heart problems
☐ High blood pressure
☐ Arthritis
☐ Claustrophobia
☐ Circulation problems
☒ Easy bleeding/bruising/anemia
☐ Sickle cell disease
☐ Liver disease/jaundice
☐ Thyroid disease
☐ Diabetes
☐ Kidney/urinary problems
☒ Gastric/abdominal pain/heartburn
☐ Hearing problems
☐ Glaucoma/eye problems
☐ Back pain
☐ Seizures
☐ Other

RISK SCREENING

Have you had a blood transfusion? ☐ Yes ☒ No
Do you smoke? ☒ Yes ☐ No
If yes, how many pack(s)? 1 pack/day
Does anyone in your household smoke? ☒ Yes ☐ No
Do you drink alcohol? ☒ Yes ☐ No
If yes, how often? daily How much? $\frac{1}{2}$-1 bottle of wine
When was your last drink? 11/22
Do you take any recreational drugs? ☐ Yes ☒ No
If yes, type:______ Route:
Frequency:______ Date last used:______/______/______

FOR WOMEN Ages 12–52

Is there any chance you could be pregnant? ☐ Yes ☒ No
If yes, expected date (EDC):
Gravida/Para:

ALL WOMEN

Date of last Pap smear: over 2 years ago
Do you perform regular breast self-exams? ☐ Yes ☒ No

ALL MEN

Do you perform regular testicular exams? ☐ Yes ☐ No

Additional comments:

✗ *Marie Schwartz, RN*
Signature/Title

Client name: Paris Marshall
DOB: 8/7
Age: 34
Sex: Female
Education: Law school graduate
Occupation: Attorney
Hours of work: 7 AM–6 PM plus weekends as needed
Household members: Self
Ethnic background: African American
Religious affiliation: Catholic
Referring physician: P. Bennett

Chief complaint:
Lost consciousness in exercise class

Patient history:
Paris Marshall is a 34-year-old attorney. When she was in high school, Paris was a straight-A student and president of the debate team and honor society. In college, she maintained a perfect 4.0/4.0 GPA majoring in engineering. Paris also spent a lot of time at the university recreation center, often swimming for an hour in the early morning before classes, walking 3 miles on the indoor track after lunch, and then doing aerobics in the afternoon for 1 hour. Her roommate taught Paris how to purge using her toothbrush. Another student on her dorm floor told Paris about laxatives and enemas. This behavior continued during law school and intensified as the stress of law school increased. In an effort to maintain her weight below 120 lbs, she resumed restricting her intake as she did in high school. The hungrier she got, the more determined she was not to give in to it. She thought this kind of self-control would help her in the practice of law. Graduating at the top of her law school class, Paris was recruited by the most prestigious law firm in town, where she is trying to make partner. She has not had a menstrual period in over 2 years.
Type of Tx: Paris tried to stop restricting her food and purging on her own, but any time she was under stress, she reverted to the only coping mechanisms she knew. Once, in law school, she was hospitalized over the weekend because she was severely dehydrated, but was released after 24 hours.
Meds: OTC laxatives every other day
Smoker: Yes, 1 pack per day
Family Hx: What? HTN *Who?* Father

Physical exam:
General appearance: Emaciated, tired-looking young woman who appears older than her stated age.
Vitals: Temp 98.1°F, BP 90/60 mm Hg, HR 50 bpm, RR 18 bpm
Heart: Bradycardia with normal rhythm
HEENT:
Eyes: PERRLA, fundi without lesions
Ears: Clear
Nose: Normal mucous membranes
Throat: Dry mucous membranes, no inflammation, tonsillar hypertrophy, scratches on posterior pharynx, erosion of dental enamel

Genitalia: Normal
Neurologic: Oriented × 3
Extremities: Fingernails and toenails brittle
Skin: Piloerection present; rough, dry skin with lanugo. Tenting of skin noted. Some bruising noted.
Chest/lungs: Clear
Abdomen: Some mild edema noted

Nutrition Hx:

General: "I love to cook! When I'm not working at the office, I'm at home cooking. I cook a lot of 'bad' food, especially when I have experienced a lot of stress at work, but I give most of it to friends and family, or take it to work. I know I have a problem with how I deal with food and eating. . . . I've always tried to help myself, but I've never passed out before. Maybe I need the help of others this time."

24-hour recall:

AM:	¼ whole-wheat bagel, 4 oz calcium-fortified orange juice, 6 oz black coffee
Lunch:	Black coffee—2–3 c
Afternoon snack:	12-oz can Diet Coke
Dinner:	6 green peas, 18 oz water
Snack:	12 oz Diet Coke

Food allergies/intolerances/aversions: Self-reported food intolerances include meats, dairy foods, and "desserts"
Previous nutrition therapy? No
Food purchase/preparation: Self
Vit/min intake: Multivitamin/mineral daily

Dx:

Anorexia nervosa with binge/purge tendencies

Tx plan:

EKG, CBC, chemistry panel, upper and lower GI series, referrals to Multidisciplinary Eating Disorder Treatment team (RD and psychologist) and dentist

UH UNIVERSITY HOSPITAL

NAME: Paris Marshall
AGE: 34
PHYSICIAN: J. Roth, MD

DOB: 8/7
SEX: F

********************CHEMISTRY********************

DAY: 1
DATE: 1/5
TIME:
LOCATION:

	NORMAL		UNITS
Albumin	3.5–5	3.0 L	g/dL
Total protein	6–8	5 L	g/dL
Prealbumin	16–35	14.5 L	mg/dL
Transferrin	250–380 (women) 215–365 (men)	170	mg/dL
Sodium	136–145	148 H	mEq/L
Potassium	3.5–5.5	3.0 L	mEq/L
Chloride	95–105	95	mEq/L
PO_4	2.3–4.7	4.6	mg/dL
Magnesium	1.8–3	1.7 L	mg/dL
Osmolality	285–295	295	mmol/kg/H_2O
Total CO_2	23–30	26	mEq/L
Glucose	70–110	115 H	mg/dL
BUN	8–18	17.5	mg/dL
Creatinine	0.6–1.2	1.1	mg/dL
Uric acid	2.8–8.8 (women) 4.0–9.0 (men)	7.2	mg/dL
Calcium	9–11	10	mg/dL
Bilirubin	≤ 0.3	0.3	mg/dL
Ammonia (NH_3)	9–33	25	µmol/L
ALT	4–36	36	U/L
AST	0–35	40	U/L
Alk phos	30–120	200	U/L
CPK	30–135 (women) 55–170 (men)	146 H	U/L
LDH	208–378	680	U/L
CHOL	120–199	195	mg/dL
HDL-C	> 55 (women) > 45 (men)	60 H	mg/dL
VLDL	7–32	9	mg/dL
LDL	< 130	129	mg/dL
LDL/HDL ratio	< 3.22 (women) < 3.55 (men)	2.39	
Apo A	101–199 (women) 94–178 (men)	180	mg/dL
Apo B	60–126 (women) 63–133 (men)	120	mg/dL
TG	35–135 (women) 40–160 (men)	189	mg/dL
T_4	4–12	10	mcg/dL
T_3	75–98	70 L	mcg/dL
HbA_{1C}	3.9–5.2	4.2	%

UH UNIVERSITY HOSPITAL

NAME: Paris Marshall DOB: 8/7
AGE: 34 SEX: F
PHYSICIAN: J. Roth, MD

HEMATOLOGY************************************

DAY: 1
DATE: 1/5
TIME:
LOCATION:

	NORMAL		UNITS
WBC	4.8-11.8	4.6 L	× $10^3/mm^3$
RBC	4.2-5.4 (women) 4.5-6.2 (men)	5.0	× $10^6/mm^3$
HGB	12-15 (women) 14-17 (men)	10	g/dL
HCT	37-47 (women) 40-54 (men)	30	%
MCV	80-96	90	μm^3
RETIC	0.8-2.8	1.0	%
MCH	26-32	26	pg
MCHC	31.5-36	31.5	g/dL
RDW	11.6-16.5	14.6	%
Plt Ct	140-440	298	× $10^3/mm^3$
Diff TYPE			
ESR	0-25 (women) 0-15 (men)	18	mm/hr
% GRANS	34.6-79.2	57.7	%
% LYM	19.6-52.7	19.2	%
SEGS	50-62	55	%
BANDS	3-6	4.5	%
LYMPHS	24-44	20 L	%
MONOS	4-8	6.0	%
EOS	0.5-4	3.2	%
Ferritin	20-120 (women) 20-300 (men)	20	mg/mL
ZPP	30-80	50	µmol/mol
Vitamin B_{12}	24.4-100		ng/dL
Folate	5-25	5	µg/dL
Total T cells	812-2,318	2,256	mm^3
T-helper cells	589-1,505	1,284	mm^3
T-suppressor cells	325-997	835	mm^3
PT	11-16	14	sec

Case Questions

I. Understanding the Disease and Pathophysiology

1. Describe the diagnostic criteria for anorexia nervosa (AN), bulimia nervosa (BN), and binge eating disorder (BED). Include all types (binging/purging AN, restrictive AN, purging BN, non-purging BN), and discuss which type of eating disorder you believe Paris presents with. Provide examples to support your rationale.

2. Describe the common psychological, socioeconomic, and environmental characteristics of an individual with AN.

3. What does research indicate about the possible role of genetics in eating disorders?

4. How does binge eating disorder (BED) differ from BN?

5. What is the long-term prognosis for AN, BN, and BED?

6. Describe the medical consequences associated with AN, BN, and BED.

7. Define *starvation*, *binge eating*, and *purging*.

8. Describe the metabolic response to voluntary starvation. Compare Paris's signs and symptoms to the metabolic response to starvation.

9. To be successful, treatment of eating disorders must include a team approach among physicians, registered dietitians, and psychologists. Describe the role of each in treatment.

10. Why might it be necessary to include a psychiatrist as a member of the treatment team?

II. Understanding the Nutrition Therapy

11. Briefly, what are the primary nutrition therapy goals for acute diagnosis of AN? How will these goals change as treatment progresses?

12. What are the primary nutrition therapy goals for BN?

13. What are the primary nutrition therapy goals for BED?

14. Describe prevention strategies that could reduce a person's risk of developing AN, BN, or BED.

III. Nutrition Assessment

A. Evaluation of Weight/Body Composition

15. What are the typical differences in body weight between someone with AN and someone with BN?

16. Calculate and interpret Paris's BMI.

17. What would be an appropriate weight for her in 1 month? In 3 months? In 1 year? Describe the rationale for choosing the weight values you did.

B. Calculation of Nutrient Requirements

18. Calculate the outpatient treatment energy requirements for Paris.

C. Intake Domain

19. Using her 24-hour recall, calculate this patient's current energy and protein intake.

20. List any nutrition problems within the intake domain using the appropriate diagnostic term.

D. Clinical Domain

21. Evaluate Paris's lab results.

22. During nutritional repletion, Paris should be monitored closely for refeeding syndrome. What are the characteristics of refeeding syndrome?

23. Why was the EKG ordered?

E. Behavioral–Environmental Domain

24. Identify a minimum of five questions that the dietitian would ask regarding Paris's purging behaviors.

25. Paris asks you for a list of "good" foods to eat and "bad" foods to avoid. What should you tell her?

26. From the information gathered, list possible nutrition problems within the behavioral–environmental domain using the appropriate diagnostic term.

IV. Nutrition Diagnosis

27. Select two high-priority nutrition problems and complete PES statements for each.

V. Nutrition Intervention

28. For each PES statement written, establish an ideal goal (based on signs and symptoms) and an appropriate intervention (based on etiology).

VI. Nutrition Monitoring and Evaluation

29. When should you schedule your next counseling session with Paris?

30. What parameters can be used to measure Paris's response to treatment?

31. What would you assess at this follow-up counseling?

32. What medical conditions warrant residential or inpatient treatment?

33. Compare three eating disorder treatment facilities (e.g., discuss treatment options, treatment model, and the facilities' professional staff).

Bibliography

American Dietetic Association. Position of the American Dietetic Association: Nutrition intervention in the treatment of anorexia nervosa, bulimia nervosa, and other eating disorders. *J Am Diet Assoc.* 2006;106:2073–2082.

American Psychiatric Association. *Diagnostic and Statistical Manual of Mental Disorders, 4th Ed. Text Revision (DSM-IV-TR).* Washington, DC: American Psychiatric Association; 2000.

American Psychiatric Association. Practice guidelines for the treatment of patients with eating disorders. *Am J Psychiatry.* 2000;157(Suppl):1–39.

Bulik CM, Reba L, Siega-Riz AM, Reichborn-Kjennerud T. Anorexia nervosa: Definition, epidemiology, and cycle of risk. *Int J Eat Disorder.* 2005;37:S2–S9.

Bulik CM, Sullivan PF, Rozzi F, Furberg H, Lichtenstein P, Pedersen NL. Prevalence, heritability, and prospective risk factors for anorexia nervosa. *Arch Gen Psychiatry.* 2006;65:305–312.

Coughlin JW, Guarda AS. Behavioral disorders affecting food intake: Eating disorders and other psychiatric conditions. In: Shills ME, Shike M, Ross AC, Caballero B, Cousins RJ, eds. *Modern Nutrition in Health and Disease.* Philadelphia, PA: Lippincott, Williams & Wilkins; 2006:1353–1361.

Fairburn CG, Harrison PJ. Eating disorders. *Lancet.* 2003;361:407–416.

Lee RD. Energy balance and body weight. In: Nelms M, Sucher K, Long S. *Nutrition Therapy and Pathophysiology.* Belmont, CA: Thomson/Brooks-Cole; 2007:323–369.

Mehler PS. Bulimia nervosa. *N Engl J Med.* 2003;349:875–881.

Mehler PS. Diagnosis and care of patients with anorexia nervosa in primary care settings. *Ann Intern Med.* 2001;134:1048–1059.

Mehler PS. Osteoporosis in anorexia nervosa: Prevention and treatment. *Int J Eat Disorder.* 2003;33:113–126.

Mitchell JE, Cook-Myers T, Wonderlich SA. Diagnostic criteria for anorexia nervosa: Looking ahead to DSM-V. *Int J Eat Disord.* 2005;37:S95–S97.

Perkins R. Medical complications of the eating disorders. In: *Student Health Services Manual.* Carbondale, IL: Southern Illinois University, unpublished manual, 2007.

Walsh BT. Eating disorders. In: Kasper DL, Braunwald E, Fauci AS, Hauser SL, Longo DL, Jameson JL, eds. *Harrison's Principles of Internal Medicine.* 16th ed. New York, NY: McGraw-Hill; 2005:430–433.

Internet Resources

Academy for Eating Disorders. http://www.aedweb.org/

Anorexia Nervosa and Related Eating Disorders, Inc.: What Causes Eating Disorders? http://www.anred.com/causes.html

Eating Disorder Referral and Information Center. http://www.edreferral.com/

Eating Disorders Mirror Mirror. http://www.mirror-mirror.org/eatdis.htm

MedlinePlus: Eating Disorders. http://www.nlm.nih.gov/medlineplus/eatingdisorders.html

National Eating Disorders Association. http://www.nationaleatingdisorders.org/p.asp?WebPage_ID=337

National Eating Disorder Information Centre (NEDIC). http://www.nedic.ca

National Institute of Mental Health. http://www.nimh.nih.gov/index.shtml

Office on Women's Health: Eating Disorders. http://www.4woman.gov/owh/pub/factsheets/eatingdis.htm

Something Fishy Website on Eating Disorders. http://www.something-fishy.org/

Weight-Control Information Network. http://win.niddk.nih.gov/index.htm

Weight-Control Information Network: Binge Eating Disorder. http://win.niddk.nih.gov/publications/binge.htm

Case 5

Polypharmacy of the Elderly: Drug–Nutrient Interactions

Objectives

After completing this case, the student will be able to:

1. Integrate knowledge of pharmacology, nutrient–nutrient, and drug–nutrient interaction(s) into the nutrition care process.
2. Describe unique nutritional needs of the elderly.
3. Interpret pertinent laboratory parameters in the elderly.
4. Assess nutritional risk factors for the elderly patient.
5. Determine nutrition diagnoses and write appropriate PES statements.
6. Determine appropriate nutrition interventions to correct drug–nutrient interactions and improve nutritional status as established by the nutrition diagnoses.

Bob Kaufman, an 85-year-old male, has been brought to the hospital emergency room because of a change in his mental status. Mr. Kaufman suffers from several chronic diseases that are currently treated with multiple medications.

ADMISSION DATABASE

Name: Bob Kaufman
DOB: 1/12 (age 85)
Physician: Curtis Martin, MD

BED #	DATE:	TIME:	TRIAGE STATUS (ER ONLY):
1	5/15	1500	☐ Red ☐ Yellow ☐ Green ☐ White

Initial Vital Signs

TEMP:	RESP:	SAO$_2$:	
97.2	17		
HT:	**WT (lb):**	**B/P:**	**PULSE:**
5′5″	196	160/82	86
LAST TETANUS		**LAST ATE**	**LAST DRANK**
unknown		12 noon	30 minutes ago

PRIMARY PERSON TO CONTACT:
Name: Megan Smith (daughter)
Home #: 555-223-4589
Work #: 555-222-3421

ORIENTATION TO UNIT: ☒ Call light ☒ Television/telephone ☒ Bathroom ☒ Visiting ☒ Smoking ☒ Meals ☒ Patient rights/responsibilities

CHIEF COMPLAINT/HX OF PRESENT ILLNESS

"We brought my father in because he is more confused. His blood sugar is normal. I thought we should make sure he's OK."

PERSONAL ARTICLES: (Check if retained/describe)
☐ Contacts ☐ R ☐ L ☒ Dentures ☒ Upper ☒ Lower
☒ Jewelry: wedding band
☒ Other: eyeglasses

ALLERGIES: Meds, Food, IVP Dye, Seafood: Type of Reaction

NKA

VALUABLES ENVELOPE:
☐ Valuables instructions

PREVIOUS HOSPITALIZATIONS/SURGERIES

Back surgery? Procedure 15 years ago
TURP 2° prostate CA–10 years ago
Lower GI bleed 2° diverticulitis–2 hospitalizations 10 and 12 years ago

INFORMATION OBTAINED FROM:
☒ Patient ☐ Previous record
☒ Family ☐ Responsible party

Signature *Megan Smith*

Home Medications (including OTC) **Codes: A=Sent home** **B=Sent to pharmacy** **C=Not brought in**

Medication	Dose	Frequency	Time of Last Dose	Code	Patient Understanding of Drug
Diovan ↑K, ↑Mg	80 mg	daily	this AM	A	no
Prilosec	20 mg	daily	this AM	A	no
Neurontin ↑wt, ↑appetite	300 mg	BID	this AM	A	no
furosemide ↑K, ↑Mg	20 mg	1–2 as needed	this AM	A	no
Zocor ↓fat, ↓chol – No GF	20 mg	daily	this AM	A	no
isosorbide mono	60 mg	daily	this AM	A	no
trazodone Avoid SJW, Caution c CaF	25 mg	at bedtime	last night	A	no
aspirin	325 mg	daily	this AM	A	no
sodium bicarbonate ↓Na	650 mg x 2	TID	this AM	A	no
NPH insulin/regular insulin	10 U/3 U	AM/before dinner	this AM	A	no
multivitamin	1	daily	this AM	A	no

avoid ginko, ginseng, or valerian root

Fe separately, Caution c calcium

Do you take all medications as prescribed? ☐ Yes ☒ No If no, why? Daughter is unclear about meds. Patient is confused.

PATIENT/FAMILY HISTORY

☐ Cold in past two weeks	☒ High blood pressure Patient	☒ Kidney/urinary problems Patient
☐ Hay fever	☒ Arthritis Patient	☐ Gastric/abdominal pain/heartburn
☐ Emphysema/lung problems	☐ Claustrophobia	☒ Hearing problems Patient
☐ TB disease/positive TB skin test	☒ Circulation problems Patient	☐ Glaucoma/eye problems
☒ Cancer Patient	☒ Easy bleeding/bruising/anemia Patient	☒ Back pain Patient
☐ Stroke/past paralysis	☐ Sickle cell disease	☐ Seizures
☐ Heart attack	☐ Liver disease/jaundice	☐ Other
☒ Angina/chest pain Patient	☐ Thyroid disease	
☐ Heart problems	☒ Diabetes Patient (Type 2)	

RISK SCREENING

Have you had a blood transfusion? ☒ Yes ☐ No
Do you smoke? ☐ Yes ☒ No
If yes, how many pack(s)?
Does anyone in your household smoke? ☐ Yes ☒ No
Do you drink alcohol? ☐ Yes ☒ No
If yes, how often?______ How much?
When was your last drink?______/______/______
Do you take any recreational drugs? ☐ Yes ☒ No
If yes, type:______ Route:
Frequency:______ Date last used:______/______/______

FOR WOMEN Ages 12–52

Is there any chance you could be pregnant? ☐ Yes ☐ No
If yes, expected date (EDC):
Gravida/Para:

ALL WOMEN

Date of last Pap smear:
Do you perform regular breast self-exams? ☐ Yes ☐ No

ALL MEN

Do you perform regular testicular exams? ☐ Yes ☒ No

Additional comments:

x *Suzanne Miller, RN, BSN*
Signature/Title

Client name: Bob Kaufman
DOB: 1/12
Age: 85
Sex: Male
Education: High school, 1 year of college
Occupation: Postal clerk
Hours of work: Retired
Household members: Daughter age 45 and son-in-law age 52, grandsons ages 16 and 11—all in good health
Ethnic background: Caucasian
Religious affiliation: Episcopalian
Referring physician: Curtis Martin, MD

Chief complaint:
"We brought my father to the hospital because he has become more confused. Sometimes he forgets little things—he is 85, you know, but he generally is not confused. I checked his blood glucose first, but that was normal. I thought I had best bring him in to make sure everything was OK."

Patient history:
Onset of disease: Sudden onset of confusion that has been increasing over the past 24 hours. Patient moved to live with daughter and her family almost 3 years ago. Daughter states that her father is responsible for his own medicine. She is really not even aware of everything that he takes. He does his own insulin injections and his own blood glucose monitoring. Her father still drives almost every day. He keeps his own doctor visits. He does volunteer work at his church and at the local elementary school. Daughter provides most of his meals except for breakfast, which he usually cooks.
Type of Tx: Currently treated for CAD, type 2 DM, peripheral neuropathy, and renal insufficiency.
PMH: CAD; type 2 DM; renal insufficiency; peripheral neuropathy, osteoarthritis, Hx of prostate CA; diverticulitis/diverticulosis
Meds: Diovan; Prilosec; Neurontin; furosemide; isosorbide mononitrate; trazodone; sodium bicarbonate; aspirin; multivitamin; and NPH and regular insulin
Smoker: No
Family Hx: What? CA *Who?* Mother

Physical exam:
General appearance: Cheerful, obese, elderly gentleman who is obviously confused and appears slightly restless
Vitals: Temp 97.2°F, BP 160/82 mm Hg, HR 86 bpm, RR 16 bpm
Heart: Regular rate and rhythm; soft systolic murmur
HEENT:
Eyes: PERRLA
Ears: Clear
Nose: Clear
Throat: No exudate
Mouth: Loose-fitting dentures; membranes dry
Neurologic: Inconsistent orientation to time, place, and person

Extremities: Significant neuropathy present
Skin: Warm to touch; numerous pinpoint hemorrhages; fragile
Chest/lungs: Lungs clear to auscultation throughout, bilaterally
Peripheral vascular: All pulses present and equal; feet are cool to touch and have slight discoloration
Abdomen: Obese; bowel sounds present

Nutrition Hx:
General: Daughter states that appetite is good—"probably too good!" Daughter states that she prepares most meals. Her father snacks between meals, but she states that she tries to have low-sugar and low-fat choices available. He weighed almost 225 lbs when he came to live with her and her family almost 3 years ago. His weight has been stable for the past year. Her biggest concern nutritionally is that her father never seems to drink fluids except at mealtime, and she is worried that he doesn't get enough. "I will pour him a glass of water between meals. He will take one sip, and then he just lets it sit there." She states that she tries to keep his calories down and limits simple sugars. That is about as far as they go with diabetic restrictions. She states, "I just don't feel my father will eat anything more restrictive. I figure at 85, we'll just do the best we can."

Usual dietary intake:
AM: Egg Beaters—12-oz carton scrambled with 1 tbsp shredded cheese, 2 slices bacon, 1 slice toast, ½ c cranberry juice, 3 c coffee with fat-free creamer. About twice a week, he has cornflakes with a banana for breakfast.
Lunch: Usually from senior center—diabetic lunch—2–3 oz meat, 1–2 vegetables ½ c each, roll, ½ c fruit, 6–8 oz iced tea
Dinner: 3–4 oz meat, rice, potato, or noodle—1 c, 1 slice bread, ½ c fresh fruit, 6–8 oz iced tea
Snacks: Usually 2–3 × daily: Sugar-free Jell-O, low-fat yogurt, microwave popcorn

24-hour recall: Not available
Food allergies/intolerances/aversions: NKA
Previous nutrition therapy? Yes—when first diagnosed with diabetes over 15 years ago
Where? He has attended diabetic classes in the past.
Food purchase/preparation: Daughter
Vit/min intake: Multivitamin daily
Anthropometric data: Ht 5′5″, Wt 196 lbs, UBW 195–225 lbs

Tx plan:
Admit to Internal Medicine: Dr. Curtis Martin
Vitals: Routine; SBGM ac q meal
Lab: CBC, SMA
Head: CT to R/O CVA
Diet: 1,800 kcal ADA diet
Activity: Bed rest with supervision
Meds: Sliding scale Humulin Regular: < 200 do nothing; 200–300 5 U SQ; 300–400 10 U SQ; > 400 call MD; Diovan 80 mg; isosorbide mononitrate 60 mg. Continue insulin prescription from home.

Hospital course:

Head CT normal. *Dx:* Metabolic alkalosis 2° to excessive intake of sodium bicarbonate; mild dehydration. Additional labs consistent with underlying diagnoses of type 2 DM, renal insufficiency. Patient received NS 40 mEq of KCl @ 75 cc/hr for 24 hours. As electrolyte abnormalities resolved, confusion resolved as well. Patient stated prior to discharge that he was confused with medications, and there appears to be a misconception on dosage of furosemide and sodium bicarbonate. Discharge medications were adjusted. Pharmacy and nutrition consult ordered prior to discharge.

UH UNIVERSITY HOSPITAL

NAME: Bob Kaufman
AGE: 85
PHYSICIAN: Curtis Martin, MD

DOB: 1/12
SEX: M

CHEMISTRY

DAY:		Admit	2	
DATE:		5/15	5/16	
TIME:				
LOCATION:				
	NORMAL			UNITS
Albumin	3.5–5	3.4 L		g/dL
Total protein	6–8	6.0		g/dL
Prealbumin	16–35	20		mg/dL
Transferrin	250–380 (women) 215–365 (men)	210		mg/dL
Sodium	136–145	145	138	mEq/L
Potassium	3.5–5.5	3.4 L	3.8	mEq/L
Chloride	95–105	98	99	mEq/L
PO_4	2.3–4.7	4.5		mg/dL
Magnesium	1.8–3	1.7		mg/dL
Osmolality	285–295	310 H	296 H	mmol/kg/H_2O
Total CO_2	23–30	30	27	mEq/L
Glucose	70–110	172 H	155 H	mg/dL
BUN	8–18	32 H	33 H	mg/dL
Creatinine	0.6–1.2	1.5 H	1.5 H	mg/dL
Uric acid	2.8–8.8 (women) 4.0–9.0 (men)	3.7		mg/dL
Calcium	9–11	8.7		mg/dL
Bilirubin	≤ 0.3	0.3		mg/dL
Ammonia (NH_3)	9–33	10		µmol/L
ALT	4–36	22		U/L
AST	0–35	14		U/L
Alk phos	30–120	101		U/L
CPK	30–135 (women) 55–170 (men)	121		U/L
LDH	208–378	356		U/L
CHOL	120–199	175		mg/dL
HDL-C	> 55 (women) > 45 (men)	41		mg/dL
VLDL	7–32			mg/dL
LDL	< 130	135		mg/dL
LDL/HDL ratio	< 3.22 (women) < 3.55 (men)	3.29		
Apo A	101–199 (women) 94–178 (men)			mg/dL
Apo B	60–126 (women) 63–133 (men)			mg/dL
TG	35–135 (women) 40–160 (men)	175 H		mg/dL
T_4	4–12			mcg/dL
T_3	75–98			mcg/dL
HbA_{1C}	3.9–5.2	8.2 H		%

UH UNIVERSITY HOSPITAL

NAME: Bob Kaufman DOB: 1/12
AGE: 85 SEX: M
PHYSICIAN: Curtis Martin, MD

********************************ARTERIAL BLOOD GASES (ABGs)********************************

DAY: Admit
DATE: 5/15
TIME:
LOCATION:

	NORMAL		UNITS
↑ pH	7.35–7.45	7.47	
↑ pCO_2	35–45	46	mm Hg
SO_2	≥ 95		%
↑ CO_2 content	23–30	31	mmol/L
O_2 content	15–22		%
pO_2	≥ 80	83	mm Hg
Base excess	> 3		mEq/L
Base deficit	< 3		mEq/L
HCO_3	24–28	32	mEq/L
HGB	12–16 (women) 13.5–17.5 (men)		g/dL
HCT	37–47 (women) 40–54 (men)		%
COHb	< 2		%
$[Na^+]$	135–148		mmol/L
$[K^+]$	3.5–5		mEq/L

UH UNIVERSITY HOSPITAL

NAME: Bob Kaufman
AGE: 85
PHYSICIAN: Curtis Martin, MD

DOB: 1/12
SEX: M

HEMATOLOGY

DAY: Admit
DATE: 5/15
TIME:
LOCATION:

	NORMAL	Admit 5/15	UNITS
WBC	4.8-11.8	5.2	$\times 10^3/mm^3$
RBC	4.2-5.4 (women) 4.5-6.2 (men)	4.5	$\times 10^6/mm^3$
HGB	12-15 (women) 14-17 (men)	13	g/dL
HCT	37-47 (women) 40-54 (men)	40	%
MCV	80-96		μm^3
RETIC	0.8-2.8		%
MCH	26-32		pg
MCHC	31.5-36		g/dL
RDW	11.6-16.5		%
Plt Ct	140-440		$\times 10^3/mm^3$
Diff TYPE			
ESR	0-25 (women) 0-15 (men)		mm/hr
% GRANS	34.6-79.2	62.8	%
% LYM	19.6-52.7	37.1	%
SEGS	50-62		%
BANDS	3-6		%
LYMPHS	24-44		%
MONOS	4-8		%
EOS	0.5-4		%
Ferritin	20-120 (women) 20-300 (men)		mg/mL
ZPP	30-80		μmol/mol
Vitamin B_{12}	24.4-100		ng/dL
Folate	5-25		μg/dL
Total T cells	812-2,318		mm^3
T-helper cells	589-1,505		mm^3
T-suppressor cells	325-997		mm^3
PT	11-16		sec

Case Questions

I. Understanding the Disease and Pathophysiology

1. Identify each of the medical diagnoses for Mr. Kaufman.

2. Identify which of these may affect cardiac function, liver function, and renal function.

3. Are there also normal changes in renal function that occur with aging?

4. Define polypharmacy. Do you think that Mr. Kaufman's medications represent polypharmacy? Why is polypharmacy a concern in the elderly?

II. Understanding the Nutrition Therapy

5. Describe the potential nutritional complications secondary to pharmacotherapy.

6. Describe the potential effect of nutrition on the action of medications.

III. Nutrition Assessment

A. Evaluation of Weight/Body Composition

7. Mr. Kaufman is 5′5″ tall and weighs 196 lbs. Calculate his body mass index. How would you interpret this value? Should any adjustments be made in the interpretation to account for his age?

8. Calculate Mr. Kaufman's percent usual body weight. Interpret the significance of this assessment.

9. In an older individual, what specific changes occur in body composition and energy requirements that may need to be taken into consideration when completing a nutritional assessment?

B. Calculation of Nutrient Requirements

10. Calculate energy and protein requirements for Mr. Kaufman. Identify the formula/calculation method you used and explain the rationale for using it. What factors should you consider when estimating his requirements?

C. Intake Domain

11. Mr. Kaufman's daughter expressed concern regarding his fluid intake. Is this a common problem in aging? Explain.

12. There are several ways to estimate fluid needs. Calculate Mr. Kaufman's fluid needs by using at least two of these methods. How do they compare? From your evaluation of his usual intake, do you think he is getting enough fluids?

13. Evaluate Mr. Kaufman's usual intake for both caloric and protein intake. How does it compare to the MyPyramid recommendations?

14. From the information gathered within the intake domain, list Mr. Kaufman's possible nutrition problems using the diagnostic term.

15. Do you think Mr. Kaufman needs to take a multivitamin? In general, do needs for vitamins and minerals change with aging? What reference would you use to determine recommended amounts of the micronutrients?

D. Clinical Domain

16. Mr. Kaufman was diagnosed with mild metabolic alkalosis and dehydration. What is metabolic alkalosis? Read Mr. Kaufman's history and physical. What signs and symptoms does the patient present with that may be consistent with metabolic alkalosis and dehydration? Explain.

17. What laboratory values support his medical history of renal insufficiency? What laboratory value(s) support this diagnosis of metabolic alkalosis? Which are consistent with dehydration? What laboratory values support his medical history of type 2 diabetes mellitus?

Laboratory	**Normal**	**Mr. Kaufman's Value**
Albumin	3.5–5 g/dL	
Potassium	3.5–5.5 mEq/L	
Osmolality	285–295 mmol/kg/H_2O	
Glucose	70–110 mg/dL	
BUN	8–18 mg/dL	
Creatinine	0.6–1.2 mg/dL	
HbA_{1c}	3.9–5.2%	
pH	7.35–7.45	
pCO_2	35–45 mm Hg	
CO_2	23–30 mmol/L	
HCO_3	24–28 mEq/L	

18. Using the following table, list all the medications that Mr. Kaufman was taking at home. Identify the function of each medication.

Medication	**Function**	**Drug–Drug Interaction**	**Drug–Nutrient Interaction**

19. Identify all drug–drug interactions and then identify any drug–nutrient interactions for the medications.

20. What medications are the most likely to have contributed to the abnormal lab values and thus this diagnosis? Why?

21. What does the HbA_{1C} measure? What can this value tell you about Mr. Kaufman's overall control over his diabetes?

22. From the information gathered within the clinical domain, list possible nutrition problems using the diagnostic term.

E. Behavioral–Environmental Domain

23. List possible behavioral–environmental nutrition problems.

IV. Nutrition Diagnosis

24. Select two high-priority nutrition problems and complete PES statements for each.

V. Nutrition Intervention

25. For each of the PES statements that you have written, establish an ideal goal (based on the signs and symptoms) and an appropriate intervention (based on the etiology).

26. Would you make diabetes education a priority in your nutrition counseling for Mr. Kaufman? What methods might you use to help maximize his glucose control? How would you assess the patient's and daughter's readiness for change?

Bibliography

American Dietetic Association. *Nutrition Diagnosis and Intervention: Standardized Language for the Nutrition Care Process.* Chicago, IL: American Dietetic Association; 2006.

American Dietetic Association. Older Adults. In: *Manual of Clinical Dietetics.* 6th ed. Chicago, IL: American Dietetic Association; 2000:141–157.

Bartlett S, Mirian M, Taren D, Muramoto M. *Geriatric Nutrition Handbook.* 5th ed. Florence, KY: International Thomson; 1998.

Beard K. Adverse reactions as a cause of hospital admissions for the aged. *Drug Aging.* 1992:2:356–363.

Blumberg J, Couris R. Pharmacology, nutrition, and the elderly: Interactions and implications. In: Chernoff R, ed. *Geriatric Nutrition.* 2nd ed. Gaithersberg, MD: Aspen, 1999.

Chang C, Liu PY, Yang Y, Wu C, Lu F. Use of the Beers criteria to predict adverse drug reactions among first-visit elderly outpatients. *Pharmacotherapy.* 2005:25(6):831–838.

Chernoff R. Thirst and fluid requirements. *Nutr Rev.* 1994;52(Suppl):S3–S5.

Chidester JC, Spangler AA. Fluid intake in the institutionalized elderly. *J Am Diet Assoc.* 1997;97:23–28.

Cook MC, Tarren DL. Nutritional implications of medication use and misuse in elderly. *J Fla Med Assoc.* 1990;77(6):606–613.

Gallo RM, Fulmer T, Paveza G, Reichel W, eds. *Handbook of Geriatric Assessment.* 3rd ed. Baltimore, MD: Aspen; 2000.

Holben DH, Hassell JT, Williams JL, Helle B. Fluid intake compared with established standards and symptoms of dehydration among elderly residents of a long-term-care facility. *J Am Diet Assoc.* 1999;99(11):1447–1450.

Kurpad AV. Protein and amino acid requirements in the elderly. *Eur J Clin Nutr.* 2000;54:S131–S142.

Lindeman RD, Romero LJ, Liang HC, Baumgartner RN, Koehler KM, Garry PJ. Do elderly persons need to be encouraged to drink more fluids? *J Gerontol A Biol Sci Med Sci.* 2000;55(7):M359–M360.

Nelms MN. Acid-base balance. In: Nelms M, Sucher K, Long S. *Nutrition Therapy and Pathophysiology.* Belmont, CA: Thomson/Brooks-Cole; 2007:203–217.

Nelms MN. Assessment of nutrition status and risk. In: Nelms M, Sucher K, Long S. *Nutrition Therapy and Pathophysiology.* Belmont, CA: Thomson/Brooks-Cole; 2007:101–135.

Nelms MN. Fluid and electrolyte balance. In: Nelms M, Sucher K, Long S. *Nutrition Therapy and Pathophysiology.* Belmont, CA: Thomson/Brooks-Cole; 2007:181–201.

Nelms MN. Pharmacology. In: Nelms M, Sucher K, Long S. *Nutrition Therapy and Pathophysiology.* Belmont, CA: Thomson/Brooks-Cole; 2007:297–322.

Quandt SA, McDonald J, Arcury TA, Bell RA, Vitolins MZ. Nutritional self-management of elderly widows in rural communities. *Gerontologist.* 2000;40(1):86–96.

Rolls BJ. Regulation of food and fluid intake in the elderly. *Ann NY Acad Sci.* 1989;561:217–255.

Rolls BJ, Phillips PA. Aging and disturbances of thirst and fluid balance. *Nutr Rev.* 1990;48:137–144.

Standing Committee on the Scientific Evaluation of Dietary Reference Intakes, Food and Nutrition Board, Institute of Medicine. *Dietary Reference Intakes for Calcium, Phosphorus, Magnesium, Vitamin D, and Fluoride.* Washington, DC: National Academy Press, 1997.

Standing Committee on the Scientific Evaluation of Dietary Reference Intakes, Food and Nutrition Board, Institute of Medicine. *Dietary Reference Intakes for Thiamin, Riboflavin, Niacin, Vitamin B6, Folate, Vitamin B12, Pantothenic Acid, Biotin, and Choline.* Washington, DC: National Academy Press, 1999.

Internet Resources

Medscape: Drug Interaction Checker. http://www.medscape.com/druginfo/druginterchecker?src=google

Merck Manuals Online Library: Drug–Nutrient Interactions. http://www.merck.com/mmpe/sec01/ch001/ch001d.html

Polypharmacy in Older Adults. http://altdirectory.com/Polypharmacy/

Unit Two

NUTRITION THERAPY FOR CARDIOVASCULAR DISORDERS

Cardiovascular disease is the leading cause of death in the United States. Risk factors for cardiovascular disease include dyslipidemia, smoking, diabetes mellitus, high blood pressure, obesity, and physical inactivity. Researchers estimate that more than 70 million Americans have one or more forms of cardiovascular disease; as a result, many patients that the health care team encounters will have conditions related to cardiovascular disease.

This section includes three of the most common diagnoses: hypertension (HTN), myocardial infarction (MI), and congestive heart failure (CHF). All these diagnoses require a significant medical nutrition therapy component for their care.

Over 65 million people in the United States have hypertension. Hypertension is defined as a systolic blood pressure of 140 mm Hg or higher and a diastolic pressure of 90 mm Hg or higher. Essential hypertension, which is the most common form of hypertension, is of unknown etiology. Case 6 focuses on lifestyle modifications as the first step in treatment of hypertension accompanied by dyslipidemia in a female patient. This case incorporates the pharmacological treatment of hypertension, and you will use the most recent information from *Dietary Approaches to Stop Hypertension* (DASH) as the center of the medical nutrition therapy intervention. Because cardiovascular disease is a complex, multifactorial condition, Case 6 provides the opportunity to evaluate these multiple risk factors through all facets of nutrition assessment. We specifically emphasize interpretation of laboratory indices for dyslipidemia. In this case, you will also determine the clinical classification and treatment of abnormal serum lipids, explore the use of drug therapy to treat dyslipidemias, and develop appropriate nutrition interventions using the Therapeutic Lifestyle Change (TLC) recommendations as the framework for these diagnoses.

Case 7 focuses on the acute care of an individual suffering a myocardial infarction (MI). Over 900,000 Americans had a new or recurrent myocardial infarction in 2005 (available at: http://www.americanheart.org/presenter.jhtml?identifier=3037327; accessed January 22, 2008). Ischemia of the vessels within the heart results in death of the affected heart tissue. This case lets you evaluate pertinent assessment measures for the individual suffering an MI and then develop an appropriate nutrition care plan that complements the medical care for prevention of further cardiac deterioration.

Case 8 addresses the long-term consequences of cardiovascular disease in a patient suffering from congestive heart failure (CHF). In CHF, the heart cannot pump effectively, and the lack of oxygen and nutrients affects the body's tissues. CHF is a major public health problem in the United States, and its incidence is increasing. Without a heart transplant, prognosis is poor. This advanced case requires you to integrate understanding of the physiology of several body systems as you address heart failure's metabolic effects. Additionally, this case allows you to explore the role of the health care team in palliative care.

Case 6

Hypertension and Cardiovascular Disease

Objectives

After completing this case, the student will be able to:

1. Describe the physiology of blood pressure regulation.
2. Apply knowledge of the pathophysiology of hypertension and dyslipidemias to identify and explain common nutritional problems associated with these diseases.
3. Understand the role of nutrition therapy as an adjunct to the pharmacotherapy of hypertension.
4. Understand the role of nutrition therapy as an adjunct to the pharmacotherapy and surgical and other medical treatment of cardiovascular disease.
5. Interpret laboratory parameters for nutritional implications and significance.
6. Analyze nutrition assessment data to evaluate nutritional status and identify specific nutrition problems.
7. Determine nutrition diagnoses and write appropriate PES statements.
8. Develop a nutrition care plan with appropriate measurable goals, interventions, and strategies for monitoring and evaluation that addresses the nutrition diagnoses of this case.

Mrs. Corey Anderson is a 54-year-old housewife. She has treated her newly diagnosed hypertension for the past year with lifestyle changes, including diet, smoking cessation, and exercise. She is in to see her physician for further evaluation and treatment for essential hypertension. Blood was drawn 2 weeks prior to this appointment and shows an abnormal lipid profile.

Name: Corey Anderson
DOB: 10/9 (age 54)
Physician: A. Thornton

ADMISSION DATABASE

BED #	DATE:	TIME:	TRIAGE STATUS (ER ONLY):
2	6/25	1419	☐ Red ☐ Yellow ☐ Green ☐ White

Initial Vital Signs

TEMP:	RESP:	SAO$_2$:	
98.6	35		
HT (in): 5′6″	WT (lb): 160	B/P: 160/100	PULSE: 80
LAST TETANUS: 10 years ago		LAST ATE: 1230	LAST DRANK: 1330

PRIMARY PERSON TO CONTACT:
Name: Ben Anderson
Home #: 555-7128
Work #: 555-2157

ORIENTATION TO UNIT: ☒ Call light ☒ Television/telephone ☒ Bathroom ☒ Visiting ☒ Smoking ☒ Meals ☒ Patient rights/responsibilities

CHIEF COMPLAINT/HX OF PRESENT ILLNESS

"I have high blood pressure, and now my cholesterol is too high."

PERSONAL ARTICLES: (Check if retained/describe)
☐ Contacts ☐ R ☐ L ☐ Dentures ☐ Upper ☐ Lower
☐ Jewelry:
☒ Other: glasses

ALLERGIES: Meds, Food, IVP Dye, Seafood: Type of Reaction

NKA

VALUABLES ENVELOPE:
☐ Valuables instructions

PREVIOUS HOSPITALIZATIONS/SURGERIES

Hernia repair 10 years ago

INFORMATION OBTAINED FROM:
☒ Patient ☐ Previous record
☐ Family ☐ Responsible party

Signature *Corey Anderson*

Home Medications (including OTC) **Codes: A=Sent home** **B=Sent to pharmacy** **C=Not brought in**

Medication	Dose	Frequency	Time of Last Dose	Code	Patient Understanding of Drug

Do you take all medications as prescribed? ☐ Yes ☐ No If no, why?

PATIENT/FAMILY HISTORY

☒ Cold in past two weeks Patient
☐ Hay fever
☐ Emphysema/lung problems
☐ TB disease/positive TB skin test
☐ Cancer
☐ Stroke/past paralysis
☒ Heart attack Mother
☐ Angina/chest pain
☒ Heart problems Mother
☒ High blood pressure Patient & mother
☒ Arthritis Patient
☐ Claustrophobia
☐ Circulation problems
☐ Easy bleeding/bruising/anemia
☐ Sickle cell disease
☐ Liver disease/jaundice
☐ Thyroid disease
☐ Diabetes
☐ Kidney/urinary problems
☐ Gastric/abdominal pain/heartburn
☐ Hearing problems
☐ Glaucoma/eye problems
☐ Back pain
☐ Seizures
☐ Other

RISK SCREENING

Have you had a blood transfusion? ☐ Yes ☒ No
Do you smoke? ☒ Yes ☐ No
If yes, how many pack(s)? 1/day for 30 yrs
Does anyone in your household smoke? ☒ Yes ☐ No
Do you drink alcohol? ☒ Yes ☐ No
If yes, how often? occ How much? 1-2 beers
When was your last drink? 6/15
Do you take any recreational drugs? ☐ Yes ☒ No
If yes, type:_______ Route:
Frequency:_______ Date last used:______/______/______

FOR WOMEN Ages 12–52

Is there any chance you could be pregnant? ☐ Yes ☒ No
If yes, expected date (EDC):
Gravida/Para:

ALL WOMEN

Date of last Pap smear: 3/3
Do you perform regular breast self-exams? ☒ Yes ☐ No

ALL MEN

Do you perform regular testicular exams? ☐ Yes ☐ No

Additional comments:

x *Connie L. Bussard, RN*
Signature/Title

Client name: Corey Anderson
DOB: 10/9
Age: 54
Sex: Female
Education: High school
Occupation: Housewife
Hours of work:
Household members: Husband age 50 in good health; children are grown and do not live at home
Ethnic background: African American
Religious affiliation: Catholic
Referring physician: Alan Thornton, MD (cardiology)

Chief complaint:
"I've tried to cut back on salt, but food just doesn't taste good without it. I want to control this high blood pressure—my mother passed away because her high blood pressure caused her to have a heart attack. And now the doctor tells me my cholesterol is high!"

Patient history:
Onset of disease: Mrs. Anderson is a 54-year-old female who is not employed outside the home. She was diagnosed 1 year ago with Stage 2 (essential) HTN. Treatment thus far has been focused on nonpharmacological measures. She began a walking program that has resulted in a 10-pound weight loss that she has been able to maintain during the past year. She walks 30 minutes 4–5 times per week, though she sometimes misses on bingo nights. She was given a diet sheet in the MD's office that outlined a 4-g Na diet. Mrs. Anderson was a 2-pack-a-day smoker but quit ("cold turkey") when she was diagnosed last year. No c/o of any symptoms related to HTN. Patient denies chest pain, SOB, syncope, palpitations, or myocardial infarction.
Type of Tx: Initiation of pharmacologic therapy with thiazide diuretics and reinforcement of lifestyle modifications to decrease fat intake. Rule out metabolic syndrome.
PMH: Not significant before Dx of HTN
Meds: Hydrochlorothiazide—25 mg daily
Smoker: No—quit 1 year ago
Family Hx: What/Who? Mother died of MI related to uncontrolled HTN

Physical exam:
General appearance: Healthy, middle-aged female who looks her age
Vitals: Temp 98.6°F, BP 160/100 mm Hg, HR 80 bpm, RR 15 bpm, Wt 160 lbs
Heart: Regular rate and rhythm, normal heart sounds—no clicks, murmurs, or gallops; no carotid bruits
HEENT:
  Eyes: No retinopathy, PERRLA
Genitalia: Normal female
Neurologic: Alert and oriented to person, place, and time
Extremities: Noncontributory
Skin: Smooth, warm, dry, excellent turgor, no edema
Chest/lungs: Lungs clear
Peripheral vascular: Pulse 4+ bilaterally, warm, no edema
Abdomen: Nontender, no guarding, normal bowel sounds

Nutrition Hx:

General: Mrs. Anderson describes her appetite as "very good." She does the majority of grocery shopping and cooking, although Mr. Anderson cooks breakfast for her on the weekends. She usually eats three meals each day, but on bingo nights, she usually skips dinner and just snacks while playing bingo. When she does this, she is really hungry when she gets home in the late evening, so she often eats a bowl of ice cream before going to bed. The Andersons usually eat out on Friday and Saturday evenings, often at pizza restaurants or steakhouses (Mrs. Anderson usually has 2 regular beers with these meals). She mentions that last year when her HTN was diagnosed, a nurse at the MD's office gave her a sheet of paper with a list of foods to avoid for a 4-g Na (no added salt) diet. She and her husband tried to comply with the diet guidelines, but they found foods bland and tasteless, and they soon abandoned the effort.

24-hour recall:

AM: 1 c coffee (black), hot (oatmeal—1 instant packet with 1 tsp margarine and 2 tsp sugar) or cold (Frosted Mini-Wheats) cereal (10 pieces), ½ c 2% milk, 1 c orange juice

Snack: 2 c coffee (black), 1 glazed donut

Lunch: 1 can Campbell's tomato bisque soup prepared with milk, 10 saltines, 1 can diet cola

PM: 6 oz baked chicken (white meat, no skin) (seasoned with salt, pepper, garlic), 1 large baked potato with 1 tbsp butter, salt, and pepper, 1 c glazed carrots (1 tsp sugar, 1 tsp butter), dinner salad with ranch-style dressing (3 tbsp)—lettuce, spinach, croutons, sliced cucumber; 2 regular beers

HS snack: 2 c butter pecan ice cream

Food allergies/intolerances/aversions: None
Previous nutrition therapy? Yes *If yes, when:* 1 year ago *Where?* MD's office
Food purchase/preparation: Self
Vit/min intake: Multivitamin/mineral daily
Current diet order: 4 g Na

Dx:

Physical exam reveals Stage 1 HTN, hypertensive heart disease, and early COPD. Complete fasting lipid profile was abnormal, EKG was WNL.

Tx plan:

Urinalysis, hematocrit, blood chemistry to include plasma glucose, potassium, BUN, creatinine, fasting lipid profile, triglycerides, calcium, uric acid
Chest X-ray
EKG
Nutrition consult
25 mg hydrochlorothiazide daily
Patient to be reassessed in 3 months

UH *UNIVERSITY HOSPITAL*

NAME: Corey Anderson DOB: 10/9
AGE: 54 SEX: F
PHYSICIAN: A. Thornton, MD

CHEMISTRY***********************************

DAY:		Admit	3 months	6 months	
DATE:					
TIME:					
LOCATION:					
	NORMAL				UNITS
Albumin	3.5–5	4.6	4.3	4.4	g/dL
Total protein	6–8	7	7	6.8	g/dL
Prealbumin	16–35	32	31	31	mg/dL
Transferrin	250–380 (women) 215–365 (men)	350	355	345	mg/dL
Sodium	136–145	136	138	137	mEq/L
Potassium	3.5–5.5	4.1	3.6	3.9	mEq/L
Chloride	95–105	102	100	101	mEq/L
PO_4	2.3–4.7	4.1	3.5	3.5	mg/dL
Magnesium	1.8–3	2.1	2.3	2.3	mg/dL
Osmolality	285–295	292	293	295	mmol/kg/H_2O
Total CO_2	23–30	30	29	29	mEq/L
Glucose	70–110	92	90	96	mg/dL
BUN	8–18	20	15	22	mg/dL
Creatinine	0.6–1.2	0.9	1.1	1.1	mg/dL
Uric acid	2.8–8.8 (women) 4.0–9.0 (men)	6.8	7.0	7.2	mg/dL
Calcium	9–11	9.2	9.0	9.1	mg/dL
Bilirubin	≤ 0.3	1.1	0.8	0.9	mg/dL
Ammonia (NH_3)	9–33	19	18	22	µmol/L
ALT	4–36	30	35	28	U/L
AST	0–35	39	34	35	U/L
Alk phos	30–120	250	115	111	U/L
CPK	30–135 (women) 55–170 (men)	100	125	134	U/L
LDH	208–378	314	323	350	U/L
CHOL	120–199	270 H	230 H	210 H	mg/dL
HDL-C	> 55 (women) > 45 (men)	30 L	35 L	38 L	mg/dL
VLDL	7–32				mg/dL
LDL	< 130	210 H	169 H	147 H	mg/dL
LDL/HDL ratio	< 3.22 (women) < 3.55 (men)	7.0 H	4.8 H	3.9 H	
Apo A	101–199 (women) 94–178 (men)	75 L	100 L	110	mg/dL
Apo B	60–126 (women) 63–133 (men)	140 H	120	115	mg/dL
TG	35–135 (women) 40–160 (men)	150 H	130	125	mg/dL
T_4	4–12	8.6	8.5	7.8	mcg/dL
T_3	75–98	95	92	93	mcg/dL
HbA_{1C}	3.9–5.2	6.2	6.3	6.5	%

UH UNIVERSITY HOSPITAL

NAME: Corey Anderson
AGE: 54
PHYSICIAN: A. Thornton, MD

DOB: 10/9
SEX: F

HEMATOLOGY**********************************

DAY:
DATE:
TIME:
LOCATION:

	NORMAL	Admit	3 months	6 months	UNITS
WBC	4.8–11.8	6.1	5.7	5.5	× $10^3/mm^3$
RBC	4.2–5.4 (women) 4.5–6.2 (men)	5.5 H	5.0	5.25	× $10^6/mm^3$
HGB	12–15 (women) 14–17 (men)	14.2	14.0	13.9	g/dL
HCT	37–47 (women) 40–54 (men)	45	44	44	%
MCV	80–96	88	90	85	μm^3
RETIC	0.8–2.8				%
MCH	26–32	29	30	29	pg
MCHC	31.5–36	35	32	33	g/dL
RDW	11.6–16.5	13.5	13.4	13.6	%
Plt Ct	140–440	430	350	366	× $10^3/mm^3$
Diff TYPE					
ESR	0–25 (women) 0–15 (men)				mm/hr
% GRANS	34.6–79.2	66.2	60.3	55.2	%
% LYM	19.6–52.7	33.8	39.9	42.1	%
SEGS	50–62	60	52.1	51	%
BANDS	3–6	1 L	2 L	2 L	%
LYMPHS	24–44	33	40	32	%
MONOS	4–8	5	3	4	%
EOS	0.5–4	1	3	2	%
Ferritin	20–120 (women) 20–300 (men)	263 H	255 H	241 H	mg/mL
ZPP	30–80				μmol/mol
Vitamin B_{12}	24.4–100	90	70.5	63.3	ng/dL
Folate	5–25	0.3	0.25	0.08	μg/dL
Total T cells	812–2,318				mm^3
T-helper cells	589–1,505				mm^3
T-suppressor cells	325–997				mm^3
PT	11–16	12.5	12.7	12.2	sec

UH UNIVERSITY HOSPITAL

NAME: Corey Anderson DOB: 10/9
AGE: 54 SEX: F
PHYSICIAN: A. Thornton, MD

URINALYSIS

DAY:		Admit	3 Months	6 Months	
DATE:					
TIME:					
LOCATION:					
	NORMAL				UNITS
Coll meth		Random specimen	Random specimen	First morning	
Color		Pale yellow	Pale yellow	Pale yellow	
Appear		Clear	Clear	Clear	
Sp grv	1.003-1.030	1.025	1.021	1.024	
pH	5-7	7.0	5.0	6.0	
Prot	NEG	NEG	NEG	NEG	mg/dL
Glu	NEG	NEG	NEG	NEG	mg/dL
Ket	NEG	Trace	1+	2+	
Occ bld	NEG	Negative	Negative	Negative	
Ubil	NEG	Negative	Negative	Negative	
Nit	NEG	Negative	Negative	Negative	
Urobil	< 1.1	0.02	0.01	Negative	EU/dL
Leu bst	NEG	Negative	Negative	Negative	
Prot chk	NEG	Negative	Negative	Negative	
WBCs	0-5	0	0	0	/HPF
RBCs	0-5	0	0	0	/HPF
EPIs	0	Rare	0	0	/LPF
Bact	0	0	0	0	
Mucus	0	0	0	0	
Crys	0	0	0	0	
Casts	0	0	0	0	/LPF
Yeast	0	0	0	0	

Case Questions

I. Understanding the Disease and Pathophysiology

1. Define blood pressure.

2. How is blood pressure normally regulated in the body?

3. What causes essential hypertension?

4. What are the symptoms of hypertension?

5. How is hypertension diagnosed?

6. List the risk factors for developing hypertension.

7. What risk factors does Mrs. Anderson currently have?

8. Hypertension is classified in stages based on the risk of developing CVD. Complete the following table of hypertension classifications.

	Blood Pressure mm Hg		
Category	**Systolic BP**		**Diastolic BP**
Normal		and	
Prehypertension		or	
Hypertension Stage 1		or	
Hypertension Stage 2		or	

9. Given these criteria, which category would Mrs. Anderson's admitting blood pressure reading place her in?

10. How is hypertension treated?

11. Dr. Thornton indicated in his admitting note that he will "rule out metabolic syndrome." What is metabolic syndrome?

12. What factors found in the medical and social history are pertinent for determining Mrs. Anderson's CHD risk category?

13. What progression of her disease might Mrs. Anderson experience?

II. Understanding the Nutrition Therapy

14. What are the most recent recommendations for nutrition therapy in hypertension? Explain the history of and rationale for the DASH diet.

15. What is the rationale for sodium restriction in treatment of hypertension? Is this controversial? Why or why not?

16. What are the Therapeutic Lifestyle Changes? Outline the major components of the nutrition therapy interventions.

17. The most recent recommendations suggest the therapeutic use of stanol esters. What are they, and what is the rationale for their use?

III. Nutrition Assessment

A. Evaluation of Weight/Body Composition

18. Calculate Mrs. Anderson's body mass index (BMI).

19. What are the health implications of this number?

B. Calculation of Nutrient Requirements

20. Calculate Mrs. Anderson's resting and total energy needs. Identify the formula/calculation method you used and explain your rationale for using it.

21. How many calories per day would you recommend for Mrs. Anderson?

22. Determine the appropriate percentages of total kilocalories from carbohydrate, protein, and lipid.

C. Intake Domain

23. Using a computer dietary analysis program or food composition table, compare Mrs. Anderson's "usual" dietary intake to her prescribed diet (DASH/TLC diet).

Food Item	Potassium (minimum 4,700 mg/ 120 mEq)	Sodium (maximum 2,400 mg/ 100 mEq)	Magnesium (500 mg)	Calcium (minimum 1,240 mg)	Total Fat (g)	Saturated Fat (g)	Cholesterol (mg)	Fiber (g)

24. What nutrients in Mrs. Anderson's diet are of major concern to you?

25. From the information gathered within the intake domain, list possible nutrition problems using the diagnostic term.

D. Clinical Domain

26. Dr. Thornton ordered the following labs: fasting glucose, cholesterol, triglycerides, creatinine, and uric acid. He also ordered an EKG. In the following table, outline the indication for these tests (tests provide information related to a disease or condition).

Parameter	Normal Value	Patient's Value	Reason for Abnormality	Nutrition Implication
Glucose	70–110 mg/dL			
BUN	8–18 mg/dL			
Creatinine	0.6–1.2 mg/dL			
Total cholesterol	120–199 mg/dL			
HDL-cholesterol	> 55 (women) mg/dL > 45 (men) mg/dL			
LDL-cholesterol	< 130 mg/dL			
Apo A	101–199 (women) mg/dL 94–178 (men) mg/dL			
Apo B	60–126 (women) mg/dL 63–133 (men) mg/dL			
Triglycerides	35–135 (women) 40–160 (men)			

27. Interpret Mrs. Anderson's risk of CAD based on her lipid profile.

28. What is the significance of apolipoprotein A and apolipoprotein B in determining a person's risk of CAD?

29. Indicate the pharmacological differences among the antihypertensive agents listed below.

Medications	Mechanism of Action	Nutritional Implications
Diuretics		
Beta-blockers		
Calcium-channel blockers		
ACE inhibitors		
Angiotensin II receptor blockers		
Alpha-adrenergic blockers		

30. What are the most common nutritional implications of taking hydrochlorothiazide?

31. Mrs. Anderson's physician has decided to prescribe an ACE inhibitor and an HMGCoA reductase inhibitor (Zocor). What changes can be expected in her lipid profile as a result of taking these medications?

32. How does an ACE inhibitor work to lower blood pressure?

33. How does a HMGCoA reductase inhibitor work to lower serum lipid?

34. What other classes of medications can be used to treat hypercholesterolemia?

35. What are the pertinent drug–nutrient interactions and medical side effects for ACE inhibitors and HMGCoA?

36. From the information gathered within the clinical domain, list possible nutrition problems using the diagnostic term.

E. Behavioral–Environmental Domain

37. What are some possible barriers to compliance?

IV. Nutrition Diagnosis

38. Select two high-priority nutrition problems and complete PES statements for each.

V. Nutrition Intervention

39. Mrs. Anderson asks you, "A lot of my friends have lost weight on that Dr. Atkins diet. Would it be best for me to follow that for awhile to get this weight off?" What can you tell Mrs. Anderson about the typical high-protein, low-carbohydrate approach to weight loss?

40. When you ask Mrs. Anderson how much weight she would like to lose, she tells you she would like to weigh 125, which is what she weighed most of her adult life. Is this reasonable? What would you suggest as a goal for weight loss for Mrs. Anderson?

41. How quickly should Mrs. Anderson lose this weight?

42. For each of the PES statements that you have written, establish an ideal goal (based on the signs and symptoms) and an appropriate intervention (based on the etiology).

43. Identify the major sources of saturated fat and cholesterol in Mrs. Anderson's diet. What suggestions would you make for substitutions and/or other changes that would help Mrs. Anderson reach her medical nutrition therapy goals?

44. Assuming that the foods in her 24-hour recall are typical of her eating pattern, outline necessary modifications you could use as a teaching tool.

Foods	Modification/Alternative(s)	Rationale
Coffee (3 c/day)		
Oatmeal (w/margarine & sugar) or Frosted Mini-Wheats		
2% low-fat milk		
Orange juice		
Glazed donut		
Canned tomato soup		
Saltine crackers		
Diet cola		
12 oz bottle regular beer		
Baked chicken		
Baked potato (w 1 tbsp butter, salt, & pepper)		
Carrots		
Salad w/ranch-style dressing		
Ice cream		

45. What would you want to reevaluate in 3 to 4 weeks at a follow-up appointment?

46. Evaluate Mrs. Anderson's labs at 6 months and then at 9 months. Have the biochemical goals been met with the current regimen?

Bibliography

American Dietetic Association. *Nutrition Diagnosis and Intervention: Standardized Language for the Nutrition Care Process.* Chicago, IL: American Dietetic Association; 2007.

American Heart Association. High blood pressure. Available at: http://www.americanheart.org/presenter.jhtml?identifier=2114 Accessed March 25, 2008.

American Heart Association. AHA Guidelines for Primary Prevention of Cardiovascular Disease and Stroke: 2002 Update. Available at: http://circ.ahajournals.org/cgi/content/full/106/3/388. Accessed March 25, 2008.

American Heart Association. For professionals: Risk factors and coronary heart disease. Available at: http://www.americanheart.org/presenter.jhtml?identifier=539. Accessed March 25, 2008.

American Heart Association. For professionals: Metabolic Syndrome. Available at: http://www.americanheart.org/presenter.jhtml?identifier=534. Accessed March 25, 2008.

Cater NB. Plant sterol ester: Review of cholesterol-efficacy and implications for coronary heart disease reduction. *PrevCardiol.* 2000;3:121–130.

Lacey K. The nutrition care process. In: Nelms M, Sucher K, Long S. *Nutrition Therapy and Pathophysiology.* Belmont, CA: Thomson/Brooks-Cole; 2007:39–64.

National Heart, Lung, and Blood Institute; National Institutes of Health. The Practical Guide: Identification, Evaluation, and Treatment of Overweight and Obesity in Adults. Available at: http://www.nhlbi.nih.gov/guidelines/obesity/practgde.htm. Accessed March 25, 2008.

National Institutes of Health, National Heart, Lung, and Blood Institute. *Your Guide to Lowering Blood Pressure with DASH.* U.S. Dept. of Health and Human Services, NIH Publication No. 06-4082. Originally printed 1998, revised April 2006. Available at: http://www.nhlbi.nih.gov/health/public/heart/hbp/dash/new_dash.pdf. Accessed March 25, 2008.

Nelms MN. Assessment of nutrition status and risk. In: Nelms M, Sucher K, Long S. *Nutrition Therapy and Pathophysiology.* Belmont, CA: Thomson/Brooks-Cole; 2007:101–135.

Pronsky ZM. *Food and Medication Interactions.* 15th ed. Birchrunville, PA: Food–Medication Interactions; 2008.

Pujol TJ, Tucker JE. Diseases of the cardiovascular system. In: Nelms M, Sucker K, Long S. *Nutrition Therapy and Pathophysiology.* Belmont, CA: Thomson/Brooks-Cole; 2007:371–420.

Seventh Report of the Joint National Committee on Detection, Evaluation, and Treatment of High Blood Pressure (JNC7). Available at: http://www.nhlbi.nih.gov/guidelines/hypertension/jnc7full.htm. Accessed June 28, 2007.

Third Report of the Expert Panel on Detection, Evaluation, and Treatment of High Blood Cholesterol in Adults (Adult Treatment Panel III Update). Bethesda, MD: National Institutes of Health (2004). Available at: http://www.nhlbi.nih.gov/guidelines/cholesterol/index.htm. Accessed March 25, 2008.

Windhauser MM, Ernst DB, Karania NM, Crawford SW, Redican SE, Swain JF, Karimbakas JM, Champagne CM, Hoben KP, Evans MA. Translating the Dietary Approach to Stop Hypertension diet from research to practice: Dietary and behavior change techniques. DASH Collaborative Research Group. *J Am Diet Assoc.* 1999;99(8 Suppl):S90–S95.

Internet Resources

MedlinePlus: National Institutes of Health: Heart Diseases. http://www.nlm.nih.gov/medlineplus/heartdiseases.html

DASH Diet Eating Plan. http://dashdiet.org/

National Heart, Lung, and Blood Institute: Joint National Committee on Prevention, Detection Evaluation and Treatment of High Blood Pressure. http://www.nhlbi.nih.gov/guidelines/hypertension/

U.S. Department of Agriculture: Nutrient Data Laboratory. http://www.ars.usda.gov/ba/bhnrc/ndl

Case 7

Myocardial Infarction

Objectives

After completing this case, the student will be able to:

1. Describe the progression of atherosclerosis and its role in developing a myocardial infarction.
2. Identify and explain common nutritional problems associated with a myocardial infarction.
3. Interpret laboratory parameters for nutritional implications and significance.
4. Analyze nutrition assessment data to evaluate nutritional status and identify specific nutrition problems.
5. Determine nutrition diagnoses and write appropriate PES statements.
6. Develop a nutrition care plan with appropriate measurable goals, interventions, and strategies for monitoring and evaluation that addresses the nutrition diagnoses of this case.

Mr. Klosterman, a 61-year-old man, is admitted through the emergency room of University Hospital after experiencing a sudden onset of severe precordial pain on the way home from work. Mr. Klosterman is found to have suffered a myocardial infarction and is treated with an emergency angioplasty of the infarct-related artery.

Name: James Klosterman
DOB: 12/1 (age 61)
Physician: Regina H. Smith, MD

ADMISSION DATABASE

BED #	DATE:	TIME:	TRIAGE STATUS (ER ONLY):
1	3/25	1354	☐ Red ☐ Yellow ☐ Green ☐ White

Initial Vital Signs

TEMP:	RESP:	SAO$_2$:	
98.4	20	80	
HT: 5′10″	WT (lb): 185	B/P: 118/78	PULSE: 92
LAST TETANUS 2005		LAST ATE 1030	LAST DRANK 1100

PRIMARY PERSON TO CONTACT:
Name: Sally Klosterman
Home #: 404-555-9214
Work #: 404-555-1822

ORIENTATION TO UNIT: ☒ Call light ☒ Television/telephone ☒ Bathroom ☒ Visiting ☒ Smoking ☒ Meals ☒ Patient rights/responsibilities

CHIEF COMPLAINT/HX OF PRESENT ILLNESS

Severe unrelenting chest pain for past 1.5 hours

PERSONAL ARTICLES: (Check if retained/describe)
☐ Contacts ☐ R ☐ L ☐ Dentures ☐ Upper ☐ Lower
☐ Jewelry:
☐ Other:

ALLERGIES: Meds, Food, IVP Dye, Seafood: Type of Reaction

Sulfa-hives

VALUABLES ENVELOPE:
☐ Valuables instructions

INFORMATION OBTAINED FROM:
☒ Patient ☐ Previous record
☒ Family ☐ Responsible party

PREVIOUS HOSPITALIZATIONS/SURGERIES

none

Signature *Sally Klosterman*

Home Medications (including OTC) **Codes: A=Sent home** **B=Sent to pharmacy** **C=Not brought in**

Medication	Dose	Frequency	Time of Last Dose	Code	Patient Understanding of Drug
none					

Do you take all medications as prescribed? ☐ Yes ☐ No If no, why?

PATIENT/FAMILY HISTORY

☐ Cold in past two weeks
☐ Hay fever
☒ Emphysema/lung problems Patient
☐ TB disease/positive TB skin test
☒ Cancer Maternal grandmother
☐ Stroke/past paralysis
☐ Heart attack
☒ Angina/chest pain Patient
☐ Heart problems

☐ High blood pressure
☐ Arthritis
☐ Claustrophobia
☐ Circulation problems
☐ Easy bleeding/bruising/anemia
☐ Sickle cell disease
☐ Liver disease/jaundice
☐ Thyroid disease
☐ Diabetes

☐ Kidney/urinary problems
☐ Gastric/abdominal pain/heartburn
☐ Hearing problems
☐ Glaucoma/eye problems
☐ Back pain
☐ Seizures
☐ Other

RISK SCREENING

Have you had a blood transfusion? ☐ Yes ☒ No
Do you smoke? ☒ Yes ☐ No
If yes, how many pack(s)? 1/day for 40 years
Does anyone in your household smoke? ☒ Yes ☐ No
Do you drink alcohol? ☒ Yes ☐ No
If yes, how often? 1 drink/day How much?
When was your last drink? 3/24
Do you take any recreational drugs? ☐ Yes ☒ No
If yes, type:______ Route:
Frequency:______ Date last used:_____/_____/_____

FOR WOMEN Ages 12–52

Is there any chance you could be pregnant? ☐ Yes ☐ No
If yes, expected date (EDC):
Gravida/Para:

ALL WOMEN

Date of last Pap smear:
Do you perform regular breast self-exams? ☐ Yes ☐ No

ALL MEN

Do you perform regular testicular exams? ☒ Yes ☐ No

Additional comments:

× *Mark Little, RN*
Signature/Title

Client name: James Klosterman
DOB: 12/1
Age: 61
Sex: Male
Education: BS degree
Occupation: Lutheran minister
Hours of work: 40/week
Household members: Wife age 61
Ethnic background: German
Religious affiliation: Lutheran
Referring physician: Regina H. Smith, MD (internal medicine)

Chief complaint:
Severe, unrelenting precordial chest pain for the past 1.5 hours

Patient history:
Onset of disease: 61-year-old male who noted the sudden onset of severe precordial pain on the way home from work. The pain is described as pressure-like pain radiating to the jaw and left arm. The patient has noted an episode of emesis and nausea. He denies palpitations or syncope. He denies prior history of pain. He admits to smoking cigarettes (1 pack/day for 40 years). He denies hypertension, diabetes, or high cholesterol. He denies SOB.
Type of Tx: Hospitalization, emergency coronary angiography with angioplasty of infarct-related artery, coronary care unit, rhythm monitoring, bed rest, sequential cardiograms, and cardiac enzymes
PMH: Surgery; cholecystectomy 10 years ago, appendectomy 30 years ago
Meds: None. *Allergies:* Sulfa drugs
Smoker: Yes—40 years, 1 pack per day
Family Hx: What? CAD *Who?* Father—MI age 59

Physical exam:
General appearance: Mildly overweight male in acute distress from chest pain
Vitals: Temp 98.4°F, BP 118/78 mm Hg, HR 92 bpm, RR 20 bpm
Heart: PMI 5 ICS MCL focal. S1 normal intensity. S2 normal intensity and split. S4 gallop at the apex. No murmurs, clicks, or rubs.
HEENT:
Head: Normocephalic
Eyes: EOMI, fundoscopic exam WNL. No evidence of atherosclerosis, diabetic retinopathy, or early hypertensive changes.
Ears: TM normal bilaterally
Nose: WNL
Throat: Tonsils not infected, uvula midline, gag normal
Genitalia: Grossly physiologic
Neurologic: No focal localizing abnormalities. DTR symmetric bilaterally.
Extremities: No C, C, E
Skin: Diaphoretic and pale
Chest/lungs: Lungs clear to auscultation and percussion
Peripheral vascular: PPP

Abdomen: RLQ scar and midline suprapubic scar. BS WNL. No hepatomegaly, splenomegaly, masses, inguinal lymph nodes, or abdominal bruits.
Height/Weight: 70″, 185 lbs

Nutrition Hx:
General: Appetite good. Has been trying to change some things in his diet. Wife indicates that she has been using "corn oil" instead of butter and has tried not to fry foods as often.

24-hour recall:

Breakfast:	None
Midmorning snack:	1 large cinnamon raisin bagel with 1 tbsp fat-free cream cheese, 8 oz orange juice, coffee
Lunch:	1 c canned vegetable beef soup,; sandwich with 4 oz roast beef, lettuce, tomato, dill pickles, 2 tsp mayonnaise; 1 small apple; 8 oz 2% milk
Dinner:	2 lean pork chops (3 oz each), 1 large baked potato, 2 tsp margarine, ½ c green beans, ½ c coleslaw (cabbage with 1 tbsp salad dressing), 1 sl apple pie
Snack:	8 oz 2% milk, 1 oz pretzels

Food allergies/intolerances/aversions: None
Previous nutrition therapy? Yes *If yes, when:* Last year *Where?* Community dietitian
Food purchase/preparation: Spouse
Vit/min intake: None

Tx plan:
IV heparin 5,000 units bolus followed by 1,000 unit/hour continuous infusion with a PTT at 2 × control
Chewable aspirin 160 mg PO and continued every day
Lopressor 50 mg twice daily
Lidocaine prn
NPO until procedure completed
Type and cross for 6 units of packed cells

Hospital course:
Patient's chest pain resolved after two sublingual NTG at 3-minute intervals and 2 mg of IV morphine. In the cath lab, the patient was found to have a totally occluded distal right coronary artery and a 70% occlusion in the left circumflex coronary artery. The left anterior descending was patent. Angioplasty of the distal right coronary artery resulted in a patent infarct-related artery with near normal flow. A stent was left in place to stabilize the patient and limit infarct size. Left ventricular ejection fraction was normal at 42%, and a posterobasilar scar was present with hypokinesis. A consult was made for nutrition counseling and for referral to cardiac rehabilitation. Patient was discharged on the following medications: Lopressor 50 mg every day; lisinopril 10 mg every day; Nitro-Bid 9 mg twice daily; NTG, 0.4 mg sublingually prn for chest pain; ASA 81 mg every day; Lipitor 10 mg every day at bedtime.

UH UNIVERSITY HOSPITAL

NAME: James Klosterman
AGE: 61
PHYSICIAN: Regina H. Smith, MD

DOB: 12/1
SEX: M

CHEMISTRY

DAY:		Day 1	Day 2	Day 3	
DATE:					
TIME:					
LOCATION:					
	NORMAL				UNITS
Albumin	3.5-5	4.2	4.3	4.2	g/dL
Total protein	6-8	6.0	5.9 L	6.1	g/dL
Prealbumin	16-35	30	32	31	mg/dL
Transferrin	250-380 (women) 215-365 (men)	250	240	260	mg/dL
Sodium	136-145	141	142	138	mEq/L
Potassium	3.5-5.5	4.2	4.1	3.9	mEq/L
Chloride	95-105	103	102	100	mEq/L
PO_4	2.3-4.7	3.1	3.2	3.0	mg/dL
Magnesium	1.8-3	2.0	2.3	2.0	mg/dL
Osmolality	285-295	292	290	291	mmol/kg/H_2O
Total CO_2	23-30	20 L	24	26	mEq/L
Glucose	70-110	136 H	106	104	mg/dL
BUN	8-18	14	15	13	mg/dL
Creatinine	0.6-1.2	1.1	1.1	1.1	mg/dL
Uric acid	2.8-8.8 (women) 4.0-9.0 (men)	7.0	6.8	6.6	mg/dL
Calcium	9-11	9.4	9.4	9.4	mg/dL
Bilirubin	≤ 0.3	0.1	0.1	0.2	mg/dL
Ammonia (NH_3)	9-33	26	22	25	μmol/L
ALT	4-36	30	215 H	185 H	U/L
AST	0-35	25	245 H	175 H	U/L
Alk phos	30-120	150	145	140	U/L
CPK	30-135 (women) 55-170 (men)	75	500 H	335 H	U/L
CPK-MB	0	0	75 H	55 H	U/L
LDH	208-378	325	685 H	365	U/L
CHOL	120-199	235 H	226 H	214 H	mg/dL
HDL-C	> 55 (women) > 45 (men)	30 L	32 L	33 L	mg/dL
VLDL	7-32	45	44	40	mg/dL
LDL	< 130	160 H	150 H	141 H	mg/dL
LDL/HDL ratio	< 3.22 (women) < 3.55 (men)	5.3 H	4.7 H	4.3 H	
Apo A	101-199 (women) 94-178 (men)	72 L	80 L	98 L	mg/dL
Apo B	60-126 (women) 63-133 (men)	115	110	105	mg/dL
Troponin I	< 0.2	2.4 H	2.8 H		ng/dL
Troponin T	< 0.03	2.1 H	2.7 H		ng/dL
TG	35-135 (women) 40-160 (men)	150	140	130	mg/dL
T_4	4-12	7.6	7.8	7.4	mcg/dL
T_3	75-98	85	87	88	mcg/dL
HbA_{1C}	3.9-5.2	6.5	6.3	6.0	%

UH UNIVERSITY HOSPITAL

NAME: James Klosterman DOB: 12/1
AGE: 61 SEX: M
PHYSICIAN: Regina H. Smith, MD

HEMATOLOGY

DAY:		Day 1	Day 2	Day 3	
DATE:					
TIME:					
LOCATION:					
	NORMAL				UNITS
WBC	4.8–11.8	11,000	9,320	8,800	× 10^3/mm^3
RBC	4.2–5.4 (women) 4.5–6.2 (men)	4.7	4.75	4.68	× 10^6/mm^3
HGB	12–15 (women) 14–17 (men)	15	14.8	14.4	g/dL
HCT	37–47 (women) 40–54 (men)	45	45	44	%
MCV	80–96	91	92	90	μm^3
RETIC	0.8–2.8				%
MCH	26–32	30	31	30	pg
MCHC	31.5–36	33	32	33	g/dL
RDW	11.6–16.5	13.2	12.8	13.0	%
Plt Ct	140–440	320	295	280	× 10^3/mm^3
Diff TYPE					
ESR	0–25 (women) 0–15 (men)				mm/hr
% GRANS	34.6–79.2	86 H	80 H	78	%
% LYM	19.6–52.7	14	20	22	%
SEGS	50–62	84 H	80 H	78 H	%
BANDS	3–6	2 L	0 L	0 L	%
LYMPHS	24–44	14 L	20 L	22 L	%
MONOS	4–8	0 L	0 L	0 L	%
EOS	0.5–4	0.5	0.5	0.5	%
Ferritin	20–120 (women) 20–300 (men)	190	208	196	mg/mL
ZPP	30–80				μmol/mol
Vitamin B_{12}	24.4–100	75	76	78	ng/dL
Folate	5–25	8	9	11	μg/dL
Total T cells	812–2,318	1,600	1,600	1,600	mm^3
T-helper cells	589–1,505	900	900	900	mm^3
T-suppressor cells	325–997	400	400	400	mm^3
PT	11–16	12.6	12.6	12.4	sec

UH UNIVERSITY HOSPITAL

NAME: James Klosterman DOB: 12/1
AGE: 61 SEX: M
PHYSICIAN: Regina H. Smith, MD

*****URINALYSIS*****

DAY:		1	2	3	
DATE:					
TIME:					
LOCATION:					
	NORMAL				UNITS
Coll meth		First morning	First morning	First morning	
Color		Pale yellow	Pale yellow	Pale yellow	
Appear		Clear	Clear	Clear	
Sp grv	1.003–1.030	1.020	1.015	1.018	
pH	5–7	5.8	5.0	6	
Prot	NEG	NEG	NEG	NEG	mg/dL
Glu	NEG	NEG	NEG	NEG	mg/dL
Ket	NEG	Trace	NEG	NEG	
Occ bld	NEG	NEG	NEG	NEG	
Ubil	NEG	NEG	NEG	NEG	
Nit	NEG	NEG	NEG	NEG	
Urobil	< 1.1	NEG	NEG	Trace	EU/dL
Leu bst	NEG	NEG	NEG	NEG	
Prot chk	NEG	NEG	NEG	NEG	
WBCs	0–5	0	0	0	/HPF
RBCs	0–5	0	0	0	/HPF
EPIs	0	0	0	0	/LPF
Bact	0	0	0	0	
Mucus	0	0	0	0	
Crys	0	0	0	0	
Casts	0	0	0	0	/LPF
Yeast	0	0	0	0	

Case Questions

I. Understanding the Disease and Pathophysiology

1. Mr. Klosterman had a myocardial infarction. Explain what happened to his heart.

2. Mr. Klosterman was treated with an angioplasty and stent placement. Explain this medical procedure and its purpose.

3. What risk factors indicated in his medical record can be addressed through nutrition therapy?

4. Mr. Klosterman and his wife are concerned about the future of his heart health. What role does cardiac rehabilitation play in his return to normal activities and in determining his future heart health?

II. Understanding the Nutrition Therapy

5. Are there any current recommendations for nutritional intake during a hospitalization following a myocardial infarction?

III. Nutrition Assessment

A. Evaluation of Weight/Body Composition

6. What is the healthy weight range for an individual of Mr. Klosterman's height?

B. Calculation of Nutrient Requirements

7. This patient is a Lutheran minister. He does get some exercise daily. He walks his dog outside for about 15 minutes at a leisurely pace.

   a. Calculate his energy need.

   b. How many grams of protein should he have daily?

C. Intake Domain

8. Using Mr. Klosterman's 24-hour recall, calculate the total number of calories he consumed as well as the energy distribution of calories for protein, carbohydrate, and fat using the exchange system.

9. From the information gathered within the intake domain, list possible nutrition problems using the diagnostic term.

D. Clinical Domain

10. Examine the chemistry results for Mr. Klosterman. Which labs are consistent with the MI diagnosis? Explain.

11. Why were the levels higher on day 2?

12. What is abnormal about his lipid profile? Indicate the abnormal levels.

13. Mr. Klosterman was prescribed the following medications on discharge. What are the food–medication interactions for this list of medications?

Medication	Possible Food–Nutrient Interactions
Lopressor 50 mg daily	
Lisinopril 10 mg daily	
Nitro-Bid 9.0 mg twice daily	
NTG 0.4 mg sl prn chest pain	
ASA 81 mg daily	

14. From the information gathered within the clinical domain, list possible nutrition problems using the diagnostic term.

E. Behavioral–Environmental Domain

15. You talk with Mr. Klosterman and his wife, a math teacher at the local high school. They are friendly and seem cooperative. They are both anxious to learn what they can do to prevent another heart attack. What questions will you ask them to assess how to best help them?

16. What other issues might you consider to support the success of his lifestyle change?

17. From the information gathered within the behavioral–environmental domain, list possible nutrition problems using the diagnostic term.

IV. Nutrition Diagnosis

18. Select two high-priority nutrition problems and complete PES statements for each.

V. Nutrition Intervention

19. For each of the PES statements that you have written, establish an ideal goal (based on the signs and symptoms) and an appropriate intervention (based on the etiology).

20. Mr. Klosterman and his wife ask about supplements. "My roommate here in the hospital told me I should be taking vitamin E and—I think it was folate along with omega-3 fatty acid supplements." What does the research say about vitamin E, folate, and omega-3 fatty acid supplementation for this patient?

VI. Nutrition Monitoring and Evaluation

21. What would you want to assess in 3 to 4 weeks when he and his wife return for additional counseling?

Bibliography

Dargie H. Myocardial infarction: Redefined or reinvented? [editorial]. *Heart.* 2002;88(1):1–3.

Hemilä H, Miller ER. Evidenced-based medicine and vitamin E supplementation. *Am J Clin Nutr.* 2007;86:261–262.

Kris-Etherton PM, Harris WS, Appel LJ; American Heart Association Nutrition Committee. Fish consumption, fish oil, omega-3 fatty acids, and cardiovascular disease. *Circulation.* 2003;23:e20–e30.

Leaf A, Albert CM, Josephson M, Steinhaus D, Kluger J, Kang JX, Cox B, Zhang H, Schoenfeld D. Prevention of fatal arrhythmias in high-risk subjects by fish oil n-3 fatty acid intake. *Circulation.* 2005;112(18):2762–2768.

Lindeboom JE, Shahin GMM, Bosker HA. Images in cardiology: Myocardial infarction and coronary thrombosis. *Heart.* 2002;88(1):60.

Meier MA, Al-Badr WH, Cooper JV, Kline-Rogers EM, Smith DE, Eagle KA, Mehta RH. The new definition of myocardial infarction: Diagnostic and prognostic implications in patients with acute coronary syndromes. *Arch Intern Med.* 2002;162(14):1585–1589.

Miller ER, Pastor-Barriuso R, Dalal D, Riemersma RA, Appel LJ, Guallar E. Meta-analysis: High-dosage vitamin E supplementation may increase all-cause mortality. *Ann Intern Med.* 2005;142:37–46.

National Heart, Lung, and Blood Institute. *Third Report of the Expert Panel on Detection, Evaluation, and Treatment of High Blood Cholesterol in Adults (Adult Treatment Panel III).* Bethesda, MD: National Institutes of Health; 2002. Available at: http://www.nhlbi.nih.gov/guidelines/cholesterol/index.htm. Accessed June 28, 2007.

Pronsky Z. *Food Medication Interactions.* 15th ed. Pottstown, PA: Food Medication Interactions, 2008.

Pujol TJ, Tucker JE. Diseases of the cardiovascular system. In: Nelms M, Sucker K, Long S. *Nutrition Therapy and Pathophysiology.* Belmont, CA: Thomson/Brooks-Cole; 2007:371–420.

SoS Investigators. Coronary artery bypass surgery versus percutaneous coronary intervention with stent implantation in patients with multivessel coronary artery disease (the Stent or Surgery trial): A randomized controlled trial. *Lancet.* 2002;360(i9338):965–970.

Wang C, Harris WS, Chung M, Lichtenstein AH, Balk EM, Kupelnick B, Jordan HS, Lau J. n-3 Fatty acids from fish or fish-oil supplements, but not alpha-linolenic acid, benefit cardiovascular disease outcomes in primary- and secondary-prevention studies: A systematic review. *Am J Clin Nutr.* 2006;84(1):5–17.

Internet Resources

American Heart Association. http://www.americanheart.org

eMedicine: Myocardial Infarction. http://www.emedicine.com/EMERG/topic327.htm

MedLinePlus: High Blood Pressure. http://www.nlm.nih.gov/medlineplus/highbloodpressure.html

Medscape: Medscape Cardiology. http://www.medscape.com/patiented/acutemi

Merck Manuals Online Library: Acute Coronary Syndromes (ACS). http://www.merck.com/mmpe/sec07/ch073/ch073c.html#sec07-ch073-ch073c-619

National Heart, Lung, and Blood Institute: The Aspirin Myocardial Infarction Study (AMIS). http://www.nhlbi.nih.gov/resources/deca/descriptions/amis.htm

Cleveland Clinic: Acute Myocardial Infarction. http://clevelandclinicmeded.com/diseasemanagement/cardiology/acutemi/acutemi.htm

Internet Pathology Laboratory Webpath: Myocardial Infarction. http://library.med.utah.edu/WebPath/TUTORIAL/MYOCARD/MYOCARD.html

U.S. Department of Agriculture: Nutrient Data Laboratory. http://www.ars.usda.gov/ba/bhnrc/ndl

WebMD. http://www.webmd.com

Case 8

Congestive Heart Failure with Resulting Cardiac Cachexia

Objectives

After completing this case, the student will be able to:

1. Use nutrition assessment information to determine baseline nutritional status.
2. Correlate a patient's signs and symptoms with the pathophysiology of congestive heart failure.
3. Evaluate laboratory indices for nutritional implications and significance.
4. Demonstrate understanding of nutrition support options for congestive heart failure.
5. Identify the roles of pharmacologic intervention and drug–nutrient interactions.
6. Determine appropriate nutritional interventions for the patient with congestive heart failure and cardiac cachexia.

Dr. Charles Peterman, an 85-year-old retired physician, is admitted with acute symptoms related to his congestive heart failure. Dr. Peterman has a long history of cardiac disease, including a previous myocardial infarction and mitral valve disease.

ADMISSION DATABASE

Name: Charles Peterman
DOB: 4/15 (age 85)
Physician: Douglas A. Schmidt, MD

BED #	DATE:	TIME:	TRIAGE STATUS (ER ONLY):
2	3/31	1600	☒ Red ☐ Yellow ☐ Green ☐ White

Initial Vital Signs

TEMP:	RESP:	SAO$_2$:	
101	25	80%	
HT:	**WT (lb):**	**B/P:**	**PULSE:**
5′10″	165	90/70	101
LAST TETANUS		**LAST ATE**	**LAST DRANK**
35 years ago		lunch	1400

PRIMARY PERSON TO CONTACT:
Name: Jean Peterman
Home #: 555-561-8556
Work #: N/A

ORIENTATION TO UNIT: ☒ Call light ☒ Television/telephone ☒ Bathroom ☒ Visiting ☒ Smoking ☒ Meals ☒ Patient rights/responsibilities

CHIEF COMPLAINT/HX OF PRESENT ILLNESS

Passed out–85-year-old male with chronic heart failure

ALLERGIES: Meds, Food, IVP Dye, Seafood: Type of Reaction

Shellfish, aspirin, ibuprofen–hives

PREVIOUS HOSPITALIZATIONS/SURGERIES

1970 acute diverticulitis

PERSONAL ARTICLES: (Check if retained/describe)
☐ Contacts ☐ R ☐ L ☐ Dentures ☐ Upper ☐ Lower
☒ Jewelry: wedding band
☐ Other:

VALUABLES ENVELOPE:
☐ Valuables instructions

INFORMATION OBTAINED FROM:
☐ Patient ☐ Previous record
☒ Family ☐ Responsible party

Signature *Jean Peterman*

Home Medications (including OTC) **Codes: A=Sent home** **B=Sent to pharmacy** **C=Not brought in**

Medication	Dose	Frequency	Time of Last Dose	Code	Patient Understanding of Drug
Lanoxin	0.125 mg	once daily	8 AM	C	yes
Lasix	80 mg	twice daily	5 PM	C	yes
lisinopril	30 mg	once daily	8 AM	C	yes
Centrum Silver	2 tablets	once daily	8 AM	C	yes
Lopressor	25 mg	once daily	8 AM	C	yes
Zocor	20 mg	once daily	9 PM	C	yes
calcium carbonate	500 mg	twice daily	5 PM	C	yes
Metamucil	1 tbsp	twice daily	6 PM	C	yes
Aldactone	25 mg	once daily	8 AM	C	yes

Do you take all medications as prescribed? ☒ Yes ☐ No If no, why?

PATIENT/FAMILY HISTORY

☐ Cold in past two weeks
☒ Hay fever Patient
☐ Emphysema/lung problems
☐ TB disease/positive TB skin test
☐ Cancer
☐ Stroke/past paralysis
☒ Heart attack Patient
☐ Angina/chest pain
☒ Heart problems Patient
☒ High blood pressure Patient
☒ Arthritis Patient
☐ Claustrophobia
☐ Circulation problems
☐ Easy bleeding/bruising/anemia
☐ Sickle cell disease
☐ Liver disease/jaundice
☐ Thyroid disease
☐ Diabetes
☒ Kidney/urinary problems Patient
☒ Gastric/abdominal pain/heartburn Patient
☒ Hearing problems Patient
☐ Glaucoma/eye problems
☒ Back pain Patient
☐ Seizures
☐ Other

RISK SCREENING

Have you had a blood transfusion? ☐ Yes ☒ No
Do you smoke? ☐ Yes ☒ No
If yes, how many pack(s)?
Does anyone in your household smoke? ☐ Yes ☒ No
Do you drink alcohol? ☐ Yes ☒ No
If yes, how often? How much?
When was your last drink? ______/______/______
Do you take any recreational drugs? ☐ Yes ☒ No
If yes, type:______ Route:
Frequency:______ Date last used:______/______/______

FOR WOMEN Ages 12–52

Is there any chance you could be pregnant? ☐ Yes ☐ No
If yes, expected date (EDC):
Gravida/Para:

ALL WOMEN

Date of last Pap smear:
Do you perform regular breast self-exams? ☐ Yes ☐ No

ALL MEN

Do you perform regular testicular exams? ☒ Yes ☐ No

Additional comments:

x *Samuel Layton, RN*
Signature/Title

Client name: Charles Peterman
DOB: 4/15
Age: 85
Sex: Male
Education: Postgraduate
Occupation: Physician
Hours of work: Retired
Household members: Wife age 82, in good health
Ethnic background: Caucasian
Religious affiliation: Presbyterian
Referring physician: Douglas A. Schmidt, MD (cardiology)

Chief complaint:
Patient collapsed at home and was brought to the emergency room by ambulance.

Patient history:
Onset of disease: CHF × 2 years
Type of Tx: Medical Tx of CAD, HTN, and CHF
PMH: Long-standing history of CAD, HTN, mitral valve insufficiency, previous anterior MI
Meds: Lanoxin 0.125 mg once daily, Lasix 80 mg twice daily, Aldactone 25 mg once daily, lisinopril 30 mg po once daily, Lopressor 25 mg once daily, Zocor 20 mg once daily, Metamucil 1 tbsp twice daily, calcium carbonate 500 mg twice daily, Centrum 2 tablets once daily
Smoker: No
Family Hx: What? HTN, CAD *Who?* Parents

Physical exam:
General appearance: Elderly male in acute distress
Vitals: Temp 98°F, pulse 110, RR 24 bpm, BP 90/70 mm Hg
Heart: Diffuse PMI in AAL in LLD; Grade II holosystolic murmur at the apex radiating to the left sternal border; first heart sound diminished, and second heart sound preserved; third heart sound present
Skin: Gray, moist
HEENT:
Eyes: Ophthalmoscopic exam reveals AV crossing changes and arteriolar spasm
Ears: WNL
Nose: WNL
Throat: Jugular venous distension in sitting position with a positive hepatojugular reflux
Chest/lungs: Rales in both bases posteriorly
Abdomen: Ascites, no masses, liver tender to A&P
Genitalia: WNL
Extremities: 4+ pedal edema
Peripheral vascular: WNL
Neurologic: WNL
Height/Weight: Admission 70″, 165 lbs

Nutrition Hx:
General: Wife reports that Dr. Peterman's appetite has been poor for the last 6 months, with no real weight loss that she can determine. "It's very hard to know the difference between his real weight and any fluid that he is retaining." She describes difficulty eating due to SOB and nausea.
Usual dietary intake: Generally likes all foods but has recently been eating only soft foods, especially ice cream. Tries to drink 2 cans of Ensure Plus each day.
24-hour recall: Has had only sips of drinks for the past 24 hours
Food allergies/intolerances/aversions: Shellfish
Previous nutrition therapy: Not specifically but has monitored salt intake for the past 2 years as well as a low-fat, low-cholesterol diet for at least the previous 10 years
Food purchase/preparation: Spouse
Vit/min intake: Centrum Silver 2 ×/day, calcium supplement 1,000 mg/day

Dx:
CHF with ascites and 4+ pedal edema

Tx plan:
Admit to CCU; parenteral dopamine and IV diuretics; 100 mg thiamin IV; telemetry, vitals every 1 hour × 8, every 2 hours × 8 for first 24 hours; daily ECG and chest X-rays; echocardiogram; Chem 24, urinalysis, strict I & Os

Hospital course:
Swan-Ganz catheter inserted. Echocardiogram indicated severe cardiomegaly secondary to end-stage congestive heart failure. Enteral feeding initiated but discontinued due to severe diarrhea. Patient had a living will that stated he wanted no other extraordinary measures taken to prolong his life.
The patient was able to express his wishes verbally: he requested oral feedings and palliative care only. Patient expired after 2-week hospitalization.

UH UNIVERSITY HOSPITAL

NAME: Charles Peterman DOB: 4/15
AGE: 85 SEX: M
PHYSICIAN: Douglas A. Schmidt, MD

CHEMISTRY

DAY:		1	3	7	
DATE:		3/31	4/2	4/6	
TIME:					
LOCATION:					
	NORMAL				UNITS
Albumin	3.5–5	2.8 L	2.7 L	2.6 L	g/dL
Total protein	6–8	5.8 L	5.6 L	5.5 L	g/dL
Prealbumin	16–35	15 L	11 L	10 L	mg/dL
Transferrin	250–380 (women) 215–365 (men)	350	355	352	mg/dL
Sodium	136–145	132 L	133 L	133 L	mEq/L
Potassium	3.5–5.5	3.7	3.6	3.8	mEq/L
Chloride	95–105	98	100	99	mEq/L
PO_4	2.3–4.7	4.0	3.8	3.6	mg/dL
Magnesium	1.8–3	2.0	1.9	1.8	mg/dL
Osmolality	285–295	292	299	290	mmol/kg/H_2O
Total CO_2	23–30	26	24	25	mEq/L
Glucose	70–110	110	106	102	mg/dL
BUN	8–18	32 H	34 H	30 H	mg/dL
Creatinine	0.6–1.2	1.6 H	1.7 H	1.5 H	mg/dL
Uric acid	2.8–8.8 (women) 4.0–9.0 (men)	6.0	6.4 H	6.7 H	mg/dL
Calcium	9–11	9.0	8.8	8.9	mg/dL
Bilirubin	≤ 0.3	1.0	1.1	0.9	mg/dL
Ammonia (NH_3)	9–33	32	30	34	µmol/L
ALT	4–36	100 H	120 H	115 H	U/L
AST	0–35	70 H	80 H	85 H	U/L
Alk phos	30–120	200	190	200	U/L
CPK	30–135 (women) 55–170 (men)	150 H	175 H	200 H	U/L
LDH	208–378	350	450	556	U/L
CHOL	120–199	150	162	149	mg/dL
HDL-C	> 55 (women) > 45 (men)	30 L	31 L	30 L	mg/dL
VLDL	7–32	40	42	39	mg/dL
LDL	< 130	180 H	160 H	152 H	mg/dL
LDL/HDL ratio	< 3.22 (women) < 3.55 (men)	6 H	5.2 H	5.1 H	
Apo A	101–199 (women) 94–178 (men)	60 L	65 L	70 L	mg/dL
Apo B	60–126 (women) 63–133 (men)	140 H	138 H	136 H	mg/dL
TG	35–135 (women) 40–160 (men)	150	145	140	mg/dL
T_4	4–12	8.0	7.8	7.6	mcg/dL
T_3	75–98	160	156	150	mcg/dL
HbA_{1C}	3.9–5.2	6.8			%

UH UNIVERSITY HOSPITAL

NAME: Charles Peterman DOB: 4/15
AGE: 85 SEX: M
PHYSICIAN: Douglas A. Schmidt, MD

HEMATOLOGY

	NORMAL				UNITS
DAY:		1	3	7	
DATE:		3/31	4/2	4/6	
TIME:					
LOCATION:					
WBC	4.8-11.8	11 H	10.5 H	9.8	$\times 10^3/mm^3$
RBC	4.2-5.4 (women) 4.5-6.2 (men)	5.5	6.5 H	6.4 H	$\times 10^6/mm^3$
HGB	12-15 (women) 14-17 (men)	14	14.3	14.5	g/dL
HCT	37-47 (women) 40-54 (men)	41	42	42	%
MCV	80-96	90	89	91	μm^3
RETIC	0.8-2.8	0.9	1.1	1.0	%
MCH	26-32	31	31	30	pg
MCHC	31.5-36	33	34	32	g/dL
RDW	11.6-16.5	12	13	12	%
Plt Ct	140-440	300	290	310	$\times 10^3/mm^3$
Diff TYPE					
ESR	0-25 (women) 0-15 (men)	11	10	11	mm/hr
% GRANS	34.6-79.2	76	82 H	72	%
% LYM	19.6-52.7	24	18 L	28	%
SEGS	50-62	65 H	73 H	66 H	%
BANDS	3-6	11 H	9 H	6	%
LYMPHS	24-44	20 L	17 L	26	%
MONOS	4-8	20 L	17 L	26	%
EOS	0.5-4	4	1 L	2 L	%
Ferritin	20-120 (women) 20-300 (men)	100	96	98	mg/mL
ZPP	30-80				μmol/mol
Vitamin B_{12}	24.4-100	32	40	41	ng/dL
Folate	5-25	10	8	12	μg/dL
Total T cells	812-2,318	1,000	1,100	1,200	mm^3
T-helper cells	589-1,505	800	860	840	mm^3
T-suppressor cells	325-997	460	440	500	mm^3
PT	11-16	12.2	12.3	12.3	sec

Case Questions

I. Understanding the Disease and Pathophysiology

1. Outline the typical pathophysiology of heart failure. The onset of heart failure usually can be traced to damage from an MI and atherosclerosis. Is this consistent with Dr. Peterman's history? Relate this to your discussion of the pathophysiology.

2. Identify the specific signs and symptoms in the patient's physical examination that are consistent with heart failure. For any three of these signs and symptoms, narratively connect them to the physiological changes that you described in question 1.

3. Heart failure is often described as R-sided failure or L-sided failure. What is the difference? How are the clinical manifestations different?

4. Dr. Peterman's admitting diagnosis was cardiac cachexia. What is cardiac cachexia? What are the characteristic symptoms? Explain the role of the underlying heart disease in the development of cardiac cachexia.

II. Understanding the Nutrition Therapy

5. Dr. Peterman's wife states that they have monitored their salt intake for several years. What is the role of sodium restriction in the treatment of heart failure? What level of sodium restriction is recommended for the outpatient with heart failure?

6. Should he be placed on a fluid restriction? If so, how would this assist with the treatment of his heart failure? What specific foods are typically "counted" as a fluid?

7. Identify any common nutrient deficiencies that are found in patients with heart failure.

III. Nutrition Assessment

A. Evaluation of Weight/Body Composition

8. Identify factors that would affect the interpretation of Dr. Peterson's weight and body composition. How does Dr. Peterson's weight change during the first week of his hospitalization?

B. Calculation of Nutrient Requirements

9. Calculate Dr. Peterman's energy and protein requirements. Explain your rationale for the weight you have used in your calculation.

10. Calculate Dr. Peterman's fluid requirements.

C. Intake Domain

11. Dr. Peterman was started on an enteral feeding when he was admitted to the hospital. Outline a nutrition therapy regimen for him that includes formula choice, total volume, and goal rate.

12. List the possible nutrition problems within the intake domain using the correct diagnostic term.

D. Clinical Domain

13. Identify any abnormal biochemical values and assess them using the following table.

Parameter	Normal Value	Patient's Value	Reason for Abnormality	Nutrition Implication

14. The following chart lists the drugs/supplements that were prescribed for Dr. Peterman. Give the rationale for the use of each. In addition, describe any nutrition implications for these medications.

Medication	Rationale for Use	Nutrition Implications
Lanoxin		
Lasix		
Dopamine		
Thiamin		

15. Identify possible nutrition problems within the clinical domain using the correct diagnostic term.

IV. Nutrition Diagnosis

16. Select two high-priority nutrition problems and complete the PES statement for each.

V. Nutrition Intervention

17. Dr. Peterman was not able to tolerate the enteral feeding because of diarrhea. What recommendations could be made to improve tolerance to the tube feeding?

18. The tube feeding was discontinued because of continued problems. Parenteral nutrition cannot be considered at this time because of the need to severely restrict fluid. What recommendations could you make to optimize Dr. Peterman's oral intake?

19. This patient had a living will that expressed his wishes regarding life support measures and requested palliative care only. What is a living will? What is palliative care?

20. Dr. Peterman is not receiving parenteral or enteral nutritional support. What is the role of the registered dietitian during palliative care?

Bibliography

Akahsi YJ, Springer J, Anker SD. Cachexia in chronic heart failure: Prognostic implications and novel therapeutic approaches. *Curr Heart Fail Rep.* 2005;2198–2203.

Berger MM, Mustafa I. Metabolic and nutritional support in acute cardiac failure. *Curr Opin Clin Nutr Metab Care.* 2003;6:195–201.

Gallagher-Allred C. *Nutritional Care of the Terminally Ill.* Rockville, MD: Aspen; 1989.

Gomberg-Maitland M, Baran DA, Fuster V. Treatment of congestive heart failure: Guidelines for the primary care physician and the health failure specialist. *Arch Intern Med.* 2001;161:342–352.

Hunt SA, Abraham WT, Chin MH, Feldman AM, Francis GS, Ganiats TG, Jessup ML, Konstam MA, Mancini DM, Michl K, Oates JA, Rahko PS, Silver MA, Stevenson LW. ACC/AHA guideline update for the diagnosis and management of chronic heart failure in the adult: A report of the American College of Cardiology/American Heart Association Task Force on Practice Guidelines. *Circulation.* 2005;105:e154–e234.

Kuehneman T, Saulsbury D, Splett P, Chapman DB. Demonstrating the impact of nutrition intervention in a heart failure program. *J Am Diet Assoc.* 2002;102:1790–1794.

Lainscak M, Cleland JG, Lenzen MJ, Keber I, Goode K, Follath F, Komajda M, Swedberg K. Nonpharmacologic measures and drug compliance in patients with heart failure: Data from the EuroHeart Failure Survey. *Am J Cardiol.* 2007;99(6B):31D–37D.

Lennie TA, Moser DK, Heo S, Chung ML, Zambroski CH. Factors influencing food intake in patients with heart failure: A comparison with healthy elders. *J Cardiovasc Nurs.* 2006;21(2):123–129.

Pasini E, Opasich C, Pastoris O, Aquilani R. Inadequate nutritional intake for daily life activity of clinically stable patients with chronic heart failure. *Am J Cardiol.* 2004;93:41A–43A.

Price RJ, Witham MD, McMurdo ME. Defining the nutritional status and dietary intake of older heart failure patients. *Eur J Cardiovasc Nurs.* 2007;6(3):178–183.

Pujol TJ, Tucker J. Diseases of the Cardiovascular System. In: Nelms M, Sucher K, Long S. *Nutrition Therapy and Pathophysiology.* Belmont, CA: Thomson/Brooks-Cole; 2007:371–420.

Riegel B, Moser DK, Powell M, Rector TS, Havranek EP. Nonpharmacologic care by heart failure experts. *J Card Fail.* 2006;12(2):149–153.

Internet Resources

American Dietetic Association: ADA Nutrition Care Manual (by subscription). http://www.nutritioncaremanual.org

American Heart Association: Heart Failure. http://www.americanheart.org/presenter.jhtml?identifier=1486

Heart Failure Online. http://www.heartfailure.org

Heart Failure Society of America. http://www.hfsa.org/

MedlinePlus, National Library of Medicine, National Institutes of Health: Heart Failure. http://www.nlm.nih.gov/medlineplus/ency/article/000158.htm

Unit Three

NUTRITION THERAPY FOR GASTROINTESTINAL DISORDERS

The six cases presented in this section cover a wide array of diagnoses that ultimately affect normal digestion and absorption. These conditions use medical nutrition therapy as a cornerstone for their treatment.

In some disorders, such as celiac disease, medical nutrition therapy is the *only* treatment. With other GI problems, it is important to understand that, because of the symptoms the patient experiences, nutritional status is often in jeopardy. Nausea, vomiting, diarrhea, constipation, and malabsorption are common with these disorders. Interventions in these cases are focused on treating such symptoms in order to restore nutritional health.

Case 9 targets gastroesophageal reflux disease (GERD). More than 20 million Americans suffer from symptoms of gastroesophageal reflux daily, and more than 100 million suffer occasional symptoms. Gastroesophageal reflux disease most frequently results from lower esophageal sphincter (LES) incompetence. Factors that influence LES competence include both physical and lifestyle factors. This case identifies the common symptoms of GERD and challenges you to develop and analyze both nutritional and medical care for this patient.

Case 10 focuses on peptic ulcer disease treated pharmacologically and surgically. Peptic ulcer disease (PUD) involves ulcerations that penetrate the submucosa, usually in the antrum of the stomach or in the duodenum. Erosion may proceed to other levels of tissue and can eventually result in perforation. The breakdown in tissue allows continued insult by the highly acidic environment of the stomach. *Helicobacter pylori* has been established to be a major cause of chronic gastritis and peptic ulcer disease. Nutrition therapy for peptic ulcer disease is highly individualized. Treatment plans should avoid foods that increase gastric secretions and restrict any particular food or beverage that the patient does not tolerate. This case describes the complications of PUD resulting in hemorrhage and perforation that require surgical intervention. Nutritional complications, such as dumping syndrome and malabsorption, often accompany gastric surgery. This case also introduces the transition from enteral nutrition support to the appropriate oral diet for postoperative use.

The next four cases target conditions affecting the small and large intestines. These conditions, whose etiologies are all different, involve the symptoms of diarrhea, constipation, and sometimes malabsorption. In all the cases, nutrition therapy is one of the major modes of treatment.

Case 11 addresses the metabolic complications of diarrhea and dehydration. This pediatric case allows you to assess fluid and electrolyte imbalances, as well as interpret nutrition assessment for children, and to plan appropriate reintroduction of solid food to help the patient recover from acute diarrhea.

Celiac disease, explored in Case 12, is an autoimmune disease that destroys the mucosa of the small intestine. This reaction is caused by exposure to gliadin, which is found in the gluten portion of grain. Treatment for this disease is total avoidance of wheat, rye, and barley. This case explores new diagnostic procedures for celiac disease, secondary malabsorption syndromes, and the use of medical nutrition therapy.

Case 13 examines diverticulosis, a condition associated with both age and low fiber intake. A diet low in fiber increases colonic intraluminal pressure as the body strives to move the small amount of stool through the colon. This increased pressure results in herniations of the colon wall, called diverticula. Long-term treatment includes a transition to a high-fiber diet. This case involves the care of acute diverticulitis and the transition to preventive care.

The final case in this section targets inflammatory bowel disease. Crohn's disease and ulcerative colitis are two conditions that fall under the diagnosis of inflammatory bowel disease. Both these conditions dramatically affect nutritional status and often require nutritional support during periods of exacerbation. This case involves the effects of Crohn's disease on digestion and absorption, the diagnosis of malnutrition, and parenteral nutrition support.

Case 9

Gastroesophageal Reflux Disease

Objectives

After completing this case, the student will be able to:

1. Apply knowledge of the pathophysiology of gastroesophageal reflux disease (GERD) in order to identify and explain common nutritional problems associated with this disease.
2. Describe basic principles of drug action required for medical treatment of GERD.
3. Discuss the rationale for nutrition recommendations to minimize the adverse symptoms of GERD.
4. Interpret pertinent laboratory parameters for nutritional implications and significance.
5. Analyze nutrition assessment data to evaluate nutritional status and identify specific nutrition problems.
6. Determine nutrition diagnoses and write appropriate PES statements.
7. Develop a nutrition care plan with appropriate measurable goals, interventions, and strategies for monitoring and evaluation that addresses the nutrition diagnoses of this case.

Jack Nelson is admitted to University Hospital for evaluation of his increasing complaints of severe indigestion. Intraesophageal pH monitoring and barium esophagram support a diagnosis of gastroesophageal reflux disease.

ADMISSION DATABASE

Name: Jack Nelson
DOB: 7/22 (age 48)
Physician: J. Phelps, MD

BED #	DATE:	TIME:	TRIAGE STATUS (ER ONLY):
1	9/22	0900	☐ Red ☐ Yellow ☐ Green ☐ White

Initial Vital Signs

TEMP:	RESP:	SAO$_2$:	
98.6	15		
HT:	WT (lb):	B/P:	PULSE:
5′9″	215	119/75	90
LAST TETANUS		LAST ATE	LAST DRANK
1 year ago		11 PM	this AM

CHIEF COMPLAINT/HX OF PRESENT ILLNESS

"My wife insisted that I come see someone. The pain was so bad that I was afraid I was having a heart attack."

ALLERGIES: Meds, Food, IVP Dye, Seafood: Type of Reaction

NKA

PREVIOUS HOSPITALIZATIONS/SURGERIES

S/p R knee arthroplasty 5 years ago

PRIMARY PERSON TO CONTACT:
Name: Mary Nelson
Home #: 555-444-5689
Work #: 555-453-5689

ORIENTATION TO UNIT: ☒ Call light ☒ Television/telephone ☒ Bathroom ☒ Visiting ☒ Smoking ☒ Meals ☒ Patient rights/responsibilities

PERSONAL ARTICLES: (Check if retained/describe)
☐ Contacts ☐ R ☐ L ☐ Dentures ☐ Upper ☐ Lower
☐ Jewelry:
☒ Other: eyeglasses

VALUABLES ENVELOPE: no
☐ Valuables instructions

INFORMATION OBTAINED FROM:
☒ Patient ☐ Previous record
☒ Family ☐ Responsible party

Signature *Jack Nelson*

Home Medications (including OTC) **Codes: A=Sent home** **B=Sent to pharmacy** **C=Not brought in**

Medication	Dose	Frequency	Time of Last Dose	Code	Patient Understanding of Drug
atenolol	50 mg	daily	this AM	C	yes
aspirin	325 mg	daily	this AM	C	yes
ibuprofen	500 mg	twice daily	this AM	C	yes

Do you take all medications as prescribed? ☒ Yes ☐ No If no, why?

PATIENT/FAMILY HISTORY

☐ Cold in past two weeks
☐ Hay fever
☐ Emphysema/lung problems
☐ TB disease/positive TB skin test
☐ Cancer
☐ Stroke/past paralysis
☒ Heart attack Father
☒ Angina/chest pain Father
☒ Heart problems Mother
☒ High blood pressure Patient
☐ Arthritis
☐ Claustrophobia
☐ Circulation problems
☐ Easy bleeding/bruising/anemia
☐ Sickle cell disease
☐ Liver disease/jaundice
☐ Thyroid disease
☐ Diabetes
☐ Kidney/urinary problems
☒ Gastric/abdominal pain/heartburn Patient
☐ Hearing problems
☐ Glaucoma/eye problems
☐ Back pain
☐ Seizures
☐ Other

RISK SCREENING

Have you had a blood transfusion? ☐ Yes ☒ No
Do you smoke? ☐ Yes ☒ No
If yes, how many pack(s)?
Does anyone in your household smoke? ☐ Yes ☒ No
Do you drink alcohol? ☒ Yes ☐ No
If yes, how often? 3-4 × week How much? 1-2 beers
When was your last drink? last PM
Do you take any recreational drugs? ☐ Yes ☒ No
If yes, type:_______ Route:
Frequency:_______ Date last used:______/______/______

FOR WOMEN Ages 12–52

Is there any chance you could be pregnant? ☐ Yes ☐ No
If yes, expected date (EDC):
Gravida/Para:

ALL WOMEN

Date of last Pap smear:
Do you perform regular breast self-exams? ☐ Yes ☐ No

ALL MEN

Do you perform regular testicular exams? ☒ Yes ☐ No

Additional comments:

× *Cathy Mosely, RN*
Signature/Title

Client name: Jack Nelson
DOB: 7/22
Age: 48
Sex: Male
Education: BA
Occupation: Retail manager of local department store
Hours of work: M–F, works consistently in evenings and on weekends as well
Household members: Wife age 42, 2 sons ages 10 and 16—all in good health
Ethnic background: Caucasian
Religious affiliation: Protestant
Referring physician: Patricia Phelps, MD (family practice)

Chief complaint:
"My wife insisted that I come see someone. I am taking Tums constantly and am really uncomfortable from this constant indigestion! It was so bad yesterday that I was afraid I was having a heart attack. I also recently hurt my shoulder when I was coaching my son's baseball team, but as long as I take Advil, I am able to cope with that pain."

Patient history:
Onset of disease: Has been experiencing increased indigestion over last year. Previously only at night but now almost constantly.
Type of Tx: Taking OTC antacids
PMH: Essential HTN—Dx 1 year ago; s/p R knee arthroplasty 5 years ago
Meds: Atenolol 50 mg daily; 325 mg aspirin daily; multivitamin daily; 500 mg ibuprofen twice daily for last month
Smoker: No
Family Hx: What? CAD *Who?* Father

Physical exam:
General appearance: Obese 48-year-old white male in mild distress
Vitals: Temp 98.6°F, BP 119/75 mm Hg, HR 90 bpm/normal, RR 16 bpm
Heart: Noncontributory
HEENT: Noncontributory
Rectal: No hemorrhoids seen or felt; prostate not enlarged or soft; stool—slight Heme +
Neurologic: Oriented × 4
Extremities: No edema; normal strength, sensations, and DTR
Skin: Warm, dry
Chest/lungs: Lungs clear to auscultation and percussion
Peripheral vascular: Pulses full—no bruits
Abdomen: No distention. BS present in all regions. Liver percusses approx 8 cm at the midclavicular line, one fingerbreadth below the right costal margin. Epigastric tenderness without rebound or guarding.

Nutrition Hx:

General: Patient relates that he has gained almost 35 lbs since his knee surgery. He attributes this to a decrease in ability to run and has not found a consistent replacement for exercise. Patient states that he plays with his children on the weekends but that is the most exercise that he receives. He states that he probably has been eating and drinking more over the last year, which he attributes to stress. He is worried about his family history of heart disease, which is why he takes an aspirin each day. He has not really followed any diet restrictions.

Usual dietary intake:

AM:	1½–2 c dry cereal (Cheerios, bran flakes, Crispix); ½–¾ c skim milk; 16–32 oz orange juice
Lunch:	1½ oz ham on whole-wheat bagel, 1 apple or other fruit, 1 c chips, diet soda
Snack when he comes home:	Handful of crackers, cookies, or chips; 1–2 16-oz beers
PM:	6–9 oz of meat (grilled, baked usually); pasta, rice, or potatoes—1–2 c; fresh fruit; salad or other vegetable; bread; iced tea
Late PM:	Ice cream, popcorn, or crackers. Drinks 5–6 12-oz diet sodas daily as well as unsweetened iced tea

Relates that his family's schedule has been increasingly busy so that they order pizza or stop for fast food 1–2 times per week instead of cooking.

24-hour recall:

(at home PTA):	Crispix—2 c, 1 c skim milk, 16 oz orange juice
At work:	3 12-oz Diet Pepsi
Lunch:	Fried chicken sandwich from McDonald's, small french fries, 32 oz unsweetened iced tea
Late afternoon:	2 c chips, 1 beer
Dinner:	1 breast, fried, from Kentucky Fried Chicken; 1½ c potato salad; ¼ c green bean casserole; ½ c fruit salad; 1 c baked beans; unsweetened iced tea
Bedtime:	2 c ice cream mixed with 1 c skim milk for milkshake

Food allergies/intolerances/aversions (specify): Fried foods seem to make the indigestion worse.
Previous nutrition therapy? No
Food purchase/preparation: Wife or eats out
Vit/min intake: One-A-Day for Men multivitamin daily

Tx plan:

Ambulatory 24-hour pH monitoring with Bravo pH Monitoring System. Barium esophagram—request radiologist to attempt to demonstrate reflux using abdominal pressure and positional changes. Endoscopy with biopsy to r/o *H. pylori* infection.

Hospital course:

pH monitoring and barium esophagram support diagnosis of gastroesophageal reflux disease with negative biopsy for *H. pylori.* Endoscopy indicates no ulcerations or lesions, but generalized gastritis present. Begin lansoprazole 30 mg every AM. Decrease aspirin to 75 mg daily. Consult to orthopedics for shoulder injury. D/C self-medication of ibuprofen daily. Nutrition consult.

UH UNIVERSITY HOSPITAL

NAME: Jack Nelson
AGE: 48
PHYSICIAN: P. Phelps, MD

DOB: 7/22
SEX: M

CHEMISTRY

	NORMAL		UNITS
DAY:		Admit	
DATE:		9/22	
TIME:			
LOCATION:			
Albumin	3.5-5	4.9	g/dL
Total protein	6-8	7.2	g/dL
Prealbumin	16-35	33	mg/dL
Transferrin	250-380 (women) 215-365 (men)	350	mg/dL
Sodium	136-145	144	mEq/L
Potassium	3.5-5.5	4.5	mEq/L
Chloride	95-105	102	mEq/L
PO_4	2.3-4.7	3.8	mg/dL
Magnesium	1.8-3	2.0	mg/dL
Osmolality	285-295	278	mmol/kg/H_2O
Total CO_2	23-30	28	mEq/L
Glucose	70-110	110	mg/dL
BUN	8-18	9	mg/dL
Creatinine	0.6-1.2	0.7	mg/dL
Uric acid	2.8-8.8 (women) 4.0-9.0 (men)		mg/dL
Calcium	9-11	9.1	mg/dL
Bilirubin	$\leq$ 0.3	0.8	mg/dL
Ammonia (NH_3)	9-33		μmol/L
ALT	4-36	30	U/L
AST	0-35	22	U/L
Alk phos	30-120	156	U/L
CPK	30-135 (women) 55-170 (men)	100	U/L
LDH	208-378	400	U/L
CHOL	120-199	220 H	mg/dL
HDL-C	> 55 (women) > 45 (men)	20 L	mg/dL
VLDL	7-32		mg/dL
LDL	< 130	165 H	mg/dL
LDL/HDL ratio	< 3.22 (women) < 3.55 (men)		
Apo A	101-199 (women) 94-178 (men)		mg/dL
Apo B	60-126 (women) 63-133 (men)		mg/dL
TG	35-135 (women) 40-160 (men)	178 H	mg/dL
T_4	4-12		mcg/dL
T_3	75-98		mcg/dL
HbA_{1C}	3.9-5.2		%

UH UNIVERSITY HOSPITAL

NAME: Jack Nelson DOB: 7/22
AGE: 48 SEX: M
PHYSICIAN: P. Phelps, MD

HEMATOLOGY

DAY: Admit
DATE: 9/22
TIME:
LOCATION:

	NORMAL		UNITS
WBC	4.8–11.8	5.6	$\times 10^3/mm^3$
RBC	4.2–5.4 (women) 4.5–6.2 (men)	5.2	$\times 10^6/mm^3$
HGB	12–15 (women) 14–17 (men)	14.0	g/dL
HCT	37–47 (women) 40–54 (men)	40	%
MCV	80–96	85	μm^3
RETIC	0.8–2.8	1.1	%
MCH	26–32	28	pg
MCHC	31.5–36	32	g/dL
RDW	11.6–16.5		%
Plt Ct	140–440		$\times 10^3/mm^3$
Diff TYPE			
ESR	0–25 (women) 0–15 (men)		mm/hr
% GRANS	34.6–79.2		%
% LYM	19.6–52.7		%
SEGS	50–62		%
BANDS	3–6		%
LYMPHS	24–44		%
MONOS	4–8		%
EOS	0.5–4		%
Ferritin	20–120 (women) 20–300 (men)		mg/mL
ZPP	30–80		μmol/mol
Vitamin B_{12}	24.4–100		ng/dL
Folate	5–25		μg/dL
Total T cells	812–2,318		mm^3
T-helper cells	589–1,505		mm^3
T-suppressor cells	325–997		mm^3
PT	11–16		sec

Case Questions

I. Understanding the Disease and Pathophysiology

1. How is acid produced and controlled within the gastrointestinal tract?

2. What role does lower esophageal sphincter (LES) pressure play in the etiology of gastroesophageal reflux disease? What factors affect LES pressure?

3. What are the complications of gastroesophageal reflux disease?

4. What is *H. pylori*, and why did the physician want to biopsy the patient for *H. pylori?*

5. Identify the patient's signs and symptoms that could suggest the diagnosis of gastroesophageal reflux disease.

6. Describe the diagnostic tests performed for this patient.

7. What risk factors does the patient present with that might contribute to his diagnosis? (Be sure to consider lifestyle, medical, and nutritional factors.)

8. The MD has decreased this patient's dose of daily aspirin and recommended discontinuing his ibuprofen. Why? How do aspirin and NSAIDs affect gastroesophageal disease?

9. The MD has prescribed lansoprazole. What class of medication is this? What is the basic mechanism of the drug? What other drugs are available in this class? What other groups of medications are used to treat GERD?

II. Understanding the Nutrition Therapy

10. Are there specific foods that may contribute to GERD? Why or why not?

11. Summarize the current recommendations for nutrition therapy in GERD.

III. Nutrition Assessment

A. Evaluation of Weight/Body Composition

12. Calculate this patient's percent UBW and BMI. What does this assessment of weight tell you? In what ways does this contribute to his diagnosis?

B. Calculation of Nutrient Requirements

13. Calculate energy and protein requirements for Mr. Nelson. Identify the formula/calculation method you used, and explain the rationale for using it.

C. Intake Domain

14. Complete a computerized nutrient analysis for this patient's usual intake and 24-hour recall. How does his caloric intake compare to your calculated requirements?

15. From the information gathered within the intake domain, list possible nutrition problems using the diagnostic term.

D. Clinical Domain

16. Are there any other abnormal labs that should be addressed to improve Mr. Nelson's overall cardiac health? Explain.

17. From the information gathered within the clinical domain, list possible nutrition problems using the diagnostic term.

E. Behavioral–Environmental Domain

18. What other components of lifestyle modification would you address in order to help in treating his disorder?

19. From the information gathered within the behavior–environmental domain, list possible nutrition problems using the diagnostic term.

IV. Nutrition Diagnosis

20. Select two high-priority nutrition problems and complete the PES statement for each.

V. Nutrition Intervention

21. For each of the PES statements that you have written, establish an ideal goal (based on the signs and symptoms) and an appropriate intervention (based on the etiology).

22. Outline necessary modifications for him within his 24-hour recall that you could use as a teaching tool.

Food Item	Modification	Rationale
Crispix		
Skim milk		
Orange juice		
Diet Pepsi		
Fried chicken sandwich		
French fries		
Iced tea		
Chips		
Beer		
Fried chicken		
Potato salad		
Green bean casserole		
Fruit salad		
Baked beans		
Milkshake		

Bibliography

American Dietetic Association. *Nutrition Diagnosis and Intervention: Standardized Language for the Nutrition Care Process.* Chicago, IL: American Dietetic Association; 2006.

American Gastroenterological Association. American Gastroenterological Association medical position statement: Guidelines on the use of esophageal pH recording. *Gastroenterology.* 1996;110(6):1981.

Armstrong D, Marshall JK, Chiba N, Enns R, Fallone CA, Fass R, Hollingworth R, Hunt RH, Kahrilas PJ, Mayrand S, Moayyedi P, Paterson WG, Sadowski D, vanZanten SJ, Canadian Association of Gastroenterolgy GERD Consensus Group. Canadian Consensus Conference on the management of gastroesophageal reflux disease in adults: Update 2004. *Can J Gastroenterol.* 2005;19(1):15–35.

Bjorkman DJ. Current status of nonsteroidal anti-inflammatory drug (NSAID) use in the United States: Risk factors and frequency of complications. *Am J Med.* 1999;107(6A):3S–10S.

DeVault KR, Castell DO. The Practice Parameters Committee of the American College of Gastroenterology. Updated guidelines for the diagnosis and treatment of gastroesophageal reflux disease. *Am J Gastroenterol.* 2005;100:190–200.

Dixon MF. Pathophysiology of *Helicobacter pylori* infection. *Scand J Gastroenterol.* 1994;201(Suppl):7–10.

Escott-Stump S. *Nutrition and Diagnosis-Related Care.* 6th ed. Baltimore, MD: Williams & Wilkins; 2007.

Myers BM, Smith JL, Graham DY. Effect of red pepper and black pepper on the stomach. *Am J Gastroenterol.* 1987;82:211–214.

National Digestive Diseases Information Clearinghouse. *Heartburn, Gastroesophageal Reflux (GER), and Gastroesophageal Reflux Disease (GERD).* Bethesda, MD: National Digestive Diseases Information Clearinghouse; NIH Publication No. 03-0082; May 2007. Available at: http://digestive.niddk.nih.gov/ddiseases/pubs/gerd/index.htm#4. Accessed June 20, 2007.

Nelms MN. Assessment of nutrition status and risk. In: Nelms M, Sucher K, Long S. *Nutrition Therapy and Pathophysiology.* Belmont, CA: Thomson/Brooks-Cole; 2007:101–135.

Nelms MN. Diseases of the upper gastrointestinal tract. In: Nelms MN, Sucher K, Long S. *Nutrition Therapy and Pathophysiology.* Belmont, CA: Thomson/Brooks-Cole; 2007:433–437.

O'Connell MB, Madden DM, Murray AM, Heaney RP, Kerzner LJ. Effects of proton pump inhibitors on calcium carbonate absorption in women: A randomized crossover trial. *Am J Med.* 2005;118(7):778–781.

O'Connor HJ. Review article: *Helicobacter pylori* and gastro-oesophageal reflux disease: Clinical implications and management. *Aliment Pharmacol Ther.* 1999;13(2):117–127.

Pronsky ZM. *Food-Medication Interactions.* 14th ed. Birchrunville, PA: Food-Medication Interactions; 2007.

Rodriquez S, Miner P, Robindson M, Greenwood B, Maton PN, Pappa K. Meal type affects heartburn severity. *Dig Dis Sci.* 1998;43:485–490.

Soll AH, Fass R. Gastroesophageal reflux disease: Presentation and assessment of a common, challenging disorder. *Clin Cornerstone.* 2003;5(4):2–14; discussion 14–17.

Tytgat GN. Review article: Treatment of mild and severe cases of GERD. *Aliment Pharmacol Ther.* 2002;16(Suppl 4):73–78.

Internet Resources

American College of Gastroenterology. http://www.gi.org/patients/gerd/word.asp

National Digestive Diseases Information Clearinghouse. http://digestive.niddk.nih.gov

National Library of Medicine and National Institutes of Health: MedlinePlus. http://www.nlm.nih.gov/medlineplus/gerd.html

U.S. Department of Agriculture: Nutrient Data Laboratory. http://www.ars.usda.gov/ba/bhnrc/ndl

Case 10

Ulcer Disease: Medical and Surgical Treatment

Objectives

After completing this case, the student will be able to:

1. Discuss the etiology and risk factors for development of ulcer disease.
2. Identify classes of medications used to treat ulcer disease and determine possible drug–nutrient interactions.
3. Describe the surgical procedures used to treat refractory ulcer disease and explain common nutritional problems associated with this treatment.
4. Apply knowledge of nutrition therapy guidelines for ulcer disease and gastric surgery.
5. Analyze nutrition assessment data to evaluate nutritional status and identify specific nutrition problems.
6. Determine nutrition diagnoses and write appropriate PES statements.
7. Calculate enteral nutrition formulations.
8. Evaluate a standard enteral nutritional regimen.
9. Develop a nutrition care plan with appropriate measurable goals, interventions, and strategies for monitoring and evaluation that addresses the nutrition diagnoses of this case.

Maria Rodriguez has been treated as an outpatient for her gastroesophageal reflux disease. Her increasing symptoms of hematemesis, vomiting, and diarrhea lead her to be admitted for further gastrointestinal workup. She undergoes a gastrojejunostomy to treat her perforated duodenal ulcer.

Name: Maria Rodriguez
DOB: 12/19 (age 38)
Physician: A. Gustaf, MD

ADMISSION DATABASE

BED #	DATE:	TIME:	TRIAGE STATUS (ER ONLY):
1	8/30	1700	☐ Red ☐ Yellow ☐ Green ☐ White

Initial Vital Signs

TEMP:	RESP:	SAO$_2$:	
102	32		
HT (in): 5′2″	WT (lb): 110 UBW 145	B/P: 78/60	PULSE: 68
LAST TETANUS: 5 years ago		LAST ATE: yesterday	LAST DRANK: water 1 hour ago

PRIMARY PERSON TO CONTACT:
Name: Emilio Santiago (brother)
Home #: 555-212-7890
Work #: 555-213-4563

ORIENTATION TO UNIT: ☒ Call light ☒ Television/telephone ☒ Bathroom ☒ Visiting ☒ Smoking ☒ Meals ☒ Patient rights/responsibilities

CHIEF COMPLAINT/HX OF PRESENT ILLNESS

"I found out I had an ulcer 2 weeks ago. Last night I seemed to have gotten worse. I have been vomiting, and I have diarrhea. My pain is terrible. I think I have blood in my vomit and my diarrhea."

PERSONAL ARTICLES: (Check if retained/describe)
☐ Contacts ☐ R ☐ L ☐ Dentures ☐ Upper ☐ Lower
☒ Jewelry: wedding band
☐ Other:

ALLERGIES: Meds, Food, IVP Dye, Seafood: Type of Reaction

Codeine causes nausea and vomiting.

VALUABLES ENVELOPE:
☐ Valuables instructions

PREVIOUS HOSPITALIZATIONS/SURGERIES

For delivery of her two daughters only

INFORMATION OBTAINED FROM:
☒ Patient ☐ Previous record
☒ Family ☐ Responsible party

Signature *Maria Rodriguez*

Home Medications (including OTC) **Codes: A = Sent home** **B = Sent to pharmacy** **C = Not brought in**

Medication	Dose	Frequency	Time of Last Dose	Code	Patient Understanding of Drug
bismuth subsalicylate	525 mg	4 × daily	this AM	C	yes
metronidazole	250 mg	4 × daily	this AM	C	yes
tetracycline	500 mg	4 × daily	this AM	C	yes
omeprazole	20 mg	2 × daily	this AM	C	yes

Do you take all medications as prescribed? ☒ Yes ☐ No If no, why?

PATIENT/FAMILY HISTORY

☐ Cold in past two weeks
☐ Hay fever
☐ Emphysema/lung problems
☐ TB disease/positive TB skin test
☐ Cancer
☐ Stroke/past paralysis
☐ Heart attack
☐ Angina/chest pain
☐ Heart problems
☐ High blood pressure
☐ Arthritis
☐ Claustrophobia
☐ Circulation problems
☐ Easy bleeding/bruising/anemia
☐ Sickle cell disease
☐ Liver disease/jaundice
☐ Thyroid disease
☒ Diabetes Maternal grandmother
☐ Kidney/urinary problems
☒ Gastric/abdominal pain/heartburn Patient
☐ Hearing problems
☐ Glaucoma/eye problems
☐ Back pain
☐ Seizures
☒ Other Father and grandfather had ulcer disease

RISK SCREENING

Have you had a blood transfusion? ☐ Yes ☒ No
Do you smoke? ☒ Yes ☐ No
If yes, how many pack(s)? 1.5/day for 15 years
Does anyone in your household smoke? ☒ Yes ☐ No
Do you drink alcohol? ☐ Yes ☒ No
If yes, how often?_______ How much?
When was your last drink? ______/______/______
Do you take any recreational drugs? ☐ Yes ☒ No
If yes, type:______ Route:
Frequency:______ Date last used:______/______/______

FOR WOMEN Ages 12–52

Is there any chance you could be pregnant? ☐ Yes ☒ No
If yes, expected date (EDC):
Gravida/Para: 2/2

ALL WOMEN

Date of last Pap smear: Feb. of this year
Do you perform regular breast self-exams? ☒ Yes ☐ No

ALL MEN

Do you perform regular testicular exams? ☐ Yes ☐ No

Additional comments:

× *Sophia McMillan, RN*
Signature/Title

Client name: Maria Rodriguez
DOB: 12/19
Age: 38
Sex: Female
Education: Associate's degree
Occupation: Works in computer programming for local firm
Hours of work: M–F 9–5
Household members: 2 daughters ages 12 and 14, in good health; widowed
Ethnic background: Hispanic
Religious affiliation: Catholic
Referring physician: Anna Gustaf, MD (gastroenterologist)

Chief complaint:
"I found out I had an ulcer a few weeks ago. Last night I got very sick. I have been vomiting, and I have diarrhea. My pain is terrible. There is blood in my vomit and in my diarrhea."

Patient history:
Onset of disease: Diagnosed with GERD approx. 11 months ago; diagnosed with duodenal ulcer 2 weeks ago
Type of Tx: 14-day course of bismuth subsalicylate 525 mg four times daily; metronidazole 250 mg four times daily; tetracycline 500 mg four times daily. omeprazole 20 mg twice daily × 10 days.
PMH: Gravida 2 para 2. No other significant history except history of GERD.
Meds: See above.
Smoker: Yes
Family Hx: What? DM, PUD *Who?* DM: maternal grandmother, PUD: father and grandfather

Physical exam:
General appearance: 38-year-old Hispanic female—thin, pale, and in acute distress
Vitals: Temp: 101.3°F, BP 78/60 mm Hg, HR 68 bpm, RR 32 bpm
Heart: Regular rate and rhythm, heart sounds normal
HEENT: Noncontributory
Genitalia: Normal
Rectal: Not performed
Neurologic: Alert and oriented
Extremities: Noncontributory
Skin: Warm and dry to touch
Chest/lungs: Rapid breath sounds, lungs clear
Abdomen: Tender with guarding, absent bowel sounds

Nutrition Hx:
Patient relates that she understands about the feeding she is receiving through her tube. She explained that she has eaten very little since her ulcer was diagnosed and wonders how long it will be before she can eat again. Her physicians have told her they might like her to try something by mouth in the next few days.

Usual dietary intake (prior to current illness):

AM: Coffee, 1 slice dry toast; on weekends, cooked large breakfasts for family, which included omelets, rice or grits, or pancakes, waffles, fruit

Lunch: Sandwich from home (2 oz turkey on whole-wheat bread with mustard), 1 piece of raw fruit, cookies (2–3 Chips Ahoy!)

Dinner: 2 c rice, some type of meat (2–3 oz chicken), fresh vegetables (steamed tomatoes, peppers, and onions—1 c), coffee

Usual intake includes 8–10 c coffee daily. 1–2 sodas each day.
24-hour recall: Has been NPO since admission.
Food allergies/intolerances/aversions: See Nutrition Hx.
Previous nutrition therapy? No
Food purchase/preparation: Self and daughters
Vit/min intake: None

Tx plan:
Two weeks ago as an outpatient, she is s/p endoscopy that revealed 2-cm duodenal ulcer with generalized gastritis with a positive biopsy for *Helicobacter pylori.* She has completed 10 days of a 14-day course of bismuth subsalicylate 525 mg 4 × daily, metronidazole 250 mg 4 × daily, tetracycline 500 mg 4 × daily, and omeprazole 20 mg 2 × daily prescribed for a total of 28 days.
Admitted through ER for a surgical consult for possible perforated duodenal ulcer.
On 8/31, a gastrojejunostomy (Billroth II) was completed. Patient is now s/p gastrojejunostomy secondary to perforated duodenal ulcer. Feeding jejunostomy was placed during surgery, and patient is receiving Vital HN @ 25 cc/hr via continuous drip. Nutrition consult with orders have been left to advance the enteral feeding to 50 cc/hr. She is receiving only ice chips by mouth.

UH UNIVERSITY HOSPITAL

NAME: Maria Rodriguez DOB: 12/19
PHYSICIAN: A. Gustaf, MD SEX: F

CHEMISTRY

DAY:		Admit	Post Op Day 3	
DATE:		8/30	9/3	
TIME:		0800	0600	
LOCATION:				
	NORMAL			UNITS
Albumin	3.5–5	3.0 L	3.3 L	g/dL
Total protein	6–8	5.5 L	6.0	g/dL
Prealbumin	16–35	15 L	14 L	mg/dL
Transferrin	250–380 (women) 215–365 (men)	425 H	419 H	mg/dL
Sodium	136–145	141	140	mEq/L
Potassium	3.5–5.5	4.5	4.2	mEq/L
Chloride	95–105	103	101	mEq/L
PO_4	2.3–4.7	3.7	3.5	mg/dL
Magnesium	1.8–3	1.9	1.7	mg/dL
Osmolality	285–295	295	292	mmol/kg/H_2O
Total CO_2	23–30	26	24	mEq/L
Glucose	70–110	80	128 H	mg/dL
BUN	8–18	24	15	mg/dL
Creatinine	0.6–1.2	1.1	0.9	mg/dL
Uric acid	2.8–8.8 (women) 4.0–9.0 (men)			mg/dL
Calcium	9–11	9.0	8.7	mg/dL
Bilirubin	≤0.3	1.3	0.6	mg/dL
Ammonia (NH_3)	9–33	11	10	µmol/L
ALT	4–36	30	24	U/L
AST	0–35	31	17	U/L
Alk phos	30–120	145	133	U/L
CPK	30–135 (women) 55–170 (men)			U/L
LDH	208–378			U/L
CHOL	120–199			mg/dL
HDL-C	>55 (women) >45 (men)			mg/dL
VLDL	7–32			mg/dL
LDL	<130			mg/dL
LDL/HDL ratio	<3.22 (women) <3.55 (men)			
Apo A	101–199 (women) 94–178 (men)			mg/dL
Apo B	60–126 (women) 63–133 (men)			mg/dL
TG	35–135 (women) 40–160 (men)			mg/dL
T_4	4–12			mcg/dL
T_3	75–98			mcg/dL
HbA_{1C}	3.9–5.2			%

UH UNIVERSITY HOSPITAL

NAME: Maria Rodriguez DOB: 12/19
Age: 38 SEX: F
PHYSICIAN: A. Gustaf, MD

***************************************HEMATOLOGY**

DAY:		Admit	Post Op Day 3	
DATE:		8/30	9/3	
TIME:		0800	0600	
LOCATION:				
	NORMAL			UNITS
WBC	4.8–11.8	16.3 H	12.5 H	$\times 10^3/mm^3$
RBC	4.2–5.4 (women) 4.5–6.2 (men)			$\times 10^6/mm^3$
HGB	12–15 (women) 14–17 (men)	11.2 L	10.2 L	g/dL
HCT	37–47 (women) 40–54 (men)	33 L	31 L	%
MCV	80–96	91	86	μm^3
RETIC	0.8–2.8	1.1	1.2	%
MCH	26–32			pg
MCHC	31.5–36	31 L	28.5 L	g/dL
RDW	11.6–16.5	19.5 H	22 H	%
Plt Ct	140–440	345	356	$\times 10^3/mm^3$
Diff TYPE				
ESR	0–25 (women) 0–15 (men)			mm/hr
% GRANS	34.6–79.2			%
% LYM	19.6–52.7			%
SEGS	50–62	87 H	78 H	%
BANDS	3–6	6	4	%
LYMPHS	24–44	12 L	22 L	%
MONOS	4–8	5	4	%
EOS	0.5–4	2	3	%
Ferritin	20–120 (women) 20–300 (men)	241 H	232 H	mg/mL
ZPP	30–80			µmol/mol
Vitamin B_{12}	24.4–100			ng/dL
Folate	5–25			µg/dL
Total T cells	812–2,318			mm^3
T-helper cells	589–1,505			mm^3
T-suppressor cells	325–997			mm^3
PT	11–16			sec

Name: Maria Rodriguez
Physician: A. Gustaf, MD

PATIENT CARE SUMMARY SHEET

Date: 9/3 Room: 1145 Wt Yesterday: 110 Today: 111

Temp °F	NIGHTS								DAYS								EVENINGS							
	00	01	02	03	04	05	06	07	08	09	10	11	12	13	14	15	16	17	18	19	20	21	22	23
105																								
104																								
103																								
102																								
101																								
100								×																
99																×								×
98																								
97																								
96																								
Pulse								82								78								80
Respiration								28								27								28
BP								109/78								110/82								115/72
Blood Glucose								122								115								82
Appetite/Assist								NPO								NPO								NPO
INTAKE																								
Oral																								
IV								380								400								400
TF Formula/Flush	25	25	25	25	25	25	25	25/50	25			25	25	25	25	25/50		25				25	25	25/50
Shift Total	630								600								550							
OUTPUT																								
Void																								
Cath.								480								550								200
Emesis																								
BM								128				200												
Drains								220								275								320
Shift Total	828								1025								520							
Gain																								
Loss	−198								−425								+30							
Signatures	S. Smith, RN								M. Taylor, RN								N. Parrish, RN							

Case Questions

I. Understanding the Disease and Pathophysiology

1. Identify this patient's risk factors for ulcer disease.

2. Is smoking related to ulcer disease?

3. How is *H. pylori* related to ulcer disease?

4. This patient was prescribed four different medications for treatment of her *H. pylori* infection. Identify the drug functions/mechanisms. (Use table below.)

Drug	Action
Metronidazole	
Tetracycline	
Bismuth subsalicylate	
Omeprazole	

5. What are the possible drug–nutrient side effects from Mrs. Rodriguez's prescribed regimen? (See table above.) Which drug–nutrient side effects are most pertinent to her current nutritional status?

6. Explain the surgical procedure that the patient received.

7. How may the normal digestive process change with this procedure?

II. Understanding the Nutrition Therapy

8. The most common physical side effects from this surgery are the development of early or late dumping syndrome. Describe each of these syndromes, including symptoms the patient might experience, the etiology of the symptoms, and the standard interventions for preventing/treating the symptoms.

9. What are other potential nutritional complications after this surgical procedure?

III. Nutrition Assessment

A. Evaluation of Body Weight/Body Composition

10. Assess this patient's available anthropometric data. Calculate percent UBW and BMI. Which of these is the most pertinent in identifying the patient's nutrition risk? Why?

11. What other anthropometric measures could be used to further confirm her nutritional status?

B. Calculation of Nutrient Requirements

12. Calculate energy and protein requirements for Mrs. Rodriguez. Identify the formula/calculation method you used and explain the rationale for using it.

C. Intake Domain

13. This patient was started on an enteral feeding postoperatively. Why do you think this decision was made?

14. What type of enteral formula is Vital HN? Is it an appropriate choice for this patient?

15. Why was the enteral formula started at 25 cc/hr?

16. Is the current enteral prescription meeting this patient's nutritional needs? Compare her energy and protein requirements to what is provided by the formula. If her needs are not met, what should be the goal for her enteral support?

17. What would the RD assess to monitor tolerance to the enteral feeding?

18. Go to the patient care summary sheet. For postoperative day 2, how much enteral nutrition did the patient receive? How does this compare to what was prescribed?

19. When evaluating the patient care summary sheet, you notice the patient has gained 1 pound in 24 hours. Should you address this in your nutrition note as an improvement in nutritional status?

20. As this patient is advanced to solid food, what modifications in diet would the RD address? Why? What would be a typical first meal for this patient?

21. What other considerations would you give to Mrs. Rodriguez to maximize her tolerance of solid food?

22. Mrs. Rodriguez asks for you to come to her room because she is concerned that she may have to follow a special diet forever. What might you tell her?

23. Should Mrs. Rodriguez be on any type of vitamin/mineral supplementation at home when she is discharged? Would you make any recommendations for specific types?

24. Why might Mrs. Rodriguez be at risk for iron-deficiency anemia, pernicious anemia, and/or megaloblastic anemia secondary to folate deficiency and/or poor vitamin B_{12} absorption?

25. Will the oral vitamin/mineral supplement be adequate to prevent the anemias discussed in question 24? Explain.

26. From the information gathered within the intake domain, list possible nutrition problems using the diagnostic term.

D. Clinical Domain

27. Using her admission chemistry and hematology values, which biochemical measures are abnormal? Explain.

a. Which values can be used to further assess her nutritional status? Explain.

b. Which laboratory measures (see lab report, pages 117–118) are related to her diagnosis of duodenal ulcer? Why would they be abnormal?

28. Do you think this patient is malnourished? If so, why? What criteria can be used to diagnose malnutrition? Within what category does this patient fit?

IV. Nutrition Diagnosis

29. Select two high-priority nutrition problems and complete the PES statement for each.

V. Nutrition Intervention

30. For each of the PES statements that you have written, establish an ideal goal (based on the signs and symptoms) and an appropriate intervention (based on the etiology).

31. What nutrition education should this patient receive prior to discharge?

32. Do any lifestyle issues need to be addressed with this patient? Explain.

Bibliography

American Dietetic Association. *Nutrition Diagnosis and Intervention: Standardized Language for the Nutrition Care Process.* Chicago, IL: American Dietetic Association; 2006.

Baron JH. Peptic ulcer. *Mount Sinai J Med.* 2000;67:58–62.

Brown LF, Wilson DE. Gastroduodenal ulcers: Causes, diagnosis, prevention and treatment. *Compr Ther.* 1999;25:30–38.

Del Valle J. Peptic ulcer disease and related disorders. In: Kasper DL, Braunwald E, Hauser S, Longo D, Jameson JL, eds. *Harrison's Principles of Internal Medicine.* 17th ed. New York, NY: McGraw-Hill; 2008. Available at: http://harrisons.accessmedicine.com. Accessed April 4, 2008.

Escott-Stump S. *Nutrition and Diagnosis-Related Care.* 6th ed. Baltimore, MD: Williams & Wilkins; 2007.

Graham DY, Rakel RE, Fendrick AM, Go MF, Marshall BJ, Peura DA, Scherger JE. Recognizing peptic ulcer disease. Keys to clinical and laboratory diagnosis. *Postgrad Med.* 1999;105(3):113–116,121–123,127–128.

Harbison SP, Dempsey DT. Peptic ulcer disease. *Curr Probl Surg.* 2005;42(6):346–454

Helicobacter pylori in peptic ulcer disease. *NIH Consensus Statement.* 1994;12(1):1–23.

Heuberger RA. Diseases of the hematological system. In: Nelms M, Sucher K, Long S. *Nutrition Therapy and Pathophysiology.* Belmont, CA: Thomson/Brooks-Cole; 2007:651–685.

Jamieson GG. Current status of indications for surgery in peptic ulcer disease. *World J Surg.* 2000;24:256–258.

Kurata JH, Nogawa AN. Meta-analysis of risk factors for peptic ulcer: Nonsteroidal antiinflammatory drugs, *Helicobacter pylori,* and smoking. *J Clin Gastroenterol.* 1997;24:2–17.

Lacy BE, Rosemore J. *Helicobacter pylori:* Ulcers and more: The beginning of an era. *J Nutr.* 2001;131(10):2789S–2793S.

Meurer LN, Bower DJ. Management of *Helicobacter pylori* infection. *Am Fam Physician.* 2002;65(7):1327–1336.

Nelms MN. Assessment of nutrition status and risk. In: Nelms M, Sucher K, Long S. *Nutrition Therapy and Pathophysiology.* Belmont, CA: Thomson/Brooks-Cole, 2007:101–135.

Nelms MN. Diseases of the upper gastrointestinal tract. In: Nelms M, Sucher K, Long S. *Nutrition Therapy and Pathophysiology.* Belmont, CA: Thomson/Brooks-Cole; 2007:445–451.

Pronsky ZM. *Food-Medication Interactions.* 14th ed. Birchrunville, PA: Food-Medication Interactions; 2007.

Qureshi WA, Graham DY. Diagnosis and management of *Helicobacter pylori* infection. *Clin Cornerstone.* 1999;1:18–28.

Ryan-Harshman M, Aldoori W. How diet and lifestyle affect duodenal ulcers. Review of the evidence. *Can Fam Physician.* 2004;50:727–732.

Skipper A, Nelms MN. Methods of nutrition support. In: Nelms M, Sucher K, Long S. *Nutrition Therapy and Pathophysiology.* Belmont, CA: Thomson/Brooks-Cole; 2007:149–179.

Internet Resources

Merck Manuals Online Library: Peptic Ulcer. http://www.merck.com/mmhe/sec09/ch121/ch121c.html

National Digestive Diseases Information Clearinghouse: NIDDK/National Institutes of Health. http://digestive.niddk.nih.gov/ddiseases/pubs/pepticulcers_ez/

National Library of Medicine/National Institutes of Health: MedlinePlus. http://www.nlm.nih.gov/medlineplus/ency/article/000206.htm

Case 11

Infectious Diarrhea with Resulting Dehydration

Objectives

After completing this case, the student will be able to:

1. Discuss the physiological effects of infection and dehydration.
2. Interpret laboratory parameters for nutritional implications and significance.
3. Determine nutrient, fluid, and electrolyte requirements for children.
4. Analyze nutrition assessment data to evaluate nutritional status and identify specific nutrition problems.
5. Determine nutrition diagnoses and write appropriate PES statements.
6. Prescribe appropriate nutrition therapy for dehydration and malabsorption resulting from diarrhea.
7. Develop a nutrition care plan with appropriate measurable goals, interventions, and strategies for monitoring and evaluation that addresses the nutrition diagnoses of this case.
8. Develop and implement transitional feeding plans.

Seth Jones is admitted to the pediatric unit of University Hospital with severe dehydration secondary to diarrhea. His medical evaluation reveals a diagnosis of *E. coli* 0157:H7.

UNIVERSITY HOSPITAL

ADMISSION DATABASE

Name: Seth Jones
DOB: 4/13 (age 8)
Physician: M. Hicks, MD

BED #	DATE:	TIME:	TRIAGE STATUS (ER ONLY):
2	7/22	1500	☐ Red ☐ Yellow ☐ Green ☐ White

Initial Vital Signs

TEMP:	RESP:	SAO$_2$:	
102.3	17		
HT: 4'1"	WT (lb): 50 UBW 54	B/P: 90/70	PULSE: 72
LAST TETANUS age 5		LAST ATE last AM	LAST DRANK today at noon

PRIMARY PERSON TO CONTACT:
Name: Violet and Philip Jones
Home #: 555-256-7892
Work #: 555-257-7721

ORIENTATION TO UNIT: ☒ Call light ☒ Television/telephone ☒ Bathroom ☒ Visiting ☒ Smoking ☒ Meals ☒ Patient rights/responsibilities

CHIEF COMPLAINT/HX OF PRESENT ILLNESS

"We thought he had the flu or some kind of virus. He has had diarrhea for over 4 days now. He just has not gotten any better."

ALLERGIES: Meds, Food, IVP Dye, Seafood: Type of Reaction

Bee stings-respiratory difficulty and large amounts of swelling

PREVIOUS HOSPITALIZATIONS/SURGERIES

PERSONAL ARTICLES: (Check if retained/describe)
☐ Contacts ☐ R ☐ L ☐ Dentures ☐ Upper ☐ Lower
☐ Jewelry:
☒ Other: glasses

VALUABLES ENVELOPE: no
☐ Valuables instructions

INFORMATION OBTAINED FROM:
☒ Patient ☐ Previous record
☒ Family ☐ Responsible party

Signature *Violet Jones (mother)*

Home Medications (including OTC) **Codes: A=Sent home** **B=Sent to pharmacy** **C=Not brought in**

Medication	Dose	Frequency	Time of Last Dose	Code	Patient Understanding of Drug

Do you take all medications as prescribed? ☐ Yes ☐ No If no, why?

PATIENT/FAMILY HISTORY

☐ Cold in past two weeks
☐ Hay fever
☐ Emphysema/lung problems
☐ TB disease/positive TB skin test
☐ Cancer
☐ Stroke/past paralysis
☐ Heart attack
☐ Angina/chest pain
☐ Heart problems
☒ High blood pressure Father
☐ Arthritis
☐ Claustrophobia
☐ Circulation problems
☐ Easy bleeding/bruising/anemia
☐ Sickle cell disease
☐ Liver disease/jaundice
☐ Thyroid disease
☐ Diabetes
☐ Kidney/urinary problems
☐ Gastric/abdominal pain/heartburn
☐ Hearing problems
☐ Glaucoma/eye problems
☐ Back pain
☐ Seizures
☐ Other

RISK SCREENING

Have you had a blood transfusion? ☐ Yes ☒ No
Do you smoke? ☐ Yes ☐ No
If yes, how many pack(s)?
Does anyone in your household smoke? ☐ Yes ☒ No
Do you drink alcohol? ☐ Yes ☐ No
If yes, how often?_______ How much?
When was your last drink?_______/_______/_______
Do you take any recreational drugs? ☐ Yes ☐ No
If yes, type:_______ Route:
Frequency:_______ Date last used:_______/_______/_______

FOR WOMEN Ages 12–52

Is there any chance you could be pregnant? ☐ Yes ☐ No
If yes, expected date (EDC):
Gravida/Para:

ALL WOMEN

Date of last Pap smear:
Do you perform regular breast self-exams? ☐ Yes ☐ No

ALL MEN

Do you perform regular testicular exams? ☐ Yes ☐ No

Additional comments:

x *Gina Miller, RN,*
Signature/Title

Client name: Seth Jones
DOB: 4/13
Age: 8
Sex: Male
Education: Less than high school *What grade/level?* 3rd grade
Occupation: Student
Hours of work: N/A
Household members: Father age 48, mother age 39, brother age 11, sister age 10—all well
Ethnic background: African American
Religious affiliation: African Methodist Episcopal church
Referring physician: M. Hicks, MD

Chief complaint:
"We thought he had the flu or some kind of virus. He has had diarrhea for over 4 days now. He just has not gotten any better. We are really worried—he just seems so weak and listless."

Patient history:
Parents describe that the family spent last weekend at an amusement water park. Seth, their 8-year-old son, began having diarrhea and running a fever Sunday morning. They decided to cut their weekend trip short, thinking that he had gotten the flu or some type of viral illness. Now, four days later, he is still running a fever, and the diarrhea has gotten worse instead of better. They have been giving him soft foods, soups, and liquids since he got sick. He has had very little to eat in the last 24 hours, and parents state that it has been difficult for him to even drink anything. They also note that there seems to be blood in the diarrhea now. His parents estimate that Seth has had anywhere from 8 to 15 diarrhea episodes in the past 24 hours. The other two children also have had diarrhea but have since improved. They have talked with their pediatrician several times, but Seth has not been seen by his MD. They have given Seth over-the-counter meds for diarrhea, including Pepto-Bismol and Kaopectate.
Onset of disease: Five days previous
Type of Tx: None at present
Meds: Pepto-Bismol and Kaopectate
Smoker: No
Family Hx: What? HTN *Who?* Father

Physical exam:
General appearance: Lethargic, 8-year-old African American male
Vitals: Temp 102.3°F, BP 90/70 mm Hg (orthostatic 75/62), HR 92 bpm, RR 17 bpm
Heart: Moderately elevated pulse
HEENT:
 Eyes: Sunken; sclera clear without evidence of tears
 Ears: Clear
 Nose: Dry mucous membranes
 Throat: Dry mucous membranes, no inflammation
Genitalia: Unremarkable
Neurologic: Alert, oriented × 3; irritable

Extremities: No joint deformity or muscle tenderness. No edema.
Skin: Warm, dry; reduced capillary refill (approximately 2 seconds)
Chest/lungs: Clear to auscultation and percussion
Abdomen: Tender, nondistended, minimal bowel sounds

Nutrition Hx:
General: Prior to admission, good appetite with consumption of a wide variety of foods except for vegetables

Usual dietary intake:
AM: Cereal, toast or bagel, juice
Lunch: Sandwich, chips, fruit, cookies, milk
Dinner: All meats, pasta or rice, fruit, milk
Snacks: Juice, fruit, cookies, crackers

24-hour recall: Parents estimate that child has had less than 6 oz of Gatorade in past 24 hours and that has had to be strongly encouraged through sips.
Food allergies/intolerances/aversions: NKA
Previous nutrition therapy? No
Food purchase/preparation: Parent(s)
Vit/min intake: Flintstones vitamin daily

Dx:
Moderate dehydration R/O bacterial vs. viral gastroenteritis

Tx plan:
D5W ½ normal saline with 40 mEq KCl/L 20 mL per kg/hr for 3 hours. Increase to 100 mL/kg over next 7 hours; then decrease to 100 mL/hr. Begin Pedialyte 30 cc q hr as tolerated. Fecal smear for RBC and leukocytes. Stool culture.

Hospital course:
Fecal smear with gross blood and leukocytes. Diagnosis of *E. coli* 0157:H7, presumably from contamination at water park. Local health department follow-up indicates additional 15 cases from visitors at park during same time period. Most likely source of contamination is playground fountain. Investigation of filtration and chlorination system is underway.

UH UNIVERSITY HOSPITAL

NAME: Seth Jones DOB: 4/13
AGE: 8 SEX: M
PHYSICIAN: M. Hicks, MD

CHEMISTRY*

DAY:		Admit	2	3	
DATE:		7/22	7/23	7/24	
TIME:					
LOCATION:					
	NORMAL				UNITS
Albumin	3.5-5	4.9	3.8		g/dL
Total protein	6-8	7.2	6.8		g/dL
Prealbumin	16-35				mg/dL
Transferrin	250-380 (women) 215-365 (men)				mg/dL
Sodium	136-145	148 H	144	138	mEq/L
Potassium	3.5-5.5	3.2 L	3.7	3.7	mEq/L
Chloride	95-105	105	101	102	mEq/L
PO_4	2.3-4.7	4.3	3.5	3.6	mg/dL
Magnesium	1.8-3	2.2	1.9	1.8	mg/dL
Osmolality	285-295	309 H	304 H	292	mmol/kg/H_2O
Total CO_2	23-30	31 H	28	27	mEq/L
Glucose	70-110	71	108	101	mg/dL
BUN	8-18	22	10	10	mg/dL
Creatinine	0.6-1.2	1.4 H	0.7	0.6	mg/dL
Uric acid	2.8-8.8 (women) 4.0-9.0 (men)	3.6			mg/dL
Calcium	9-11	9.1			mg/dL
Bilirubin	≤ 0.3	1.1			mg/dL
Ammonia (NH_3)	9-33				μmol/L
ALT	4-36				U/L
AST	0-35				U/L
Alk phos	30-120				U/L
CPK	30-135 (women) 55-170 (men)				U/L
LDH	208-378				U/L
CHOL	120-199				mg/dL
HDL-C	> 55 (women) > 45 (men)				mg/dL
VLDL	7-32				mg/dL
LDL	< 130				mg/dL
LDL/HDL ratio	< 3.22 (women) < 3.55 (men)				
Apo A	101-199 (women) 94-178 (men)				mg/dL
Apo B	60-126 (women) 63-133 (men)				mg/dL
TG	35-135 (women) 40-160 (men)				mg/dL
T_4	4-12				mcg/dL
T_3	75-98				mcg/dL
HbA_{1C}	3.9-5.2				%

UH UNIVERSITY HOSPITAL

NAME: Seth Jones DOB: 4/13
AGE: 8 SEX: M
PHYSICIAN: M. Hicks, MD

URINALYSIS

DAY:		Admit	2	3	
DATE:		7/22	7/23	7/24	
TIME:					
LOCATION:					
	NORMAL				UNITS
Coll meth	Random specimen	First morning	First Morning	First morning	
Color	Pale yellow	Amber	Straw	Pale Yellow	
Appear	Clear	Cloudy	Slightly Hazy	Clear	
Sp grv	1.003-1.030	1.039	1.020	1.008	
pH	5-7	4.8	5.2	5.6	
Prot	NEG	Neg	Neg	Neg	mg/dL
Glu	NEG	Neg	Neg	Neg	mg/dL
Ket	NEG	+1	Neg	Neg	
Occ bld	NEG	Neg	Neg	Neg	
Ubil	NEG	Neg	Neg	Neg	
Nit	NEG	Neg	Neg	Neg	
Urobil	< 1.1	0.5	0.7	0.9	EU/dL
Leu bst	NEG	Neg	Neg	Neg	
Prot chk	NEG	Neg	Neg	Neg	
WBCs	0-5	2	1	0	/HPF
RBCs	0-5	1	0	0	/HPF
EPIs	0	0	0	0	/LPF
Bact	0	0	0	0	
Mucus	0	0	0	0	
Crys	0	0	0	0	
Casts	0	0	0	0	/LPF
Yeast	0	0	0	0	

Case Questions

I. Understanding the Disease and Pathophysiology

1. Define *diarrhea.* How do osmotic and secretory diarrhea differ? Which does Seth have? What criteria did you use to make this decision?

2. What are the physiological consequences of prolonged diarrhea?

3. What are electrolytes?

II. Understanding the Nutrition Therapy

4. Outline the specific modifications that are recommended for an individual with diarrhea.

5. Clear liquids are often recommended for someone with diarrhea. Why may this be contraindicated? What are the controversies surrounding the use of a clear liquid diet?

6. What is the BRAT diet? Why are these foods recommended?

7. What is the potential role of pro- and prebiotics in treating diarrhea? What specific foods could you recommend?

III. Nutrition Assessment

A. Evaluation of Weight/Body Composition

8. Assess Seth's height and weight. Which weight would you use—admission weight or usual body weight? What is the most appropriate tool to assess height and weight for an 8-year-old child?

B. Calculation of Nutrient Requirements

9. What are Seth's energy and protein needs?

10. What are Seth's fluid requirements?

C. Intake Domain

11. The physician ordered D5W ½ NS with 40 mEq KCL @100 mL/hr.

    a. What is D5W?

    b. What is NS?

    c. How much sodium does this solution provide in 1 liter? In 24 hours?

    d. How much energy does this solution provide?

    e. How much potassium does it provide?

12. The physician also ordered Pedialyte 30 cc q hour. What is Pedialyte?

13. From the information gathered within the intake domain, list possible nutrition problems using the diagnostic term.

D. Clinical Domain

14. What signs and symptoms in the physician's interview and physical examination indicate that Seth may be dehydrated?

15. Evaluate Seth's laboratory values on day 1 of his admission. Use the following table to organize your information.

    a. Which values are consistent with diagnosing dehydration?

    b. Does Seth have an electrolyte imbalance?

c. Does Seth have an acid–base imbalance? Explain. Are there other tests needed to confirm an acid–base imbalance?

Chemistry	Normal Value	Seth's Value	Reason for Abnormality	Nutritional Implications

16. Assess Seth's urinalysis report. Which measures are consistent with his diagnosis?

17. From the information gathered within the clinical domain, list possible nutrition problems using the diagnostic term.

IV. Nutrition Diagnosis

18. Select two high-priority nutrition problems and complete the PES statement for each.

V. Nutrition Intervention

19. What other solutions are similar to Pedialyte?

20. Should Seth have been made NPO? Why or why not?

21. Once Seth's electrolyte imbalances are corrected and diarrhea begins to decrease, what type of diet would you recommend for transition from Pedialyte to solid food?

22. For each of the PES statements that you have written, establish an ideal goal (based on the signs and symptoms) and an appropriate intervention (based on the etiology).

Bibliography

American Academy of Pediatrics. Practice parameter: The management of acute gastroenteritis in young children. *Pediatrics.* 1996;97(3):424–435.

American Dietetic Association. Nutrition management of diarrhea in childhood. In: American Dietetic Association. *Manual of Clinical Dietetics.* 6th ed. Chicago, IL: American Dietetic Association; 1996:237–246.

Broussard EK, Surawicz CM. Probiotics and prebiotics in clinical practice. *Nutr Clin Care.* 2004;7(3):104–113.

Burkhart DM. Management of acute gastroenteritis in children. *Am Fam Physician.* 1999;60(9):2555–2566.

Donowitz M, Kokke FT, Saidi R. Evaluation of patients with chronic diarrhea. *N Engl J Med.* 1995;332(11):725–729.

Duggan C, Santosham M, Glass RI. The management of acute diarrhea in children: Oral rehydration, maintenance and nutritional therapy. *MMWR Recomm Rep.* 1992;41:1–20.

Fayad IM, Hashem M, Duggan C, Refat M, Bakir M, Fontaine O, Santosham M. Comparative efficacy of rice-based and glucose-based oral rehydration salts plus early reintroduction of food. *Lancet.* 1993;342:772–775.

Gavin N, Merrick N, Davidson B. Efficacy of glucose based oral rehydration therapy. *Pediatrics.* 1996;98:45–51.

Hoghton MA, Mittal NK, Sandhu BK, Mahdi G. Effects of immediate modified feeding on infantile gastroenteritis. *Br J Gen Pract.* 1996;46:173–175.

Khan AM, Sarker SA, Alam NH, Hossain MS, Guchs GJ, Salam MA. Low osmolar oral rehydration salts solution in the treatment of acute watery diarrhea in neonates and young infants: A randomized, controlled clinical trial. *J Health Popul Nutr.* 2005:23:52–57.

Mackenzie A., Barnes G, Shann F. Clinical signs of dehydration in children. *Lancet.* 1989;2(8663):605–607.

Murphy C, Hahn S, Volmink J. Reduced osmolarity oral rehydration solution for treating cholera. *Cochrane Database Syst Rev.* 2004;18(4):CD003754.

Nelms MN. Acid-base balance. In: Nelms M, Sucher K, Long S. *Nutrition Therapy and Pathophysiology.* Belmont, CA: Thomson/Brooks-Cole; 2007:203–217.

Nelms MN. Assessment of nutrition status and risk. In: Nelms M, Sucher K, Long S. *Nutrition Therapy and Pathophysiology.* Belmont, CA: Thomson/Brooks-Cole; 2007:101–135.

Nelms MN. Diseases of the lower gastrointestinal tract. In: Nelms M, Sucher K, Long S. *Nutrition Therapy and Pathophysiology.* Belmont, CA: Thomson/Brooks-Cole; 2007:457–474.

Nelms MN. Fluid and electrolyte balance. In: Nelms M, Sucher K, Long S. *Nutrition Therapy and Pathophysiology.* Belmont, CA: Thomson/Brooks-Cole; 2007:181–201.

Northrup RS, Flanigan TP. Gastroenteritis. *Pediatr Rev.* 1994;15:461–472.

O'Sullivan GC, Kelly P, O'Halloran S, Collins C, Collins JK, Dunne C, Shanahan F. Probiotics: An emerging therapy. *Curr Pharm Des.* 2005;11:3–10.

Sabol VK, Carlson KK. Diarrhea: Applying research to bedside practice. *AACN Adv Crit Care.* 2007;18(1):32–44.

Vila J, Ruiz J, Gallardo F, Vargas M, Soler L, Figueras MJ, Gascon J. *Aeromonas* spp. and traveler's diarrhea: Clinical features and antimicrobial resistance. *Emerg Infect Dis.* 2003;9(5):552–555. Available at: http://www.cdc.gov/ncidod/EID/vol9no5/02-0451.htm. Accessed March 31, 2008.

Internet Resources

Centers for Disease Control: *Escherichia coli.* http://www.cdc.gov/nczved/dfbmd/disease_listing/stec_gi.html

Merck Manuals Online Library: Dehydration. http://www.merck.com/mmpe/sec19/ch276/ch276b.html

National Guideline Clearinghouse: Dehydration and Fluid Maintenance. http://www.guidelines.gov/summary/summary.aspx?doc_id=3305&nbr=002531&string=staffing

Case 12

Celiac Disease

Objectives

After completing this case, the student will be able to:

1. Apply knowledge of the pathophysiology of celiac disease to identify and explain common nutritional problems associated with the disease.
2. Apply knowledge of nutrition therapy for celiac disease.
3. Analyze nutrition assessment data to evaluate nutritional status and identify specific nutrition problems.
4. Determine nutrition diagnoses and write appropriate PES statements.
5. Develop a nutrition care plan with appropriate measurable goals, interventions, and strategies for monitoring and evaluation that addresses the nutrition diagnoses of this case.

Mrs. Melissa Gaines is admitted to University Hospital with severe weight loss, extreme fatigue, and diarrhea. A small bowel biopsy reveals a diagnosis of celiac disease with secondary malabsorption and anemia.

ADMISSION DATABASE

Name: Melissa Gaines
DOB: 3/14 (age 36)
Physician: Roger Smith, MD

BED #	DATE:	TIME:	TRIAGE STATUS (ER ONLY):
1	11/12	0800	☐ Red ☐ Yellow ☐ Green ☐ White

Initial Vital Signs

TEMP:	RESP:	SAO$_2$:	
98.4	17		
HT:	**WT (lb):**	**B/P:**	**PULSE:**
5′3″	92	110/74	71
LAST TETANUS		**LAST ATE**	**LAST DRANK**
8 years ago		yesterday lunch	last PM–bedtime

PRIMARY PERSON TO CONTACT:
Name: Michael Gaines
Home #: 555-256-7894
Work #: 555-254-9900

ORIENTATION TO UNIT: ☒ Call light ☒ Television/telephone ☒ Bathroom ☒ Visiting ☒ Smoking ☒ Meals ☒ Patient rights/responsibilities

CHIEF COMPLAINT/HX OF PRESENT ILLNESS

"I have lost a tremendous amount of weight, and I have been having terrible diarrhea for awhile now. I don't even have the energy to get off the couch right now."

ALLERGIES: Meds, Food, IVP Dye, Seafood: Type of Reaction

Maybe NutraSweet?

PREVIOUS HOSPITALIZATIONS/SURGERIES

2 live births; last child born in September of this year–5 lb 2 oz full term

PERSONAL ARTICLES: (Check if retained/describe)
☒ Contacts ☒ R ☒ L ☐ Dentures ☐ Upper ☐ Lower
☒ Jewelry: wedding band
☐ Other:

VALUABLES ENVELOPE:
☐ Valuables instructions

INFORMATION OBTAINED FROM:
☒ Patient ☐ Previous record
☒ Family ☐ Responsible party

Signature *Melissa Gaines*

Home Medications (including OTC) **Codes: A=Sent home** **B=Sent to pharmacy** **C=Not brought in**

Medication	Dose	Frequency	Time of Last Dose	Code	Patient Understanding of Drug
prenatal vitamins	1	daily	this AM	C	yes
Kaopectate	2 tbsp	every 3–4 hours	this PM	C	yes

Do you take all medications as prescribed? ☒ Yes ☐ No If no, why?

PATIENT/FAMILY HISTORY

☐ Cold in past two weeks
☐ Hay fever
☐ Emphysema/lung problems
☐ TB disease/positive TB skin test
☐ Cancer
☐ Stroke/past paralysis
☒ Heart attack Father
☒ Angina/chest pain Father
☒ Heart problems Father
☐ High blood pressure
☐ Arthritis
☐ Claustrophobia
☐ Circulation problems
☐ Easy bleeding/bruising/anemia
☐ Sickle cell disease
☐ Liver disease/jaundice
☐ Thyroid disease
☐ Diabetes
☐ Kidney/urinary problems
☐ Gastric/abdominal pain/heartburn
☐ Hearing problems
☐ Glaucoma/eye problems
☐ Back pain
☐ Seizures
☐ Other

RISK SCREENING

Have you had a blood transfusion? ☐ Yes ☒ No
Do you smoke? ☐ Yes ☐ No
If yes, how many pack(s)?
Does anyone in your household smoke? ☐ Yes ☐ No
Do you drink alcohol? ☐ Yes ☐ No
If yes, how often?_______ How much?
When was your last drink?_______/_______/_______
Do you take any recreational drugs? ☐ Yes ☐ No
If yes, type:_______ Route:
Frequency:_______ Date last used:_______/_______/_______

FOR WOMEN Ages 12–52

Is there any chance you could be pregnant? ☐ Yes ☒ No
If yes, expected date (EDC):
Gravida/Para: 3/2

ALL WOMEN

Date of last Pap smear: 10/26
Do you perform regular breast self-exams? ☒ Yes ☐ No

ALL MEN

Do you perform regular testicular exams? ☐ Yes ☐ No

Additional comments:

✗ *Lynnette Hall, RN, BSN*
Signature/Title

Client name: Melissa Gaines
DOB: 3/14
Age: 36
Sex: Female
Education: Bachelor's degree
Occupation: Previously, secretary for hospital administrator—now at home since recent delivery of son
Hours of work: N/A
Household members: Husband age 42, son age 4, son age 3 months—all well
Ethnic background: Caucasian
Religious affiliation: None
Referring physician: Roger Smith, MD (gastroenterology)

Chief complaint:
"I have lost a tremendous amount of weight, and I have been having terrible diarrhea for awhile now. I don't even have the energy to get off the couch right now."

Patient history:
Onset of disease: Patient relates having diarrhea on and off for most of her adult life that she can remember. "It seems that a lot of people in my family have 'funny' stomachs. My mom and grandmother both have problems with diarrhea off and on. Mine got much worse during my most recent pregnancy, and now it is debilitating." She recently delivered a 5 lb 2 oz healthy son at 39 weeks gestation.
She states that she gained 11 lbs during this pregnancy but has since lost about 30 lbs. She now weighs 92 lbs. Her greatest nonpregnant weight was 112 lbs just prior to this pregnancy. She describes the diarrhea as foul smelling and indicates symptoms are not affected by what she eats in that she generally has diarrhea no matter what she eats. "I started out breastfeeding my son but stopped about 3 weeks ago because I felt so bad."
Type of Tx: None at present
PMH: 3 pregnancies—2 live births, 1 miscarriage at 22 weeks. No other significant medical history.
Meds: Prenatal vitamins, Kaopectate
Smoker: Yes
Family Hx: What? CAD *Who?* Father

Physical exam:
General appearance: Thin, pale woman who complains of fatigue, weakness, and diarrhea
Vitals: Temp 98.2°F, BP 108/72 mm Hg, HR 78 bpm/normal, RR 17 bpm
Heart: Regular rate and rhythm. Heart sounds normal.
HEENT:
Eyes: PERRLA sclera pale; fundi benign
Throat: Pharynx clear without postnasal drainage
Genitalia: Deferred
Neurologic: Intact; alert and oriented
Extremities: No edema, strength 5/5
Skin: Pale without lesions
Chest/lungs: Lungs clear to percussion and auscultation
Abdomen: Not distended; diminished bowel sounds

Nutrition Hx:
General: Patient states that she is hungry all the time. "I do eat, but it seems that every time I eat in any large amount that I almost immediately have diarrhea. I do not have nausea or vomiting." Foods that are fried and meat—especially beef—tend to make the diarrhea worse. Relates that she has been relying on chicken noodle soup, crackers, and Sprite for the last several days. Patient states that her greatest nonpregnant weight was prior to her last pregnancy, when she weighed 112 lbs. She gained 11 lbs with her pregnancy, and her full-term son weighed 5 lbs 2 oz.
Usual dietary intake: Likes all foods but has found that she avoids eating because it causes her diarrhea to start.

24-hour recall (prior to admission):
AM: 1 slice whole-wheat toast, 1 tsp butter; hot tea with 2 tsp sugar
Lunch: 1 c chicken noodle soup, 2–3 saltine crackers, ½ c applesauce, 12 oz Sprite; throughout rest of day, sips of Sprite.
Dinner: none

Food allergies/intolerances/aversions: NKA
Previous nutrition therapy? No
Food purchase/preparation: Self
Vit/min intake: Still taking prenatal vitamins
Anthropometric measures: TSF 7.5 mm, MAC 180 mm

Dx:
Celiac disease with secondary malabsorption and anemia

Tx plan:
24-hour stool collection for direct visual examination; white blood cells; occult blood; Sudan black B fat stain; ova and parasites; electrolytes and osmolality; pH; alkalization; 72-hour fecal fat. Upper gastrointestinal endoscopy for small bowel biopsy and possible duodenal aspirate.
Diet: 100-g fat diet × 3 days

Hospital course:
Small bowel biopsy indicates flat mucosa with villus atrophy and hyperplastic crypts—inflammatory infiltrate in lamina propria. Fecal fat indicates steatorrhea and malabsorption. Positive AGA, EMA antibodies.

UH UNIVERSITY HOSPITAL

NAME: Melissa Gaines DOB: 3/14
AGE: 36 SEX: F
PHYSICIAN: Roger Smith, MD

CHEMISTRY

	NORMAL			UNITS
DAY:		Admit	3	
DATE:		11/12	11/15	
TIME:				
LOCATION:				
Albumin	3.5–5	2.9 L		g/dL
Total protein	6–8	5.5 L		g/dL
Prealbumin	16–35	13 L		mg/dL
Transferrin	250–380 (women) 215–365 (men)	350		mg/dL
Sodium	136–145	138		mEq/L
Potassium	3.5–5.5	3.7		mEq/L
Chloride	95–105	101		mEq/L
PO_4	2.3–4.7	2.8		mg/dL
Magnesium	1.8–3	1.6		mg/dL
Osmolality	285–295	275		mmol/kg/H_2O
Total CO_2	23–30	27		mEq/L
Glucose	70–110	72		mg/dL
BUN	8–18	9		mg/dL
Creatinine	0.6–1.2	0.7		mg/dL
Uric acid	2.8–8.8 (women) 4.0–9.0 (men)	2.8		mg/dL
Calcium	9–11	9.1		mg/dL
Bilirubin	≤ 0.3	0.2		mg/dL
Ammonia (NH_3)	9–33	10		μmol/L
ALT	4–36	12		U/L
AST	0–35	8		U/L
Alk phos	30–120	111		U/L
CPK	30–135 (women) 55–170 (men)			U/L
LDH	208–378			U/L
CHOL	120–199	119		mg/dL
HDL-C	> 55 (women) > 45 (men)			mg/dL
VLDL	7–32			mg/dL
LDL	< 130			mg/dL
LDL/HDL ratio	< 3.22 (women) < 3.55 (men)			
Apo A	101–199 (women) 94–178 (men)			mg/dL
Apo B	60–126 (women) 63–133 (men)			mg/dL
TG	35–135 (women) 40–160 (men)			mg/dL
T_4	4–12			mcg/dL
T_3	75–98			mcg/dL
HbA_{1C}	3.9–5.2			%
AGA antibodies	0	+		–
EMA antibodies	0	+		–
Fecal Fat			11.5	

UH *UNIVERSITY HOSPITAL*

NAME: Melissa Gaines DOB: 3/14
AGE: 36 SEX: F
PHYSICIAN: Roger Smith, MD

HEMATOLOGY

DAY:		Admit	
DATE:		11/12	
TIME:			
LOCATION:			
	NORMAL		UNITS
WBC	4.8-11.8	5.2	× 10^3/mm^3
RBC	4.2-5.4 (women) 4.5-6.2 (men)	4.9	× 10^6/mm^3
HGB	12-15 (women) 14-17 (men)	9.5 L	g/dL
HCT	37-47 (women) 40-54 (men)	34 L	%
MCV	80-96	90	μm^3
RETIC	0.8-2.8	0.9	%
MCH	26-32	27	pg
MCHC	31.5-36	30 L	g/dL
RDW	11.6-16.5	11.9	%
Plt Ct	140-440	220	× 10^3/mm^3
Diff TYPE			
ESR	0-25 (women) 0-15 (men)		mm/hr
% GRANS	34.6-79.2	38.6	%
% LYM	19.6-52.7	21.4	%
SEGS	50-62	55	%
BANDS	3-6	4	%
LYMPHS	24-44	28	%
MONOS	4-8	5	%
EOS	0.5-4	1	%
Ferritin	20-120 (women) 20-300 (men)	12 L	mg/mL
ZPP	30-80	85	μmol/mol
Vitamin B_{12}	24.4-100	21.2	ng/dL
Folate	5-25	3	μg/dL
Total T cells	812-2,318		mm^3
T-helper cells	589-1,505		mm^3
T-suppressor cells	325-997		mm^3
PT	11-16		sec

Case Questions

I. Understanding the Disease and Pathophysiology

1. The small bowel biopsy results state, "flat mucosa with villus atrophy and hyperplastic crypts—inflammatory infiltrate in lamina propria." What do these results tell you about the change in the anatomy of the small intestine?

2. What is the etiology of celiac disease? Is anything in Mrs. Gaines's history typical of patients with celiac disease? Explain.

3. How is celiac disease related to the damage to the small intestine that the endoscopy and biopsy results indicate?

4. What are AGA and EMA antibodies? Explain the connection between the presence of antibodies and the etiology of celiac disease.

5. What is a 72-hour fecal fat test? What are the normal results for this test?

6. Mrs. Gaines's laboratory report shows that her fecal fat was 11.5 g fat/24 hours. What does this mean?

7. Why was the patient placed on a 100-g fat diet when her diet history indicates that her symptoms are much worse with fried foods?

II. Understanding the Nutrition Therapy

8. Gluten restriction is the major component of the medical nutrition therapy for celiac disease. What is gluten? Where is it found?

9. Can patients on a gluten-free diet tolerate oats?

10. What sources other than foods might introduce gluten to the patient?

11. Can patients with celiac disease also be lactose intolerant?

III. Nutrition Assessment

A. Evaluation of Weight/Body Composition

12. Calculate the patient's percent UBW and BMI, and explain the nutritional risk associated with each value.

B. Calculation of Nutrient Requirements

13. Calculate this patient's total energy and protein needs using the Harris-Benedict equation or Mifflin-St. Jeor equation.

C. Intake Domain

14. Evaluate Mrs. Gaine's 24-hour recall for adequacy.

15. From the information gathered within the intake domain, list possible nutrition problems using the diagnostic term.

D. Clinical Domain

16. Evaluate Mrs. Gaines's laboratory measures for nutritional significance. Identify all laboratory values that support a nutrition problem.

17. Are the abnormalities identified in question 16 related to the consequences of celiac disease? Explain.

18. Are any symptoms from Mrs. Gaines's physical examination consistent with her laboratory values? Explain.

19. Evaluate Mrs. Gaines's other anthropometric measurements. Using the available data, calculate her arm muscle area.

$$AMA = \left(\frac{MAC}{4\pi}(\pi \times TSF)\right)^2$$

Interpret this information for nutritional significance.

20. From the information gathered within the clinical domain, list possible nutrition problems using the diagnostic term.

IV. Nutrition Diagnosis

21. Can you diagnose Mrs. Gaines with malnutrition? If so, what type? What is your rationale?

22. Select two high-priority nutrition problems and complete the PES statement for each.

V. Nutrition Intervention

23. For each of the PES statements that you have written, establish an ideal goal (based on the signs and symptoms) and an appropriate intervention (based on the etiology).

24. What type of diet would you initially begin when you consider the potential intestinal damage that Mrs. Gaines has?

25. Mrs. Gaines's nutritional status is so compromised that she might benefit from high-calorie, high-protein supplementation. What would you recommend?

26. Would glutamine supplementation help Mrs. Gaines during the healing process? What form of glutamine supplementation would you recommend?

27. What result can Mrs. Gaines expect from restricting all foods with gluten? Will she have to follow this diet for very long?

VI. Nutrition Monitoring and Evaluation

28. Evaluate the following excerpt from Mrs. Gaines's food diary. Identify the foods that might not be tolerated on a gluten/gliadin-free diet. For each food identified, provide an appropriate substitute.

Cornflakes	______________________
Bologna slices	______________________
Lean Cuisine—Ginger Garlic Stir Fry with Chicken	______________________
Skim milk	______________________
Cheddar cheese spread	______________________
Green bean casserole (mushroom soup, onions, green beans)	______________________
Coffee	______________________
Rice crackers	______________________
Fruit cocktail	______________________
Sugar	______________________
Pudding	______________________
V8 juice	______________________
Banana	______________________
Cola	______________________

Bibliography

Biagi F, Campanella J, Martucci S, Pezzimenti D, Ciclitira PJ, Ellis HJ, Corazza GR. A milligram of gluten a day keeps the mucosal recovery away: A case report. *Nutr Rev.* 2004;62(9):360–363.

Dickey W, Hughes DF, McMillan SA. Disappearance of endomysial antibodies in treated celiac disease does not indicate histological recovery. *Am J Gastroenterol.* 2000;95(3):712–714.

Holm K, Mäki M, Vuolteenaho N, Mustalahti K, Ashorn M, Ruuska T, Kaukinen K. Oats in the treatment of childhood coeliac disease: A 2-year controlled trial and a long-term clinical follow-up study. *Aliment Pharmacol Ther.* 2006;23(10):1463–1472.

Kemppainen TA, Kosma VM, Janatuinen EK, Julkunen RJ, Pikkarainen PH, Uusitupa MI. Nutritional status of newly diagnosed celiac disease patients before and after the institution of a celiac disease diet: Association with the grade of mucosal villous atrophy. *Am J Clin Nutr.* 1998;67:482–487.

Lee SK, Lo W, Memeo L, Rotterdam H, Green PH. Duodenal histology in patients with celiac disease after treatment with a gluten-free diet. *Gastrointest Endosc.* 2003;57(2):187–191.

Nelms MN. Assessment of nutrition status and risk. In: Nelms M, Sucher K, Long S. *Nutrition Therapy and Pathophysiology.* Belmont, CA: Thomson/Brooks-Cole; 2007:101–135.

Nelms MN. Diseases of the lower gastrointestinal tract. In: Nelms M, Sucher K, Long S. *Nutrition Therapy and Pathophysiology.* Belmont, CA: Thomson/Brooks-Cole; 2007:457–507.

Russo PA, Chartrand LJ, Seidman E. Comparative analysis of serologic screening tests for initial diagnosis of celiac disease. *Pediatrics.* 1999;104:75–78.

Thompson T. Gluten contamination of commercial oats in the United States. *N Engl J Med.* 2004;351:2021–2022.

Thompson T. Oats and the gluten-free diet. *J Am Diet Assoc.* 2003;103(3):376–379.

Internet Resources

Celiac Foundation. http://www.celiac.org

Celiac Sprue Association. http://www.csaceliacs.org

National Digestive Disease Information Clearinghouse. http://digestive.niddk.nih.gov/ddiseases/pubs/celiac/

National Library of Medicine/National Institutes of Health: MedlinePlus. http://www.nlm.nih.gov/medlineplus/celiacdisease.html

University of Chicago: Celiac Disease Center. http://www.celiacdisease.net

U.S. Department of Agriculture: Nutrient Data Laboratory. http://www.ars.usda.gov/ba/bhnrc/ndl

Case 13

Diverticulosis with Incidence of Diverticulitis

Objectives

After completing this case, the student will be able to:

1. Apply knowledge of the pathophysiology of diverticular diseases to identify and explain common nutritional problems associated with these conditions.
2. Explain the rationale of nutrition therapy in the management of diverticulosis and diverticulitis.
3. Evaluate the nutrient composition of a dietary history.
4. Modify recipes for individual nutrient needs.
5. Analyze nutrition assessment data to evaluate nutritional status and identify specific nutrition problems.
6. Determine nutrition diagnoses and write appropriate PES statements.
7. Develop a nutrition care plan with appropriate measurable goals, interventions, and strategies for monitoring and evaluation that addresses the nutrition diagnoses of this case.

Dr. Greer admitted Mrs. Edna Meyer after she experienced rectal bleeding. An upper GI source of bleeding was ruled out, but the colonoscopy noted numerous diverticula. With additional evidence of a lower GI bleed in the sigmoid colon, a diagnosis of diverticulosis was determined.

Name: Edna Meyer
DOB: 1/17 (age 62)
Physician: Boyd Greer, MD

ADMISSION DATABASE

BED #	DATE:	TIME:	TRIAGE STATUS (ER ONLY):
2	4/22	1400	☐ Red ☐ Yellow ☐ Green ☐ White

Initial Vital Signs

TEMP:	RESP:	SAO$_2$:	
98.8	15		
HT:	**WT (lb):**	**B/P:**	**PULSE:**
5′1″	155	120/82	72
LAST TETANUS		**LAST ATE**	**LAST DRANK**
2 years ago		this AM	this AM

PRIMARY PERSON TO CONTACT:
Name: Leonard Meyer
Home #: 555-225-7855
Work #:

ORIENTATION TO UNIT: ☐ Call light ☐ Television/telephone ☐ Bathroom ☐ Visiting ☐ Smoking ☐ Meals ☐ Patient rights/responsibilities

CHIEF COMPLAINT/HX OF PRESENT ILLNESS

"I had a lot of bright red blood in my bowel movement yesterday morning."

PERSONAL ARTICLES: (Check if retained/describe)
☐ Contacts ☐ R ☐ L ☒ Dentures ☒ Upper ☒ Lower
☒ Jewelry: wedding band
☒ Other: glasses

ALLERGIES: Meds, Food, IVP Dye, Seafood: Type of Reaction

NKA

VALUABLES ENVELOPE:
☐ Valuables instructions

PREVIOUS HOSPITALIZATIONS/SURGERIES

3 live births

INFORMATION OBTAINED FROM:
☒ Patient ☐ Previous record
☒ Family ☐ Responsible party

Signature *Edna Meyer*

Home Medications (including OTC) **Codes: A=Sent home** **B=Sent to pharmacy** **C=Not brought in**

Medication	Dose	Frequency	Time of Last Dose	Code	Patient Understanding of Drug
Prinivil	5 mg	daily	0800	C	yes

Do you take all medications as prescribed? ☒ Yes ☐ No If no, why?

PATIENT/FAMILY HISTORY

☐ Cold in past two weeks
☐ Hay fever
☐ Emphysema/lung problems
☐ TB disease/positive TB skin test
☒ Cancer Mother
☐ Stroke/past paralysis
☐ Heart attack
☐ Angina/chest pain
☐ Heart problems
☒ High blood pressure Patient
☐ Arthritis
☐ Claustrophobia
☐ Circulation problems
☐ Easy bleeding/bruising/anemia
☐ Sickle cell disease
☐ Liver disease/jaundice
☐ Thyroid disease
☐ Diabetes
☐ Kidney/urinary problems
☐ Gastric/abdominal pain/heartburn
☐ Hearing problems
☐ Glaucoma/eye problems
☐ Back pain
☐ Seizures
☐ Other

RISK SCREENING

Have you had a blood transfusion? ☐ Yes ☒ No
Do you smoke? ☐ Yes ☒ No
If yes, how many pack(s)?
Does anyone in your household smoke? ☐ Yes ☒ No
Do you drink alcohol? ☐ Yes ☒ No
If yes, how often?_______ How much?
When was your last drink?______/______/______
Do you take any recreational drugs? ☐ Yes ☒ No
If yes, type:______ Route:
Frequency:______ Date last used:______/______/______

FOR WOMEN Ages 12–52

Is there any chance you could be pregnant? ☐ Yes ☐ No
If yes, expected date (EDC):
Gravida/Para:

ALL WOMEN

Date of last Pap smear: 1/02 this year
Do you perform regular breast self-exams? ☒ Yes ☐ No

ALL MEN

Do you perform regular testicular exams? ☐ Yes ☐ No

Additional comments:

x *Betsy Temple, LPN*
Signature/Title

Client name: Edna Meyer
DOB: 1/17
Age: 62
Sex: Female
Education: High school diploma
Occupation: Homemaker and works at home as a seamstress
Hours of work: Varies
Household members: Husband age 66, well; granddaughters ages 13, 15
Ethnic background: African American
Religious affiliation: Baptist
Referring physician: Dr. Boyd Greer, family practice

Chief complaint:
"I had a lot of bright red blood in my bowel movement yesterday morning."

Patient history:
Onset of disease: History of constipation off and on for most of adult life. But recently has also had some episodes of diarrhea with crampy LLQ pain. Presented to MD's office with complaint of blood expelled with bowel movement that morning. Has had 2 other episodes of bleeding in past 24 hours.
Type of Tx: None at present
PMH: Hypertension Dx 3 years previous
Meds: Prinivil (lisinopril) 5 mg daily
Smoker: No
Family Hx: What? CA *Who?* Mother died of ovarian cancer; father died of colon cancer

Physical exam:
General appearance: Slightly overweight, 62-year-old African American woman in no acute distress; somewhat anxious
Vitals: Temp 98.8°F, BP 120/82 mm Hg, HR 72 bpm, RR 15 bpm
Heart: S1 and S2 clear; no rub, gallop, or murmur; regular rate
HEENT: Unremarkable—normocephalic
Eyes: PERRLA, fundi without lesions
Ears: Clear
Nose: Clear
Throat: Supple, no adenopathy or thyromegaly, no bruits
Genitalia: Deferred
Neurologic: Alert and oriented × 4; strength 5/5 throughout, DTRs 2+ and symmetrical, sensation intact
Extremities: No edema
Skin: Warm, dry to touch
Chest/lungs: Clear to auscultation and percussion
Peripheral vascular: Peripheral pulses palpable
Abdomen: Positive bowel sounds throughout, nontender, nondistended

Nutrition Hx:

General: "My appetite is pretty good—well, probably too good. I like to cook and bake—especially for my granddaughters. I have a problem with being regular, though, and it seems to have gotten worse. I try to drink prune juice or eat prunes, and that helps a little."

Usual dietary intake:
Breakfast: White toast with butter and jam, fried egg, coffee
Lunch: Soup or sandwich—sometimes leftovers from previous day, coffee
Dinner: Meat, 1–2 vegetables, rice or potatoes, bread or biscuits, iced tea or coffee

24-hour recall: 2 slices white toast with 3 tsp margarine, 2 tsp jelly, ½ c sliced prunes, black coffee; 2 slices white bread with 1 oz ham, 1 tbsp mayonnaise, 2 oz potato chips, black coffee; 2–3 oz pork chop (trimmed and fried in 2 tbsp corn oil), 1 c macaroni and cheese, 1 biscuit, water, 2 (1″) slices pound cake with ½ c vanilla ice cream
Food allergies/intolerances/aversions: None
Previous nutrition therapy? No
Food purchase/preparation: Self
Vit/min intake: None

Dx:

Diverticulosis with evidence of lower GI bleed in sigmoid colon

Tx plan:

NPO NG to low wall suction D5NS @ 50 cc/hr; metronidazole 1 g loading dose, then 500 mg q 6 h; ciprofloxacin 400 mg q 12 h; strict I/O; schedule for colonoscopy

Hospital course:

NG aspirate heme negative—upper GI source of bleed ruled out. Colonoscopy negative for active bleeding, but numerous diverticula noted.

UH UNIVERSITY HOSPITAL

NAME: Edna Meyer DOB: 1/17
AGE: 62 SEX: F
PHYSICIAN: B. Greer, MD

CHEMISTRY

DAY: 1
DATE: 4/22
TIME:
LOCATION:

	NORMAL	UNITS	
Albumin	3.5-5	3.8	g/dL
Total protein	6-8	6.9	g/dL
Prealbumin	16-35		mg/dL
Transferrin	250-380 (women) 215-365 (men)		mg/dL
Sodium	136-145	138	mEq/L
Potassium	3.5-5.5	4.2	mEq/L
Chloride	95-105	101	mEq/L
PO_4	2.3-4.7	3.2	mg/dL
Magnesium	1.8-3	1.9	mg/dL
Osmolality	285-295	285	mmol/kg/H_2O
Total CO_2	23-30	25	mEq/L
Glucose	70-110	101	mg/dL
BUN	8-18	12	mg/dL
Creatinine	0.6-1.2	1.1	mg/dL
Uric acid	2.8-8.8 (women) 4.0-9.0 (men)	3.5	mg/dL
Calcium	9-11	9.1	mg/dL
Bilirubin	≤0.3	0.2	mg/dL
Ammonia (NH_3)	9-33	9	μmol/L
ALT	4-36	14	U/L
AST	0-35	8	U/L
Alk phos	30-120	244	U/L
CPK	30-135 (women) 55-170 (men)		U/L
LDH	208-378		U/L
CHOL	120-199	175	mg/dL
HDL-C	> 55 (women) > 45 (men)	62	mg/dL
VLDL	7-32		mg/dL
LDL	< 130	111	mg/dL
LDL/HDL ratio	< 3.22 (women) < 3.55 (men)	1.79	
Apo A	101-199 (women) 94-178 (men)		mg/dL
Apo B	60-126 (women) 63-133 (men)		mg/dL
TG	35-135 (women) 40-160 (men)	155	mg/dL
T_4	4-12		mcg/dL
T_3	75-98		mcg/dL
HbA_{1C}	3.9-5.2		%

UH UNIVERSITY HOSPITAL

NAME: Edna Meyer
AGE: 62
PHYSICIAN: B. Greer, MD

DOB: 1/17
SEX: F

HEMATOLOGY

DAY: 1
DATE: 4/22
TIME:
LOCATION:

	NORMAL		UNITS
WBC	4.8–11.8	8.5	$\times 10^3/mm^3$
RBC	4.2–5.4 (women) 4.5–6.2 (men)	4.2	$\times 10^6/mm^3$
HGB	12–15 (women) 14–17 (men)	12	g/dL
HCT	37–47 (women) 40–54 (men)	37	%
MCV	80–96	85	μm^3
RETIC	0.8–2.8	1.1	%
MCH	26–32	28	pg
MCHC	31.5–36	32.5	g/dL
RDW	11.6–16.5	12.2	%
Plt Ct	140–440		$\times 10^3/mm^3$
Diff TYPE			
ESR	0–25 (women) 0–15 (men)	18	mm/hr
% GRANS	34.6–79.2	55.2	%
% LYM	19.6–52.7	44.6	%
SEGS	50–62		%
BANDS	3–6		%
LYMPHS	24–44		%
MONOS	4–8		%
EOS	0.5–4		%
Ferritin	20–120 (women) 20–300 (men)		mg/mL
ZPP	30–80		µmol/mol
Vitamin B_{12}	24.4–100		ng/dL
Folate	5–25		µg/dL
Total T cells	812–2,318		mm^3
T-helper cells	589–1,505		mm^3
T-suppressor cells	325–997		mm^3
PT	11–16		sec

Case Questions

I. Understanding the Disease and Pathophysiology

1. Define *diverticulosis.*

2. What are the possible factors that contribute to the etiology of diverticulosis? Does Mrs. Meyer present with any of these factors?

3. What are the possible complications of diverticulosis?

4. What symptoms did Mrs. Meyer indicate in the physician's H & P that are consistent with diverticulosis?

II. Understanding the Nutrition Therapy

5. Research indicates that low fiber intake may be related to the risk for the development of diverticulosis. What is the optimal fiber intake for Mrs. Meyer? What guideline would you use to determine this optimal fiber intake?

6. Historically, patients with a history of diverticulitis have been told to avoid nuts, seeds, and hulls from foods. What foods (if any) would you suggest that she avoid in the future?

III. Nutrition Assessment

A. Evaluation of Weight/Body Composition

7. Evaluate Mrs. Meyer's anthropometric data by determining UBW, percent UBW, and BMI. Interpret your assessment.

B. Calculation of Nutrient Requirements

8. What methods could be used to estimate her nutrient requirements? Select one of these methods and calculate Mrs. Meyer's energy and protein requirements. Explain your rationale for choosing the method you used.

C. Intake Domain

9. Analyze Mrs. Meyer's 24-hour recall using a computerized dietary analysis program.

10. What are the recommendations for percentage of calories from carbohydrate, protein, and fat for Mrs. Meyer? What guideline did you use and why?

11. How does her dietary intake compare to recommendations for kcal, protein, fat, and fiber? Complete the following table.

Nutrient	Recommended Amount	Mrs. Meyer's Intake per 24-Hour Recall
Energy		
Protein		
Carbohydrate		
Fat		
Fiber		

12. From the information gathered within the intake domain, list possible nutrition problems using the diagnostic term.

D. Behavioral–Environmental Domain

13. After reading Mrs. Meyer's medical and nutritional history, list any nutrition problems within the behavioral–environmental domain using the diagnostic term.

IV. Nutrition Diagnosis

14. Select two high-priority nutrition problems and complete the PES statement for each.

V. Nutrition Intervention

15. Mrs. Meyer is currently on clear liquids. What diet would you recommend for her to progress to while she is in the hospital?

16. What should her nutrition therapy goal(s) be as the inflammation decreases?

17. For each of the PES statements that you have written, establish an ideal goal (based on the signs and symptoms) and an appropriate intervention (based on the etiology).

VI. Nutrition Monitoring and Evaluation

18. Using your previous nutrient analysis and Mrs. Meyer's medical and nutritional history, determine whether she should be concerned about any of her vitamin or mineral intakes. Explain your rationale. Would you recommend any supplementation?

19. Mrs. Meyer tells you she has several recipes for homemade quick breads that she likes to prepare. She wonders if there is a way to increase the amount of fiber in the banana bread she makes. Analyze the following recipe for fiber and fat content, and then make recommendations to increase the fiber content in the recipe. Could you make any recommendations to decrease fat content as well?

Recipe: Edna's Banana Bread

Ingredient	Fiber (g)	Fat (g)	Substitution	Fiber (g)	Fat (g)
½ c margarine					
1 c sugar					
2 eggs					
1¾ c all-purpose flour					
1 tsp baking soda					
1 tsp baking powder					
½ tsp salt					
3 jars of banana baby food					
1 tsp vanilla					
Total for recipe (1 loaf)					
Total per serving—12 slices per loaf					

20. Dr. Greer has suggested that Mrs. Meyer take Fiberall, Benefiber, or Metamucil to increase her fiber intake. What are these supplements, and how might they help?

21. What types of fiber should she increase? Identify two suggestions to increase her fiber intake using her 24-hour recall as a guideline.

Bibliography

Beitz JM. Diverticulosis and diverticulitis spectrum of a modern malady. *J Wound Ostomy Continence Nurs.* 2004;31:75–82.

Bogardus ST. What do we know about diverticular disease? A brief overview. *J Clin Gastroenterol.* 2006;40(7 Suppl 3):S108–S111.

Diverticular disease. The importance of getting enough fiber. *Mayo Clin Health Lett.* 2005;23(2):1–3.

Eglash A, Lane CH, Schneider DM. Clinical inquiries. What is the most beneficial diet for patients with diverticulosis? *J Fam Pract.* 2006;55(9):813–815.

Escott-Stump S. *Nutrition and Diagnosis-Related Care.* 6th ed. Baltimore, MD: Williams & Wilkins; 2007.

Korzenik JR. Case closed? Diverticulitis: Epidemiology and fiber. *J Clin Gastroenterol.* 2006;40(7 Suppl 3):S112–S116.

Nelms MN. Assessment of nutrition status and risk. In: Nelms M, Sucher K, Long S. *Nutrition Therapy and Pathophysiology.* Belmont, CA: Thomson/Brooks-Cole; 2007:101–135.

Nelms MN. Diseases of the lower gastrointestinal tract. In: Nelms M, Sucher K, Long S. *Nutrition Therapy and Pathophysiology.* Belmont, CA: Thomson/Brooks-Cole; 2007:457–507.

Internet Resources

American Society of Colorectal Surgeons. http://www.fascrs.org

National Digestive Diseases Information Clearinghouse: Diverticulosis and Diverticulitis. http://digestive.niddk.nih.gov/ddiseases/pubs/diverticulosis/

National Library of Medicine/National Institutes of Health: MedlinePlus. http://www.nlm.nih.gov/medlineplus/diverticulosisanddiverticulitis.html

U.S. Department of Agriculture: Nutrient Data Laboratory. http://www.ars.usda.gov/ba/bhnrc/ndl

Case 14

Inflammatory Bowel Disease: Crohn's Disease

Objectives

After completing this case, the student will be able to:

1. Apply knowledge of the pathophysiology of Crohn's disease to identify and explain common nutritional problems associated with this disease.
   - **a.** Describe the physiological changes resulting from Crohn's disease.
   - **b.** Identify the nutritional consequences of Crohn's disease.
   - **c.** Identify the nutritional consequences of surgical resection of the small intestine.
2. Describe the current medical care for Crohn's disease.
3. Identify potential drug–nutrient interactions.
4. Analyze nutrition assessment data to evaluate nutritional status and identify specific nutrition problems.
5. Determine nutrition diagnoses and write appropriate PES statements.
6. Calculate parenteral nutrition formulations.
7. Evaluate a parenteral nutrition regimen.
8. Develop a nutrition care plan with appropriate measurable goals, interventions, and strategies for monitoring and evaluation that addresses the nutrition diagnoses of this case.

Matt Sims was diagnosed with Crohn's disease 2½ years ago. He is now admitted with an acute exacerbation of that disease.

Name: Matt Sims
DOB: 7/22 (age 35)
Physician: David Tucker, MD

ADMISSION DATABASE

BED #	DATE:	TIME:	TRIAGE STATUS (ER ONLY):
2	12/15	1500	☐ Red ☐ Yellow ☐ Green ☐ White

Initial Vital Signs

TEMP:	RESP:	SAO$_2$:	
101.5	18		
HT:	**WT (lb):**	**B/P:**	**PULSE:**
5′9″	140	125/82	81
LAST TETANUS		**LAST ATE**	**LAST DRANK**
unknown		this AM	30 minutes ago

PRIMARY PERSON TO CONTACT:
Name: Mary Sims
Home #: 555-447-1476
Work #: 555-447-2322

ORIENTATION TO UNIT: ☒ Call light ☒ Television/telephone ☒ Bathroom ☒ Visiting ☒ Smoking ☒ Meals ☒ Patient rights/responsibilities

CHIEF COMPLAINT/HX OF PRESENT ILLNESS

"I was diagnosed with Crohn's disease almost $2\frac{1}{2}$ years ago. I have really done OK until this year. My doctor has mentioned that if I don't respond quickly to treatment, I might have to have surgery."

ALLERGIES: Meds, Food, IVP Dye, Seafood: Type of Reaction

Maybe milk; otherwise none

PREVIOUS HOSPITALIZATIONS/SURGERIES

$2\frac{1}{2}$ years ago–Dx. Crohn's disease

This past September hospitalized with abcess and acute exacerbation of Crohn's

PERSONAL ARTICLES: (Check if retained/describe)
☐ Contacts ☐ R ☐ L ☐ Dentures ☐ Upper ☐ Lower
☐ Jewelry:
☐ Other:

VALUABLES ENVELOPE:
☐ Valuables instructions

INFORMATION OBTAINED FROM:
☒ Patient ☐ Previous record
☐ Family ☐ Responsible party

Signature *Matt Sims*

Home Medications (including OTC) **Codes: A=Sent home** **B=Sent to pharmacy** **C=Not brought in**

Medication	Dose	Frequency	Time of Last Dose	Code	Patient Understanding of Drug
6-mercaptopurine	60 mg	daily	this AM	C	yes
multivitamin	1	daily	yesterday	C	yes

Do you take all medications as prescribed? ☒ Yes ☐ No If no, why?

PATIENT/FAMILY HISTORY

☐ Cold in past two weeks
☐ Hay fever
☐ Emphysema/lung problems
☐ TB disease/positive TB skin test
☐ Cancer
☐ Stroke/past paralysis
☐ Heart attack
☐ Angina/chest pain
☐ Heart problems
☐ High blood pressure
☐ Arthritis
☐ Claustrophobia
☐ Circulation problems
☐ Easy bleeding/bruising/anemia
☐ Sickle cell disease
☐ Liver disease/jaundice
☐ Thyroid disease
☐ Diabetes
☐ Kidney/urinary problems
☒ Gastric/abdominal pain/heartburn Patient
☐ Hearing problems
☐ Glaucoma/eye problems
☐ Back pain
☐ Seizures
☐ Other

RISK SCREENING

Have you had a blood transfusion? ☐ Yes ☒ No
Do you smoke? ☐ Yes ☒ No
If yes, how many pack(s)?
Does anyone in your household smoke? ☐ Yes ☒ No
Do you drink alcohol? ☐ Yes ☒ No
If yes, how often? ______ How much? ______/______/______
When was your last drink? ______/______/______
Do you take any recreational drugs? ☐ Yes ☒ No
If yes, type:______ Route:
Frequency:______ Date last used:______/______/______

FOR WOMEN Ages 12–52

Is there any chance you could be pregnant? ☐ Yes ☐ No
If yes, expected date (EDC):
Gravida/Para:

ALL WOMEN

Date of last Pap smear:
Do you perform regular breast self-exams? ☐ Yes ☐ No

ALL MEN

Do you perform regular testicular exams? ☒ Yes ☐ No

Additional comments:

x *Rosie Martin, RN*
Signature/Title

Client name: Matthew Sims
DOB: 7/22
Age: 35
Sex: Male
Education: Bachelor's degree
Occupation: High school math teacher
Hours of work: 8–4:30, some after-school meetings and responsibilities as advisor for school clubs
Household members: Wife age 32, well; son age 5, well
Ethnic background: Caucasian
Religious affiliation: Episcopalian
Referring physician: David Tucker, MD (gastroenterology)

Chief complaint:
"I was diagnosed with inflammatory bowel disease almost 3 years ago. At first, they thought I had ulcerative colitis, but 6 months later, it was identified as Crohn's disease. I was really sick at that time and was in the hospital for more than 2 weeks. I have done OK until school started this fall. I've noticed more diarrhea and abdominal pain, but I tried to keep going since school just started. I was here in September, and we switched medicine. I was a little better, and I went back to work. Now my abdominal pain is unbearable—I seem to have diarrhea constantly, and now I am running a fever."

Patient history:
Onset of disease: Dx Crohn's disease 2½ years ago. Initial diagnostic workup indicated acute disease within last 5–7 cm of jejunum and first 5 cm of ileum. Regimens have included corticosteroids, Azulfidine, and most recently 6-mercaptopurine.
Type of Tx: 6-mercaptopurine
PMH: Noncontributory
Meds: 6-mercaptopurine
Smoker: No
Family Hx: Noncontributory

Physical exam:
General appearance: Thin, 36-year-old white male in apparent distress
Vitals: Temp 101.5°F, BP 125/82 mm Hg, HR 81 bpm/normal, RR 18 bpm
Heart: RRR without murmurs or gallops
HEENT:
Eyes: PERRLA, normal fundi
Ears: Noncontributory
Nose: Noncontributory
Throat: Pharynx clear
Rectal: No evidence of perianal disease
Neurologic: Oriented × 4
Extremities: No edema; pulses full; no bruits; normal strength, sensation, and DTR
Skin: Warm, dry
Chest/lungs: Lungs clear to auscultation and percussion
Abdomen: Distension, extreme tenderness with rebound and guarding; minimal bowel sounds

Nutrition Hx:

General: Patient states he has been eating fairly normally for the last year. After hospitalization at initial diagnosis, he had lost almost 25 lbs, which he regained. He initially ate a low-fiber diet and worked hard to regain the weight he had lost. He drank Boost between meals for several months. His usual weight before his illness was 166–168 lbs. He was at his highest weight (168 lbs) about 6 months ago, but now states he has lost most of what he regained and has even lost more since his last hospitalization when he was at 140 lbs.

Recent dietary intake:

AM:	Cereal, small amount of skim milk, toast or bagel, juice
AM snack:	Cola, sometimes crackers or pastry
Lunch:	Sandwich (ham or turkey) from home, fruit, chips, cola
Dinner:	Meat, pasta or rice, some type of bread; rarely eats vegetables
Bedtime snack:	Cheese and crackers, cookies, cola

24-hour recall: Has been on clear liquids for past 24 hours since admission
Food allergies/intolerances/aversions: Mr. Sims says he has never liked milk but purposefully avoided it after his diagnosis. He does consume milk products, such as cheese, usually without any difficulty.
Previous nutrition therapy? Yes *If yes, when:* Last hospitalization
What? "The dietitian talked to me about ways to decrease my diarrhea—ways to keep from being dehydrated—and then we worked out a plan to help me regain weight. I know what to do—it is just that the pain and diarrhea make my appetite so bad. It is really hard for me to eat."
Food purchase/preparation: Self and wife
Vit/min intake: Multivitamin daily

Tx plan:

R/O acute exacerbation of Crohn's disease vs. infection vs. small bowel obstruction.
CBC/Chem 24
ASCA
CT scan of abdomen and possible esophagogastroduodenoscopy
D5W NS @ 75 cc/hr; Clear liquids
Surgical consult
Nutrition support consult

Hospital course:

CT scan indicated bowel obstruction; Crohn's disease classified as severe-fulminant disease. CDAI score of 400. Mr. Sims underwent resection of 200 cm of jejunum and proximal ileum with placement of jejunostomy. The ileocecal valve was preserved. Mr. Sims did not have an ileostomy, and his entire colon remains intact. Mr. Sims was placed on parenteral nutrition support immediately postoperatively, and a nutrition support consult was ordered.

UH UNIVERSITY HOSPITAL

NAME: Matt Sims
DOB: 7/22
AGE: 35
SEX: M
PHYSICIAN: D. Tucker, MD

CHEMISTRY

DAY:		Admit	
DATE:		12/15	
TIME:			
LOCATION:			
	NORMAL		UNITS
Albumin	3.5–5	3.2 L	g/dL
Total protein	6–8	5.5 L	g/dL
Prealbumin	16–35	11 L	mg/dL
Transferrin	250–380 (women) 215–365 (men)	180 L	mg/dL
C-reactive protein	< 1.0	2.8 H	mg/dL
Sodium	136–145	136	mEq/L
Potassium	3.5–5.5	3.7	mEq/L
Chloride	95–105	101	mEq/L
PO_4	2.3–4.7	2.9	mg/dL
Magnesium	1.8–3	1.8	mg/dL
Osmolality	285–295	280	mmol/kg/H_2O
Total CO_2	23–30	26	mEq/L
Glucose	70–110	82	mg/dL
BUN	8–18	11	mg/dL
Creatinine	0.6–1.2	0.8	mg/dL
Uric acid	2.8–8.8 (women) 4.0–9.0 (men)	5.1	mg/dL
Calcium	9–11	9.1	mg/dL
Bilirubin	≤ 0.3	0.2	mg/dL
Ammonia (NH_3)	9–33	11	μmol/L
ALT	4–36	35	U/L
AST	0–35	22	U/L
Alk phos	30–120	123	U/L
CPK	30–135 (women) 55–170 (men)		U/L
LDH	208–378		U/L
CHOL	120–199	149	mg/dL
HDL-C	> 55 (women) > 45 (men)	38	mg/dL
VLDL	7–32		mg/dL
LDL	< 130	111	mg/dL
LDL/HDL ratio	< 3.22 (women) < 3.55 (men)	2.92	
Apo A	101–199 (women) 94–178 (men)		mg/dL
Apo B	60–126 (women) 63–133 (men)		mg/dL
TG	35–135 (women) 40–160 (men)	85	mg/dL
T_4	4–12		mcg/dL
T_3	75–98		mcg/dL
HbA_{1C}	3.9–5.2		%

UH UNIVERSITY HOSPITAL

NAME: Matt Sims DOB: 7/22
AGE: 35 SEX: M
PHYSICIAN: D. Tucker, MD

HEMATOLOGY

DAY:		1	
DATE:		12/15	
TIME:			
LOCATION:			
	NORMAL		UNITS
WBC	4.8–11.8	11.1	$\times 10^3/mm^3$
RBC	4.2–5.4 (women) 4.5–6.2 (men)	4.9	$\times 10^6/mm^3$
HGB	12–15 (women) 14–17 (men)	12.9	g/dL
HCT	37–47 (women) 40–54 (men)	38	%
MCV	80–96	87	μm^3
RETIC	0.8–2.8	0.9	%
MCH	26–32	30	pg
MCHC	31.5–36	33	g/dL
RDW	11.6–16.5	13.2	%
Plt Ct	140–440	422	$\times 10^3/mm^3$
Diff TYPE			
ESR	0–25 (women) 0–15 (men)	35	mm/hr
% GRANS	34.6–79.2		%
% LYM	19.6–52.7		%
SEGS	50–62		%
BANDS	3–6		%
LYMPHS	24–44		%
MONOS	4–8		%
EOS	0.5–4		%
Ferritin	20–120 (women) 20–300 (men)	16 L	mg/mL
ZPP	30–80	85 H	µmol/mol
Vitamin B_{12}	24.4–100	75	ng/dL
Folate	5–25	6	µg/dL
Total T cells	812–2,318		mm^3
T-helper cells	589–1,505		mm^3
T-suppressor cells	325–997		mm^3
PT	11–16	15	sec
ASCA	neg	+	
ANCA	neg		

Name: Matt Sims
Physician: D. Tucker, MD

PATIENT CARE SUMMARY SHEET

Date: 12/20			Room: 315						Wt Yesterday: 138						Today: 138.25									
Temp °F	NIGHTS								DAYS								EVENINGS							
	00	01	02	03	04	05	06	07	08	09	10	11	12	13	14	15	16	17	18	19	20	21	22	23
105																								
104																								
103																								
102																								
101																								
100		×																						
99																								
98																								
97																								
96																								
Pulse	77																							
Respiration	18																							
BP	101/73																							
Blood Glucose	142																							
Appetite/Assist	NPO																							
INTAKE																								
Oral																								
IV	85	85	85	85	85	85	85	85	85	85	85	85	85	85	85	85	85	85	85	85	85	85	85	85
TF Formula/Flush																								
Shift Total	680								680								680							
OUTPUT																								
Void																								
Cath.		520								250				250				350			220			
Emesis																								
BM																								
Drains																								
Shift Total	520								500								570							
Gain	+160								+180								+110							
Loss																								
Signatures	Angela Phelps, RD								Leslie Snyder, RN								D. Magee, RN							

Case Questions

I. Understanding the Disease and Pathophysiology

1. What is inflammatory bowel disease? What does current medical literature indicate regarding its etiology?

2. Mr. Sims was initially diagnosed with ulcerative colitis and then diagnosed with Crohn's. How could this happen? What are the similarities and differences between Crohn's disease and ulcerative colitis?

3. What does a CDAI score of 400 indicate? What does a classification of severe-fulminant disease indicate?

4. What did you find in Mr. Sims's history and physical that is consistent with his diagnosis of Crohn's? Explain.

5. Crohn's patients often have extraintestinal symptoms of the disease. What are some examples of these symptoms? Is there evidence of these in his history and physical?

6. Which laboratory values are consistent with an exacerbation of his Crohn's disease? Identify and explain.

7. Is Mr. Sims a likely candidate for short bowel syndrome? Define *short bowel syndrome*, and provide a rationale for your answer.

8. What type of adaptation can the small intestine make after resection?

9. For what classic symptoms of short bowel syndrome should Mr. Sims's health care team monitor?

10. Mr. Sims is being evaluated for participation in a clinical trial with the new drug called Teduglutide (ALX-0600). What is this drug, and how might it help Mr. Sims?

II. Understanding the Nutrition Therapy

11. What are the potential nutritional consequences of Crohn's disease?

12. Mr. Sims has had a 200-cm resection of his jejunum and proximal ileum. How long is the small intestine, and how significant is this resection?

13. What nutrients are normally digested and absorbed in the portion of the small intestine that has been resected?

III. Nutrition Assessment

A. Evaluation of Weight/Body Composition

14. Evaluate Mr. Sims's anthropometric data by evaluating UBW and BMI. Interpret your calculations.

B. Calculation of Nutrient Requirements

15. Calculate Mr. Sims's energy requirements. Compare the Harris-Benedict, Mifflin-St. Jeor, and Ireton-Jones equations.

16. Which numbers would you use as a goal for Mr. Sims's nutrition support? Explain.

17. What would you estimate Mr. Sims's protein requirements to be?

C. Intake Domain

18. Based on your evaluation of Mr. Sims's nutritional history, taking into consideration his current hospital course, and from all information gathered within the intake domain, list possible nutrition problems.

D. Clinical Domain

19. Identify any significant laboratory measurements from both his hematology and his chemistry labs.

20. Based on your evaluation of Mr. Sims's clinical data, and taking into consideration his current hospital course, list the nutrition problems within the clinical domain.

E. Behavioral–Environmental Domain

21. From the information gathered within the behavioral–environmental domain, list possible nutrition problems using the diagnostic term.

IV. Nutrition Diagnosis

22. Select two high-priority nutrition problems and complete the PES statement for each.

V. Nutrition Intervention

23. The surgeon notes that Mr. Sims probably will not resume eating by mouth for at least 7–10 days. What information would the nutrition support team evaluate in deciding the route for nutrition support?

24. The members of the nutrition support team note that his serum phosphorus and serum magnesium are at the low end of the normal range. Why might that be of concern?

25. What is refeeding syndrome? Is Mr. Sims at risk for this syndrome? How can it be prevented?

26. Mr. Sims was started on parenteral nutrition postoperatively. Initially, he was prescribed to receive 200-g dextrose/L, 42.5-g amino acids/L, 30-g lipid/L. His parenteral nutrition was initiated at 50 cc/hr with a goal rate of 85 cc/hr. Do you agree with the team's decision to initiate parenteral nutrition? Will this meet his estimated nutritional needs? Explain. Calculate pro (g); CHO (g); lipid (g); and total kcal from his PN.

27. For each of the PES statements that you have written, establish an ideal goal (based on the signs and symptoms) and an appropriate intervention (based on the etiology).

VI Nutrition Monitoring and Evaluation

28. Indirect calorimetry revealed the following information.

Measure	Mr. Sims's Data
Oxygen consumption (mL/min)	295
CO_2 production (mL/min)	261
RQ	0.88
RMR	2022

What does this information tell you about Mr. Sims?

29. Would you make any changes in his prescribed nutrition support? What should be monitored to ensure adequacy of his nutrition support? Explain.

30. What should the nutrition support team monitor daily? What should be monitored weekly? Explain your answers.

31. Mr. Sims's serum glucose increased to 145 mg/dL. Why do you think this level is now abnormal? What should be done about it?

32. Evaluate the following 24-hour urine data: 24-hour urinary nitrogen for 12/20: 18.4 grams. By using the daily nursing record that records the amount of PN received, calculate Mr. Sims's nitrogen balance on postoperative day 4. How would you interpret this information? Should you be concerned? Are there problems with the accuracy of nitrogen balance studies? Explain.

33. On post-op day 10, Mr. Sims's team notes that he has had bowel sounds for the previous 48 hours and had his first bowel movement. The nutrition support team recommends consideration of an oral diet. What should Mr. Sims be allowed to try first? What would you monitor for tolerance? If successful, when can the parenteral nutrition be weaned?

34. What would be the primary nutrition concerns as Mr. Sims prepares for rehabilitation after his discharge? Be sure to address his need for supplementation of any vitamins and minerals. Identify two nutritional outcomes with specific measures for evaluation.

Bibliography

Bayless T, Talamini M, Kaufman H, Norwitz L, Kalloo AN. Crohn's disease. The Johns Hopkins Medical Institutions Gastroenterology and Hepatology Resource Center. Available at: http://hopkins-gi.nts.jhu.edu/pages/latin/templates/index.cfm?pg=disease1&organ=6&disease=21&lang_id=1. Accessed April 4, 2008.

Eiden KA. Nutritional considerations in inflammatory bowel disease. *Pract Gastroenterol.* 2003;May:33–54.

Galvez J, Rodriguez-Cabezas ME, Zarzuelo A. Effects of dietary fiber on inflammatory bowel disease. *Mol Nutr Food Res.* 2005;49(6):601–608.

Gottschlich M, ed. *The Science and Practice of Nutrition Support: A Case-Based Core Curriculum.* Dubuque, IA: Kendall/Hunt; 2001.

Jeejeebhoy KN. Clinical nutrition: 6. Management of nutritional problems of patients with Crohn's disease. *CMAJ.* 2002;166(7):913–918.

Jeppesen PB, Sanguinetti EL, Buchman A, Howard L, Scolapio JS, Ziegler TR, Gregory J, Tappenden KA, Holst J, Mortensen PB. Teduglutide (ALX-0600), a dipeptidyl peptidase IV resistant glucagon-like peptide 2 analogue, improves intestinal function in short bowel syndrome patients. *Gut.* 2005;54(9):1224–1231.

Kornbluth A, Sachar DB. Ulcerative colitis practice guidelines in adults (update). American College of Gastroenterology. Practice Parameters Committee. *Am J Gastroenterol.* 2004;99(7):1371–1385.

Krok KL, Lichtenstein GR. Nutrition in Crohn disease. *Curr Opin Gastroenterol.* 2003;19(2):148–153.

Lashner BA. Inflammatory bowel disease. The Cleveland Clinic Disease Management Project. 2005. Available at: http://www.clevelandclinicmeded.com/diseasemanagement/gastro/inflammatory_bowel/inflammatory_bowel.htm. Accessed June 29, 2007.

Nelms MN. Assessment of nutrition status and risk. In: Nelms M, Sucher K, Long S. *Nutrition Therapy and Pathophysiology.* Belmont, CA: Thomson/Brooks-Cole; 2007:101–135.

Nelms MN. Diseases of the lower gastrointestinal tract. In: Nelms M, Sucher K, Long S. *Nutrition Therapy and Pathophysiology.* Belmont, CA: Thomson/Brooks-Cole; 2007:457–507.

Nelms MN. Metabolic stress. In: Nelms M, Sucher K, Long S. *Nutrition Therapy and Pathophysiology.* Belmont, CA: Thomson/Brooks-Cole; 2007:785–804.

Rees-Parrish C. The clinician's guide to short bowel syndrome. *Pract Gastroenterol.* 2005;Sept:67–106.

Skipper A, Nelms MN. Methods of nutrition support. In: Nelms M, Sucher K, Long S. *Nutrition Therapy and Pathophysiology.* Belmont, CA: Thomson/Brooks-Cole; 2007:149–179.

Sundaram A, Koutkia P, Apovian CM. Nutritional management of short bowel syndrome in adults. *J Clin Gastroenterol.* 2002;34:207–220.

Internet Resources

Crohn's Colitis Foundation. http://www.ccfa.org

Mayo Clinic: Chrohn's Disease. http://www.mayoclinic.com/health/crohns-disease/DS00104

National Digestive Diseases Information Clearinghouse: Crohn's Disease. http://digestive.niddk.nih.gov/ddiseases/pubs/crohns/index.htm

Unit Four

NUTRITION THERAPY FOR PANCREATIC AND HEPATOBILIARY DISORDERS

The liver and pancreas are often called "ancillary" organs of digestion; however, the term *ancillary* does little to describe their importance in digestion, absorption, and metabolism of carbohydrate, protein, and lipid. The cases in this section portray common conditions affecting these organs and outline their effects on nutritional status. The incidence of hepatobiliary disease has significantly increased over the last several decades, with cirrhosis being the most frequent diagnosis. The most common cause of cirrhosis is chronic alcohol ingestion; the second most common cause is viral hepatitis. The cases in this section focus on those etiologies.

The incidence of malnutrition is very high in these disease states. It is often difficult for the practitioner to simultaneously treat the disease with appropriate nutrition therapy and prevent malnutrition. Generalized symptoms of these diseases center around interruption of normal metabolism in these organs. Jaundice, anorexia, fatigue, abdominal pain, steatorrhea, and malabsorption are signs or symptoms of hepatobiliary disease. These symptoms may be responsive to nutrition therapy but also interfere with maintenance of an adequate nutritional status.

Case 15 places the nutritional concerns of chronic alcoholism in the context of chronic pancreatitis. Case 16 focuses on infection with the hepatitis C virus (HCV), which is the most common hepatic viral infection in the United States. It is diagnosed in approximately 300,000 Americans each year. Almost 4 million Americans have the disease today; more than 90 percent of them have chronically infected livers, and of those, over 1 million will progress to cirrhosis or hepatic carcinoma. Researchers estimate that between the years 2010 and 2019, health care costs for HCV will approach $81.8 billion. The medical profession hopes that new treatments using interferon and ribavirin will prevent the natural progression of this disease.

In Case 17, the complications of end-stage cirrhosis are explored. Treatment of cirrhosis is primarily supportive. The only cure is a liver transplant. Therefore, nutrition therapy is crucial for preventing protein–calorie malnutrition, minimizing the symptoms of the disease, and maintaining quality of life.

Case 15

Chronic Pancreatitis Secondary to Chronic Alcoholism

Objectives

After completing this case, the student will be able to:

1. Describe the anatomic features and physiologic function of the pancreas.
2. Explain etiology and risk factors for development of chronic pancreatitis.
3. Apply working knowledge of the pathophysiology of chronic pancreatitis.
4. Collect pertinent information, and use nutrition assessment techniques to determine baseline nutritional status.
5. Calculate parenteral nutrition formulations (develop an appropriate parenteral nutrition regimen).
6. Evaluate standard parenteral nutrition formulations.
7. Identify appropriate nutrition goals.
8. Design nutrition education for the patient with alcoholism and chronic pancreatitis.
9. Demonstrate the ability to communicate in the medical record.

Ms. Elena Jordan is admitted for evaluation of her recurring epigastric pain accompanied by nausea, vomiting, diarrhea, and weight loss. Dr. Paula Bennett diagnoses Ms. Jordan with pancreatitis, probably secondary to chronic alcohol ingestion.

Name: Elena Jordan
DOB: 10/7 (age 30)
Physician: P. Bennett, MD

ADMISSION DATABASE

BED #	DATE:	TIME:	TRIAGE STATUS (ER ONLY):
2	2/22	1140	☐ Red ☐ Yellow ☐ Green ☐ White

Initial Vital Signs

TEMP:	RESP:	SAO$_2$:	
100.8	80		
HT:	WT (lb):	B/P:	PULSE:
5′8″	112	125/76	114
LAST TETANUS		LAST ATE	LAST DRANK
10+ years		over 24 hrs ago	5 hrs ago

PRIMARY PERSON TO CONTACT:
Name: Michele Jordan (mother)
Home #: 555-3847
Work #: same

ORIENTATION TO UNIT: ☒ Call light ☒ Television/telephone ☒ Bathroom ☒ Visiting ☒ Smoking ☒ Meals ☒ Patient rights/responsibilities

CHIEF COMPLAINT/HX OF PRESENT ILLNESS

severe abdominal pain

PERSONAL ARTICLES: (Check if retained/describe)
☒ Contacts ☒ R ☒ L ☐ Dentures ☐ Upper ☐ Lower
☒ Jewelry: watch
☐ Other:

ALLERGIES: Meds, Food, IVP Dye, Seafood: Type of Reaction

none known

VALUABLES ENVELOPE:
☐ Valuables instructions

PREVIOUS HOSPITALIZATIONS/SURGERIES

none

INFORMATION OBTAINED FROM:
☒ Patient ☐ Previous record
☐ Family ☐ Responsible party

Signature *Elena Jordan*

Home Medications (including OTC) **Codes: A=Sent home** **B=Sent to pharmacy** **C=Not brought in**

Medication	Dose	Frequency	Time of Last Dose	Code	Patient Understanding of Drug

Do you take all medications as prescribed? ☐ Yes ☐ No If no, why?

PATIENT/FAMILY HISTORY

☐ Cold in past two weeks
☒ Hay fever Patient
☐ Emphysema/lung problems
☐ TB disease/positive TB skin test
☐ Cancer
☐ Stroke/past paralysis
☐ Heart attack
☐ Angina/chest pain
☐ Heart problems

☐ High blood pressure
☐ Arthritis
☒ Claustrophobia Patient
☐ Circulation problems
☐ Easy bleeding/bruising/anemia
☒ Sickle cell disease Maternal grandmother
☐ Liver disease/jaundice
☐ Thyroid disease
☐ Diabetes

☐ Kidney/urinary problems
☐ Gastric/abdominal pain/heartburn
☐ Hearing problems
☐ Glaucoma/eye problems
☐ Back pain
☐ Seizures
☐ Other

RISK SCREENING

Have you had a blood transfusion? ☐ Yes ☒ No
Do you smoke? ☐ Yes ☒ No
If yes, how many pack(s)?
Does anyone in your household smoke? ☐ Yes ☐ No
Do you drink alcohol? ☒ Yes ☐ No
If yes, how often? daily How much? 2-3 drinks
When was your last drink? 2/18
Do you take any recreational drugs? ☒ Yes ☐ No
If yes, type: marijuana Route: inhale
Frequency: <1/month Date last used: doesn't remember

FOR WOMEN Ages 12–52

Is there any chance you could be pregnant? ☐ Yes ☒ No
If yes, expected date (EDC):
Gravida/Para: 0/0

ALL WOMEN

Date of last Pap smear: 2/15
Do you perform regular breast self-exams? ☒ Yes ☐ No

ALL MEN

Do you perform regular testicular exams? ☐ Yes ☐ No

Additional comments:

x *Miriam Link, RN*
Signature/Title

Client name: Elena Jordan
DOB: 10/7
Age: 30
Sex: Female
Education: Bachelor's degree
Occupation: Pharmaceutical sales rep
Hours of work: Varies—usually 50+ hours/week
Household members: Lives alone
Ethnic background: Biracial
Religious affiliation: Agnostic
Referring physician: Paula Bennett, MD (gastroenterology)

Chief complaint:
"I'm tired of hurting so much. I've had this terrible pain in my stomach for the past 2 days. I took a client out to dinner the other night, but I couldn't eat much. This has been happening off and on for the past 9 months, but the pain has never gone around to my back before."

Patient history:
Ms. Jordan is a 30-year-old woman who has been well until 12 months ago when she began to experience bouts of epigastric pain. Most recently, the pain has started to radiate to her back and lasts from 4 hours to 3 days. She c/o poor appetite and a recent, unintentional weight loss of 10 lbs. She reports two loose stools per day for the past 4 months. She says they are foul smelling. As of late, she c/o anorexia and nausea.
Onset of disease: 12 months ago
Type of Tx: Antacids
PMH: Currently weighs 112 lbs; weighed 140 a year ago
Meds: Ortho Tri-Cyclen, 28-day cycle
Smoker: No
Family Hx: No family history of GI disease

Physical exam:
General appearance: Thin, 30-year-old female with temporal muscle wasting who appears to be in a moderate amount of discomfort
Vitals: Temp 100.8°F, BP 125/76 mm Hg, HR 114 bpm, RR 80 bpm
Heart: Regular rate and rhythm, heart sounds normal
HEENT:
  Eyes: PERRLA
  Ears: Noncontributory
  Nose: Noncontributory
  Throat: Noncontributory
Genitalia: Normal female
Neurologic: Alert, oriented × 3
Extremities: Noncontributory
Skin: Smooth, warm, and dry, slightly tented, no edema
Chest/lungs: Lungs are clear

Peripheral vascular: Pulse 4+ bilaterally, warm, no edema
Abdomen: Flat, bowel sounds normal, tenderness in epigastric region; liver and spleen not enlarged

Nutrition Hx:
General: Patient reports that her appetite has usually been good, but for the last 6–9 months, she's had difficulty eating due to nausea. Ms. Jordan works in pharmaceutical sales and travels outside the area two weeks every month. Because a large part of her job entails meeting with clients, she eats many of her meals in restaurants and consumes 2–3 alcoholic beverages per night. When questioned about her history of alcohol intake, Ms. Jordan stated that she started drinking in high school on the weekends when her friends had parties. She drank only beer and had only 1–2 cans per night (total of 2–4 beers per weekend). When she entered college, Ms. Jordan continued to drink on the weekends with her friends. She would drink beer in the college bars and at house parties—often consuming 5 or more drinks in one evening. After graduation from college, she was hired as a pharmaceutical representative, which entailed eating most of her meals (while on the road) in restaurants. When she entertains clients, she feels she has to match them drink for drink to help land business. Ms. Jordan usually drinks wine or beer when out with clients because she believes only people with alcohol problems drink "hard liquor."

Usual dietary intake:
At home:

Breakfast:	Dry bagel, 1 c black coffee
Lunch:	Diet Coke, Lean Cuisine—usually Swedish meatballs (with noodles)
Dinner:	5 oz white wine while preparing dinner; grilled salmon—usually 2–3 oz, seasoned with salt and pepper; baked potato—medium sized, with butter, sour cream, and chives; 2 stalks steamed broccoli with cheese sauce (made from Cheez Whiz); 2 glasses (5 oz) white wine with dinner

On the road:

Breakfast:	¾ c dry cereal (varies) with 1½ c 2% milk, 1 c orange juice, 1 c black coffee
Lunch:	(Often doesn't eat lunch, but when she does) McDonald's fruit and yogurt parfait, medium Diet Coke
Dinner:	Usually some type of appetizer—most likely fried mushrooms; spinach salad with hot bacon dressing; fettuccine Alfredo or small (6 oz) filet mignon with garlic mashed potatoes; 2–3 glasses of wine (6-oz glasses)
After-dinner drink:	Usually sherry (3 oz)

Food allergies/intolerances/aversions: None
Previous nutrition therapy? No
Food purchase/preparation: Self, eats in restaurants often (2 weeks of each month)
Vit/min intake: None
Current diet order: NPO

Tx plan:
Pregnancy test
CBC
Chemistry with liver and pancreatic enzymes

Urinalysis
Upper GI w/small bowel follow-through
CT scan of abdomen and pelvis
1 liter NS bolus, then D5NS @ 150 cc/h
Demerol 25 mg IM q 4-6 h
72-hr stool collection for fecal fat
NPO
Chlordiazepoxide 25 mg IV q 6h × 3d
Thiamin 100 mg IV daily × 3d
Folic acid 1 mg IV daily × 3d
Multivitamins 1 amp in first liter of IV fluids

UH UNIVERSITY HOSPITAL

NAME: Elena Jordan DOB: 10/7
AGE: 54 SEX: F
PHYSICIAN: P. Bennett, MD

CHEMISTRY

DAY: Admit
DATE:
TIME:
LOCATION:

	NORMAL		UNITS
Albumin	3.5–5	3.6	g/dL
Total protein	6–8	6	g/dL
Prealbumin	16–35	20.5	mg/dL
Transferrin	250–380 (women) 215–365 (men)	155 L	mg/dL
Sodium	136–145	145	mEq/L
Potassium	3.5–5.5	4.6	mEq/L
Chloride	95–105	105	mEq/L
PO_4	2.3–4.7	3.3	mg/dL
Magnesium	1.8–3	2.1	mg/dL
Osmolality	285–295	295	mmol/kg/H_2O
Total CO_2	23–30	27	mEq/L
Glucose	70–110	130 H	mg/dL
BUN	8–18	18	mg/dL
Creatinine	0.6–1.2	0.75	mg/dL
Uric acid	2.8–8.8 (women) 4.0–9.0 (men)	4.7	mg/dL
Calcium	9–11	9.3	mg/dL
Bilirubin	≤ 0.3	1.5 H	mg/dL
Ammonia (NH_3)	9–33	27	μmol/L
ALT	4–36	45 H	U/L
AST	0–35	50 H	U/L
Alk phos	30–120	178	U/L
CPK	30–135 (women) 55–170 (men)	145	U/L
LDH	208–378	323	U/L
CHOL	120–199	225 H	mg/dL
HDL-C	> 55 (women) > 45 (men)	40 L	mg/dL
VLDL	7–32	56	mg/dL
LDL	< 130	129	mg/dL
LDL/HDL ratio	< 3.22 (women) < 3.55 (men)		
Apo A	101–199 (women) 94–178 (men)		mg/dL
Apo B	60–126 (women) 63–133 (men)		mg/dL
TG	35–135 (women) 40–160 (men)	250 H	mg/dL
T_4	4–12		mcg/dL
T_3	75–98		mcg/dL
HbA_{1C}	3.9–5.2	6.5	%

UH UNIVERSITY HOSPITAL

NAME: Elena Jordan
AGE: 30
PHYSICIAN: P. Bennett, MD

DOB: 10/7
SEX: F

***HEMATOLOGY**

DAY:		Admit	
DATE:			
TIME:			
LOCATION:			
	NORMAL		UNITS
WBC	4.8–11.8	14.5 H	$\times 10^3$/mm^3
RBC	4.2–5.4 (women) 4.5–6.2 (men)	4.8	$\times 10^6$/mm^3
HGB	12–15 (women) 14–17 (men)	11.6 L	g/dL
HCT	37–47 (women) 40–54 (men)	35.7 L	%
MCV	80–96	101.5 H	µm^3
RETIC	0.8–2.8		%
MCH	26–32	29	pg
MCHC	31.5–36	33.4	g/dL
RDW	11.6–16.5	13.6	%
Plt Ct	140–440	359	$\times 10^3$/mm^3
Diff TYPE			
ESR	0–25 (women) 0–15 (men)		mm/hr
% GRANS	34.6–79.2	84.2 L	%
% LYM	19.6–52.7	51	%
SEGS	50–62	59	%
BANDS	3–6	4	%
LYMPHS	24–44	34	%
MONOS	4–8	5	%
EOS	0.5–4	2	%
Ferritin	20–120 (women) 20–300 (men)	19.5 L	mg/mL
ZPP	30–80		µmol/mol
Vitamin B_{12}	24.4–100		ng/dL
Folate	5–25		µg/dL
Total T cells	812–2,318		mm^3
T-helper cells	589–1,505		mm^3
T-suppressor cells	325–997		mm^3
PT	11–16		sec

Case Questions

I. Understanding the Disease and Pathophysiology

1. The pancreas is an exocrine and endocrine gland. Describe the exocrine and endocrine functions of the pancreas.

2. Factors that influence pancreatic secretion during a meal can be subdivided into three phases (cephalic, gastric, and intestinal). Describe the action of the pancreas within each phase.

3. Dr. Bennett makes a diagnosis of chronic pancreatitis. Define *chronic pancreatitis.*

4. What physical symptoms in the physician's H & P are consistent with Ms. Jordan's diagnosis?

5. What is the most common etiology for pancreatitis? Explain the physiological consequences of pancreatitis.

6. Explain how alcohol is metabolized. Women absorb and metabolize alcohol differently from men, making them more vulnerable than men to alcohol-related organ damage. Describe how alcohol is metabolized differently in women than in men.

7. Describe the major health consequences associated with chronic alcohol consumption.

II. Nutrition Assessment

A. Evaluation of Weight/Body Composition

8. One year ago, Ms. Jordan weighed 140 lbs. On admission, she weighed 112 lbs. Calculate her percent weight loss.

9. Calculate her BMI.

10. After assessing her weight status, identify nutrition problems using the correct diagnostic term.

B. Calculation of Nutrient Requirements

11. Using the Mifflin-St. Jeor equation, estimate Ms. Jordan's energy needs at her current weight.

12. Calculate Ms. Jordan's protein needs.

C. Intake Domain

13. What do the U.S. Dietary Guidelines indicate regarding alcohol intake?

14. How is a "drink" (beer, wine, liquor) defined by the U.S. Dietary Guidelines?

15. Estimate Ms. Jordan's usual dietary intake for the following (show your calculations):

	At Home	On the Road
Alcohol	g	g
Alcohol kcal	kcal	kcal
Total energy	kcal	kcal
% energy as alcohol	%	%
Protein	g	g

Calculations:

16. Hospital day 2: Patient remains stable on IV fluid. Her pain has been somewhat controlled with parenteral analgesics, but she is still unable to eat. Dr. Bennett consults you to evaluate the parenteral nutrition she has suggested: a dextrose-based parenteral solution with 4.25% amino acids, 25% dextrose, electrolytes, vitamins, and trace elements at a rate of 85 cc/hr with 500 cc/day of 10% lipids. Will this meet the patient's energy and protein needs?

17. When developing parenteral regimens during pancreatitis, you may find that patients have difficulty with high-dextrose solutions as well as lipid emulsions. What guidelines exist for handling these situations?

18. Would you recommend any changes in the prescribed parenteral regimen? Explain.

19. Why do you think Dr. Bennett ordered parenteral nutrition rather than enteral nutrition support? What is the current standard of practice?

20. Identify potential nutrition problems within the intake domain using the correct diagnostic term.

D. Clinical Domain

21. Examine Ms. Jordan's lab reports. Which tests are important in diagnosing pancreatitis? What are her values?

22. What other labs are consistent with her diagnosis?

23. Dr. Bennett specifically wanted to see Ms. Jordan's blood glucose level. Why?

24. When Dr. Bennett admitted Ms. Jordan, she evaluated the severity of her pancreatitis using Ranson's criteria. What are Ranson's criteria, and how is this test scored?

25. Why were thiamin, folic acid, and a multivitamin supplement ordered on admission?

26. Ms. Jordan's mean corpuscular volume (MCV) was elevated on admission. What might cause this?

27. Identify potential nutrition problems within the clinical domain using the correct diagnostic term.

E. Behavioral–Environmental Domain

28. How would you respond to Ms. Jordan's comment that she mostly drinks wine and beer because only people who drink "hard liquor" develop alcohol problems?

29. Identify potential nutrition problems within the behavioral–environmental domain using the correct diagnostic term.

III. Nutrition Diagnosis

30. Select two high-priority nutrition problems and complete the PES statement for each.

IV. Nutrition Intervention

31. Outline nutritional management for an underweight patient with pancreatitis when the patient is able to eat again. What will be the key factors in preventing exacerbations of Ms. Jordan's pancreatitis in the future?

Bibliography

Behrman SW, Fowler ES. Pathophysiology of chronic pancreatitis. *Surg Clin North Am.* 2007;87(6):1309–1324.

Bengmark S. Bio-ecological control of acute pancreatitis: The role of enteral nutrition, pro and synbiotics. *Curr Opin Clin Nutr Metab Care.* 2005;8(5):557–561.

Binnekade JM. Review: Enteral nutrition reduces infections, need for surgical intervention, and length of hospital stay more than parenteral nutrition in acute. *Evid Based Nurs.* 2005;8(1):19.

Center for Science in the Public Interest. Fact sheet: Women and alcohol. Available at: http://www.cspinet.org/booze/women.htm. Accessed February 21, 2008.

Chick J, Kemppainen E. Estimating alcohol consumption. *Pancreatology.* 2007;7(2–3):157–161.

Clancy TE, Benoit EP, Ashley SW. Current management of acute pancreatitis. *J Gastrointest Surg.* 2005;9(3):440–452.

Davis MA. Acute pancreatitis. Library of the National Medical Society. Available at: http://www.medical-library.org. Accessed February 21, 2008.

Escott-Stump S. *Nutrition and Diagnosis-Related Care.* 6th ed. Baltimore, MD: Williams & Wilkins; 2007.

Jeejeebhoy KN. Enteral feeding. *Curr Opin Gastroenterol.* 2005;21(2):187–191.

Lasztity N, Hamvas J, Biro L, Nemeth E, Marosvolgyi T, Decsi T, Pap A, Antal M. Effect of enterally administered n-3 polyunsaturated fatty acids in acute pancreatitis—A prospective randomized clinical trial. *Clin Nutr.* 2005;24(2):198–205.

Lobo DN, Memon MA, Allison SP, Rowlands BJ. Evolution of nutritional support in acute pancreatitis. *Br J Surg.* 2000;87(6):695–707.

Mattfeldt-Beman M. Diseases of the hepatobiliary: Liver, gallbladder, exocrine pancreas. In: Nelms M, Sucher K, Long S. *Nutrition Therapy and Pathophysiology.* Belmont, CA: Thomson/Brooks-Cole; 2007:509–547.

Nagar AB, Gorelick FS. Acute pancreatitis. *Curr Opin Gastroenterol.* 2004;20(5):439–443.

Nair RJ, Lawler L, Miller MR. Chronic pancreatitis. *Am Fam Physician.* 2007;76(11):1679–1688.

National Institute on Alcohol Abuse and Alcoholism. Alcohol and the liver. Alcohol Alert No. 42. Rockville, MD: U.S. Department of Health and Human Services; 1998. Available at: http://pubs.niaaa.nih.gov/publications/aa42.htm. Accessed February 21, 2008.

National Institute on Alcohol Abuse and Alcoholism. Are women more vulnerable to alcohol effects? Alcohol Alert No. 46. Rockville, MD: U.S. Department of Health and Human Services; 1999. Available at: http://pubs.niaaa.nih.gov/publications/aa46.htmv. Accessed February 21, 2008.

Sand J, Lankisch PG, Nordback I. Alcohol consumption in patients with acute or chronic pancreatitis. *Pancreatology.* 2007;7(2-3):147–156.

Stanga Z, Giger U, Marx A, DeLegge MH. Effect of jejunal long-term feeding in chronic pancreatitis. *JPEN J Parenter Enteral Nutr.* 2005;29(1):12–20.

Vonlaufen A, Wilson JS, Pirola RC, Apte MV. Role of alcohol metabolism in chronic pancreatitis. *Alcohol Res Health.* 2007;30(1):48–54.

Internet Resources

American Academy of Family Physicians: Diagnosis and Management of Acute Pancreatitis. http://www.aafp.org/afp/20000701/164.html

American Gastroentrological Association: Pancreatitis. http://www.gastro.org/wmspage.cfm?parm1=855

eMedicine: Pancreatitis. http://www.emedicine.com/emerg/topic354.htm

Hardin MD: Pancreatitis. http://www.lib.uiowa.edu/hardin/md/pancreatitis.html

MedlinePlus: Pancreatitis. http://www.nlm.nih.gov/medlineplus/ency/article/001144.htm

Merck Manuals Online Library: Chronic pancreatitis. http://www.merck.com/mmpe/sec02/ch015/ch015c.html

National Digestive Diseases Information Clearinghouse (NDDIC): Pancreatitis. http://digestive.niddk.nih.gov/ddiseases/pubs/pancreatitis/

WebMD: Digestive Diseases: Pancreatitis. http://www.webmd.com/digestive-disorders/digestive-diseases-pancreatitis

Case 16

Acute Hepatitis

Objectives

After completing this case, the student will be able to:

1. Discuss the etiology and risk factors for development of hepatitis.
2. Apply knowledge of the pathophysiology of hepatitis to identify symptoms associated with this disease.
3. Demonstrate understanding of the role of nutrition therapy as an adjunct to pharmacotherapy and other medical treatments for this disease.
4. Interpret laboratory parameters for nutritional implications and significance.
5. Analyze nutrition assessment data to evaluate nutritional status and identify specific nutrition problems.
6. Determine nutrition diagnoses and write appropriate PES statements.
7. Develop a nutrition care plan with appropriate measurable goals, interventions, and strategies for monitoring and evaluation that addresses the nutrition diagnoses of this case.

Teresa Wilcox is a 22-year-old woman who is admitted with complaints of fatigue, vague upper quadrant pain, nausea, and anorexia. Elevated liver enzymes lead to further evaluation for possible infection with hepatitis C.

Name: Teresa Wilcox
DOB: 3/5 (age 22)
Physician: P. Horowitz, MD

ADMISSION DATABASE

BED #	DATE:	TIME:	TRIAGE STATUS (ER ONLY):
1	5/7	1400	☐ Red ☐ Yellow ☐ Green ☐ White

Initial Vital Signs

TEMP:	RESP:	SAO$_2$:	
99.6	20		
HT: 5'9"	WT (lb): 130	B/P: 100/60	PULSE: 75
LAST TETANUS: 5 years ago		LAST ATE: yesterday	LAST DRANK: this AM

PRIMARY PERSON TO CONTACT:
Name: Kevin Gustat
Home #: 555-3947
Work #: same

ORIENTATION TO UNIT: ☒ Call light ☒ Television/telephone ☒ Bathroom ☒ Visiting ☒ Smoking ☒ Meals ☒ Patient rights/responsibilities

CHIEF COMPLAINT/HX OF PRESENT ILLNESS

Fatigue, malaise
Anorexia, N/V

PERSONAL ARTICLES: (Check if retained/describe)
☒ Contacts ☒ R ☒ L ☐ Dentures ☐ Upper ☐ Lower
☒ Jewelry: ring
☒ Other: watch

ALLERGIES: Meds, Food, IVP Dye, Seafood: Type of Reaction

penicillin

VALUABLES ENVELOPE:
☐ Valuables instructions

PREVIOUS HOSPITALIZATIONS/SURGERIES

none

INFORMATION OBTAINED FROM:
☒ Patient ☐ Previous record
☐ Family ☐ Responsible party

Signature *Terri Wilcox*

Home Medications (including OTC) **Codes: A=Sent home** **B=Sent to pharmacy** **C=Not brought in**

Medication	Dose	Frequency	Time of Last Dose	Code	Patient Understanding of Drug
YAZ	1	qd	this AM	C	yes
Allegra	60 mg	qd	this AM	C	yes

Do you take all medications as prescribed? ☒ Yes ☐ No If no, why?

PATIENT/FAMILY HISTORY

☐ Cold in past two weeks
☐ Hay fever
☐ Emphysema/lung problems
☐ TB disease/positive TB skin test
☒ Cancer Maternal grandmother
☐ Stroke/past paralysis
☐ Heart attack
☐ Angina/chest pain
☐ Heart problems
☒ High blood pressure Mother
☒ Arthritis Mother
☐ Claustrophobia
☐ Circulation problems
☐ Easy bleeding/bruising/anemia
☐ Sickle cell disease
☒ Liver disease/jaundice Paternal grandfather
☐ Thyroid disease
☒ Diabetes Father
☐ Kidney/urinary problems
☒ Gastric/abdominal pain/heartburn Father
☐ Hearing problems
☒ Glaucoma/eye problems Mother
☐ Back pain
☐ Seizures
☒ Other ALS-Paternal grandmother

RISK SCREENING

Have you had a blood transfusion? ☐ Yes ☒ No
Do you smoke? ☐ Yes ☒ No
If yes, how many pack(s)?
Does anyone in your household smoke? ☐ Yes ☒ No
Do you drink alcohol? ☒ Yes ☐ No
If yes, how often? socially How much? 1-2 glasses of wine/week
When was your last drink? 5/15
Do you take any recreational drugs? ☐ Yes ☒ No
If yes, type:_______ Route:
Frequency:_______ Date last used:_______/_______/_______

FOR WOMEN Ages 12–52

Is there any chance you could be pregnant? ☐ Yes ☒ No
If yes, expected date (EDC):
Gravida/Para: 0/0

ALL WOMEN

Date of last Pap smear:
Do you perform regular breast self-exams? ☐ Yes ☐ No

ALL MEN

Do you perform regular testicular exams? ☐ Yes ☐ No

Additional comments:

x *Shannon Whitney, RN*
Signature/Title

Client name: Teresa (Terri) Wilcox
DOB: 3/5
Age: 22
Sex: Female
Education: College student
Occupation: Exotic dancer
Hours of work: Works evenings 6 PM–2 AM; takes graduate classes during the day (full course load)
Household members: Roommate who is a law school student
Ethnic background: European American
Religious affiliation: Unitarian
Referring physician: Phillip Horowitz, MD (gastroenterology)

Chief complaint:
"I just feel so tired. I can hardly move, my joints ache so much, and my muscles feel sore."

Patient history:
Onset of disease: Terri Wilcox is a 22-year-old architecture graduate student. She works full time as an exotic dancer to pay graduate school and living expenses. In addition to c/o fatigue, aches, and pains, she complains of vague right upper quadrant pain, nausea, and anorexia. She has been in relatively good health all of her life.
Gravida 0/para 0. On admission to the hospital, all laboratory tests proved negative except for liver enzymes.
Type of Tx: Rule out hepatitis C; test for anti-HCV and HCV RNA; nutrition consult to determine appropriate nutrition therapy; abstain from alcohol
PMH: Seasonal allergies treated with antihistamines
Meds: YAZ, 1 tab po daily; Allegra, 60 mg po qd
Smoker: No
Family Hx: Mother (living)—HTN, diverticulitis, cholecystitis; father (deceased)—diabetes mellitus, peptic ulcer disease; maternal grandmother—cholecystitis, bilateral breast cancer; maternal grandfather—leukemia; paternal grandfather—cirrhosis; paternal grandmother—amyotrophic lateral sclerosis

Physical exam:
General appearance: Tired-looking, college-aged female
Vitals: Temp 99.6°F, BP 100/60 mm Hg, HR 75 bpm, RR 20 bpm
Heart: Regular rate and rhythm, no gallops or rubs, point of maximal impulse at the fifth intercostal space in the midclavicular line
HEENT:
Head: Normocephalic
Eyes: Wears contact lenses to correct myopia, PERRLA
Ears: Tympanic membranes w/out lesions
Nose: Dry mucous membranes w/out lesions
Throat: Normal mucosa w/out exudates or lesions
Genitalia: Normal female
Neurologic: Alert and oriented × 3

Extremities: Normal muscular tone, normal ROM
Skin: Warm and dry
Chest/lungs: Respirations normal; no crackles, rhonchi, wheezes, or rubs noted
Peripheral vascular: Pulse 3+ bilaterally
Abdomen: Pierced umbilicus; upper right abdomen guarding

Nutrition Hx:
General: Appetite is usually good, but has not had an appetite for the past few weeks. She eats cereal and orange juice for breakfast most mornings (orange juice every AM). Takes lunch to eat on campus, or fast food at Student Union. Dinner at work, usually carryout. If carryout, it's usually Chinese food.

Usual dietary intake:
Breakfast: 1½ c Sugar Frosted Flakes or Frosted Mini-Wheats, about ½ c 2% milk; occasionally a banana sliced on top of cereal; strawberries or raspberries in season; 8 oz calcium-fortified orange juice
AM: Unsweetened, flavored hot or iced tea during morning at Student Center
Lunch: Cheeseburger—1 oz American cheese, 3 oz beef patty, bun, lettuce, tomato slice, dill pickle spear; 12 oz Diet Coke; half-order waffle-cut french fries with ketchup
Dinner: Cashew shrimp: 3 oz shrimp, 1½ c vegetables (baby corn, water chestnuts, sliced carrots, pea pods), ¼ c cashews, 1 c steamed rice; 12 oz Diet Coke
HS snack: 3–4 small Famous Amos cookies (chocolate chip with pecans), ice water

24-hour recall: Sips of orange juice, hot tea; 4 saltine crackers; 5 c Jell-O; 12 oz Sprite; ¾ c cream of chicken soup; hot tea
Food allergies/intolerances/aversions: Does not like liver or lima beans; NKA
Previous Medical Nutrition Therapy? No
Food purchase/preparation: Self and/or roommate
Vit/min intake: 400 mg vitamin E, 500 mg calcium multivitamin/mineral q d
Current diet order: High-kcal, high-protein

Tx plan:
Anti-HCV and HCV RNA tests
Chemistry, CBC I
Vitamin B-complex supplement
High-kcal, high-protein diet (per dietitian)
Bed rest
Allegra, 60 mg po daily
Alternative contraception planning

UH UNIVERSITY HOSPITAL

NAME: Teresa Wilcox DOB: 3/5
AGE: 22 SEX: F
PHYSICIAN: P. Horowitz, MD

CHEMISTRY********************************

DAY: Admit
DATE:
TIME:
LOCATION:

	NORMAL		UNITS
Albumin	3.5–5	4.2	g/dL
Total protein	6–8	7.1	g/dL
Prealbumin	16–35	36	mg/dL
Transferrin	250–380 (women) 215–365 (men)	325	mg/dL
Sodium	136–145	136	mEq/L
Potassium	3.5–5.5	4.0	mEq/L
Chloride	95–105	102	mEq/L
PO_4	2.3–4.7	3.6	mg/dL
Magnesium	1.8–3	2.1	mg/dL
Osmolality	285–295	289	mmol/kg/H_2O
Total CO_2	23–30	28	mEq/L
Glucose	70–110	105	mg/dL
BUN	8–18	11	mg/dL
Creatinine	0.6–1.2	0.9	mg/dL
Uric acid	2.8–8.8 (women) 4.0–9.0 (men)	5.9	mg/dL
Calcium	9–11	9.3	mg/dL
Bilirubin	≤0.3	1.5 H	mg/dL
Ammonia (NH_3)	9–33	28	μmol/L
ALT	4–36	340 H	U/L
AST	0–35	500 H	U/L
Alk phos	30–120	302 H	U/L
CPK	30–135 (women) 55–170 (men)	138	U/L
LDH	208–378	695 H	U/L
CHOL	120–199	199	mg/dL
HDL-C	>55 (women) >45 (men)	50 L	mg/dL
VLDL	7–32	24	mg/dL
LDL	<130	125	mg/dL
LDL/HDL ratio	<3.22 (women) <3.55 (men)		
Apo A	101–199 (women) 94–178 (men)		mg/dL
Apo B	60–126 (women) 63–133 (men)		mg/dL
TG	35–135 (women) 40–160 (men)	152	mg/dL
T_4	4–12		mcg/dL
T_3	75–98		mcg/dL
HbA_{1C}	3.9–5.2	4.9	%

UH UNIVERSITY HOSPITAL

NAME: Teresa Wilcox DOB: 3/5
AGE: 22 SEX: F
PHYSICIAN: P. Horowitz, MD

HEMATOLOGY

DAY: Admit
DATE:
TIME:
LOCATION:

	NORMAL		UNITS
WBC	4.8-11.8	12.6 H	× $10^3/mm^3$
RBC	4.2-5.4 (women) 4.5-6.2 (men)	4.2	× $10^6/mm^3$
HGB	12-15 (women) 14-17 (men)	11.5 L	g/dL
HCT	37-47 (women) 40-54 (men)	36 L	%
MCV	80-96	85	μm^3
RETIC	0.8-2.8	1.3	%
MCH	26-32	38	pg
MCHC	31.5-36	33.6	g/dL
RDW	11.6-16.5	14.7	%
Plt Ct	140-440	140	× $10^3/mm^3$
Diff TYPE			
ESR	0-25 (women) 0-15 (men)	20	mm/hr
% GRANS	34.6-79.2	56.9	%
% LYM	19.6-52.7	49.7	%
SEGS	50-62	55	%
BANDS	3-6	4.5	%
LYMPHS	24-44	30	%
MONOS	4-8	6	%
EOS	0.5-4	2.9	%
Ferritin	20-120 (women) 20-300 (men)	105	mg/mL
ZPP	30-80	65.2	μmol/mol
Vitamin B_{12}	24.4-100	50.6	ng/dL
Folate	5-25	18.7	μg/dL
Total T cells	812-2,318	1,998	mm^3
T-helper cells	589-1,505	1,487	mm^3
T-suppressor cells	325-997	652	mm^3
PT	11-16	17 H	sec

UH UNIVERSITY HOSPITAL

NAME: Teresa Wilcox
AGE: 22
PHYSICIAN: P. Horowitz, MD

DOB: 3/5
SEX: F

URINALYSIS

DAY: Admit
DATE:
TIME:
LOCATION:

	NORMAL	Admit	UNITS
Coll meth		First morning	
Color		Dark yellow	
Appear		Clear	
Sp grv	1.003–1.030	1.020	
pH	5–7	5.9	
Prot	NEG	1+	mg/dL
Glu	NEG	NEG	mg/dL
Ket	NEG	NEG	
Occ bld	NEG	NEG	
Ubil	NEG	NEG	
Nit	NEG	NEG	
Urobil	<1.1	1.8	EU/dL
Leu bst	NEG	NEG	
Prot chk	NEG	NEG	
WBCs	0–5	3.6	/HPF
RBCs	0–5	4.2	/HPF
EPIs	0	0	/LPF
Bact	0	0	
Mucus	0	0	
Crys	0	0	
Casts	0	0	/LPF
Yeast	0	0	

Case Questions

I. Understanding the Disease and Pathophysiology

1. Several specific viruses that can cause hepatitis have been identified. Describe the following characteristics of each virus.

Characteristic	Hepatitis A HAV	Hepatitis B HBV	Hepatitis C HCV	Hepatitis D HDV	Hepatitis E HEV
Likely mode of transmission					
Symptoms					
Population most often affected					
Means of reducing exposure					
Treatment					

2. Describe hepatitis C to Ms. Wilcox as you would to a patient.

3. What signs and symptoms does Ms. Wilcox have?

4. Are there any other typical signs or symptoms of hepatitis that Ms. Wilcox does not have?

5. Describe how the symptoms of hepatitis are related to the pathophysiology of this disease. Include at least five symptoms in your discussion.

6. Teresa Wilcox is devastated by the diagnosis. She tells Dr. Horowitz that she's never had a blood transfusion or been exposed to blood products. She has never used IV drugs, but did inhale cocaine once or twice at parties in college. She has had several sexual partners, and the only time she has come into contact with any kind of needles was when she had her naval pierced 6 months ago. How, most likely, did Terri contract hepatitis C?

II. Nutrition Assessment

A. Evaluation of Weight/Body Composition

7. Calculate the patient's percent UBW and BMI, and explain the nutritional risk associated with each value.

B. Calculation of Nutrient Requirements

8. Because resting energy expenditure varies in liver disease, indirect calorimetry is recommended. However, you do not have access to this means of measurement. How would you estimate Ms. Wilcox's energy and protein requirements?

9. Identify any potential nutrition problems regarding micronutrient requirements. Provide the rationale for why these micronutrients are of concern.

10. Calculate this patient's total energy and protein needs using the Mifflin-St. Jeor equation.

C. Intake Domain

11. Using the patient's usual dietary intake, help her plan a menu.

Usual Diet	Suggested Substitutions	kcal	Pro (g)
1½ c Sugar Frosted Flakes			
½ c 2% milk			
1 banana			
1 c calcium-fortified orange juice			
Iced tea			
Cheeseburger			
12 oz Diet Coke			
French fries			
3 oz shrimp			
1½ c vegetables			
1 c steamed rice			
Water			
4 cookies			

12. Identify nutrition problems within the intake domain using the correct diagnostic term.

D. Clinical Domain

13. Examine the patient's chemistry report. What values would steer Dr. Horowitz toward the patient's diagnosis?

14. What do these values measure, and what is their relationship to liver disease?

15. The results of the anti-HCV and HCV RNA tests that Dr. Horowitz ordered were positive. What does this mean?

16. Once the diagnosis of hepatitis C is made, the physician orders 3-MU interferon alfa-2b sq qd and Rebetol 200 mg po bid. What are these medications, and what do they do?

17. What are the nutritional side effects of interferon and ribavirin?

18. Given these side effects, what can the dietetic professional do to help the patient maintain positive nutritional status?

19. List nutrition problems within the clinical domain using the correct diagnostic term.

E. Behavioral–Environmental Domain

20. Ms. Wilcox tells you that a friend suggested she use milk thistle to help fight the hepatitis virus. What would you tell her?

21. List any potential nutrition problems with the behavioral–environmental domain.

III. Nutrition Diagnosis

22. Select two high-priority nutrition problems and complete the PES statement for each.

IV. Nutrition Intervention

23. As you assess Terri Wilcox's nutritional status, what are your concerns?

24. Dr. Horowitz requested your consultation to order the patient's diet. What do you recommend?

25. How will you be able to determine whether this diet prescription is appropriate?

26. For each of the PES statements that you have written, establish an ideal goal (based on the signs and symptoms) and an appropriate intervention (based on the etiology).

Bibliography

Agency for Healthcare Research and Quality. Milk thistle: Effects on liver disease and cirrhosis and clinical adverse effects. Evidence Report/Technology Assessment: Number 21. 2000. Available at: http://www.ahrq.gov/clinic/epcsums/milktsum.htm. Accessed January 28, 2008.

Escott-Stump S. *Nutrition in Diagnosis-Related Care.* 6th ed. Baltimore, MD: Lippincott Williams & Wilkins; 2007.

Matarese LE, Gottschlich MM. *Contemporary Nutrition Support Practice: A Clinical Guide.* Philadelphia, PA: WB Saunders Company; 1996.

Mattfeldt-Beman M. Diseases of the hepatobiliary: Liver, gallbladder, exocrine pancreas. In: Nelms M, Sucher K, Long S. *Nutrition therapy and pathophysiology.* Belmont, CA: Thomson/Brooks-Cole; 2007:509–547.

Internet Resources

Centers for Disease Control: Viral Hepatitis C. http://www.cdc.gov/ncidod/diseases/hepatitis/c/index.htm

eMedicine: Hepatitis C. http://www.emedicine.com/med/topic993.htm

Hepatitis Foundation International. http://www.hepfi.org

Mayo Clinic: Hepatitis C. http://www.mayoclinic.com/health/hepatitis-c/DS00097

MedlinePlus: Hepatitis C. http://www.nlm.nih.gov/medlineplus/hepatitisc.html

National Digestive Diseases Information Clearinghouse (NDDC): Chronic Hepatitis C: Current Disease Management. http://digestive.niddk.nih.gov/ddiseases/pubs/chronichepc/

National Hepatitis C Program. http://www.hepatitis.va.gov/

World Health Organization: Hepatitis C. http://www.who.int/mediacentre/factsheets/fs164/en/

Case 17

Cirrhosis of the Liver with Resulting Hepatic Encephalopathy

Objectives

After completing this case, the student will be able to:

1. Integrate knowledge of the pathophysiology of cirrhosis with the nutrition care process.
2. Research and discuss the current role of nutrition in the development and treatment of hepatic encephalopathy.
3. Identify and apply pertinent nutrition assessment indices for the patient with cirrhosis.
4. Develop nutrition diagnoses for the patient.
5. Identify appropriate nutrition therapy goals.
6. Determine key components of nutrition education for the patient with cirrhosis.

Teresa Wilcox, introduced in Case 16, is admitted to University Hospital with increasing symptoms of liver disease 3½ years after being diagnosed with acute hepatitis. A liver biopsy and CT scan confirm her diagnosis of cirrhosis of the liver secondary to chronic hepatitis C infection.

Name: Teresa Wilcox
DOB: 3/5 (age 26)
Physician: P. Horowitz, MD

ADMISSION DATABASE

BED #	DATE:	TIME:	TRIAGE STATUS (ER ONLY):
2	12/19	1400	☐ Red ☐ Yellow ☐ Green ☐ White

Initial Vital Signs

TEMP:	RESP:	SAO$_2$:	
96.9	19		
HT: 5'9"	WT (lb): 125	B/P: 102/65	PULSE: 72
LAST TETANUS: 8 yrs ago		LAST ATE: AM	LAST DRANK: AM

PRIMARY PERSON TO CONTACT:
Name: Kevin Gustat
Home #: 555-3947
Work #: same

ORIENTATION TO UNIT: ☒ Call light ☒ Television/telephone ☒ Bathroom ☒ Visiting ☒ Smoking ☒ Meals ☒ Patient rights/responsibilities

CHIEF COMPLAINT/HX OF PRESENT ILLNESS

N/V, anorexia, fatigue, weakness

ALLERGIES: Meds, Food, IVP Dye, Seafood: Type of Reaction

penicillin

PREVIOUS HOSPITALIZATIONS/SURGERIES

3 years ago for hepatitis C

PERSONAL ARTICLES: (Check if retained/describe)
☒ Contacts ☒ R ☒ L ☐ Dentures ☐ Upper ☐ Lower
☒ Jewelry:
☒ Other: cell phone

VALUABLES ENVELOPE:
☐ Valuables instructions

INFORMATION OBTAINED FROM:
☒ Patient ☐ Previous record
☐ Family ☐ Responsible party

Signature *Terri Wilcox*

Home Medications (including OTC) **Codes: A=Sent home** **B=Sent to pharmacy** **C=Not brought in**

Medication	Dose	Frequency	Time of Last Dose	Code	Patient Understanding of Drug
YAZ	1	qd	this AM	C	yes
Allegra	60 mg	qd	this AM	C	yes

Do you take all medications as prescribed? ☒ Yes ☐ No If no, why?

PATIENT/FAMILY HISTORY

☐ Cold in past two weeks
☐ Hay fever
☐ Emphysema/lung problems
☐ TB disease/positive TB skin test
☒ Cancer Maternal grandmother
☐ Stroke/past paralysis
☐ Heart attack
☐ Angina/chest pain
☐ Heart problems
☒ High blood pressure Mother
☒ Arthritis Mother
☐ Claustrophobia
☐ Circulation problems
☐ Easy bleeding/bruising/anemia
☐ Sickle cell disease
☒ Liver disease/jaundice Patient
☐ Thyroid disease
☒ Diabetes Father
☐ Kidney/urinary problems
☒ Gastric/abdominal pain/heartburn Father
☐ Hearing problems
☒ Glaucoma/eye problems Mother
☐ Back pain
☐ Seizures
☒ Other ALS-Paternal grandmother

RISK SCREENING

Have you had a blood transfusion? ☐ Yes ☒ No
Do you smoke? ☐ Yes ☒ No
If yes, how many pack(s)?
Does anyone in your household smoke? ☐ Yes ☒ No
Do you drink alcohol? ☒ Yes ☐ No
If yes, how often? socially How much? 1-2 glasses
When was your last drink? 5/15
Do you take any recreational drugs? ☐ Yes ☒ No
If yes, type:______ Route:
Frequency:______ Date last used:_____/_____/_____

FOR WOMEN Ages 12–52

Is there any chance you could be pregnant? ☐ Yes ☒ No
If yes, expected date (EDC):
Gravida/Para: 0/0

ALL WOMEN

Date of last Pap smear:
Do you perform regular breast self-exams? ☒ Yes ☐ No

ALL MEN

Do you perform regular testicular exams? ☐ Yes ☐ No

Additional comments:

x *Shannon Whitney, RN*
Signature/Title

Client name: Teresa (Terri) Wilcox
DOB: 3/5
Age: 26
Sex: Female
Education: Doctoral graduate student
Occupation: Graduate teaching assistant
Hours of work: Teaches late morning and late afternoon; takes classes and conducts research during most evenings
Household members: Roommate who is a law student
Ethnic background: European American
Religious affiliation: Unitarian
Referring physician: Phillip Horowitz, MD (gastroenterology)

Chief complaint:
"It just seems as if I can't get enough rest. I feel so weak. Sometimes I'm so tired I can't go to campus to teach my classes. Does my skin look yellow to you?"

Patient history:
Terri Wilcox is a 26-year-old architecture doctoral student who was in relatively good health until 3 years ago when she was Dx with hepatitis C. Currently, she c/o fatigue, anorexia, N/V, and weakness. She has lost 10 lbs since her last office visit, which was 6 months ago. She also reports that she has bruising of her skin that did not happen previously and does not appear to be related to injury.
Type of Tx: Rule out cirrhosis
PMH: Hepatitis C Dx 3 years ago—previously treated with alpha-interferon and ribavirin; seasonal allergies treated with antihistamines
Meds: YAZ, 1 tab po daily; Allegra 60 mg po qd
Smoker: No
Family Hx: What? Mother (living)—HTN, diverticulitis, cholecystitis, carpal tunnel syndrome; father (deceased)—diabetes mellitus, peptic ulcer disease; maternal grandmother—cholecystitis, bilateral breast cancer; maternal grandfather—leukemia; paternal grandfather—cirrhosis; paternal grandmother—amyotrophic lateral sclerosis

Physical exam:
General appearance: Tired-looking young female
Vitals: Temp 96.9°F, BP 102/65 mm Hg, HR 72 bpm, RR 19 bpm
Heart: Regular rate and rhythm, no gallops or rubs, point of maximal impulse at the fifth intercostal space in the midclavicular line
HEENT:
  Head: Normocephalic
  Eyes: Wears contact lenses to correct myopia, PERRLA
  Ears: Tympanic membranes w/out lesions
  Nose: Dry mucous membranes w/out lesions
  Throat: Enlarged esophageal veins
Genitalia: Normal female
Neurologic: Alert and oriented × 3

Extremities: Normal muscular tone, normal ROM; no edema; no asterixis noted
Skin: Warm and dry; bruising noted on lower arms and legs; telangiectasias noted on chest
Chest/lungs: Respirations normal; no crackles, rhonchi, wheezes, or rubs noted
Peripheral vascular: Pulse 3+ bilaterally
Abdomen: Pierced umbilicus, mild distension RUQ, splenomegaly w/out hepatomegaly; no ascites

Nutrition Hx:
General: Has not had an appetite for the past few weeks. She states that she drinks calcium-fortified orange juice for breakfast most mornings. Lunch is usually soup and crackers with a Diet Coke. Dinner at home, but may be carry-out. If carry-out, it's usually Chinese or Italian food.
Usual dietary intake: Sips of water, juice, and Diet Coke only. Has not eaten for the past 2 days.
Food allergies/intolerances/aversions: Does not like liver or lima beans
Previous nutrition therapy? 3 years ago: small, frequent meals, plenty of liquids.
Food purchase/preparation: Self and/or significant other
Vit/min intake: 400 mg vitamin E, 600 mg calcium with 400 IU vitamin D, multivitamin/mineral daily, 200 mg milk thistle twice daily, chicory 3 grams daily, 500 mg ginger at least twice daily.
Current diet order: Soft, 4 gram sodium, high-kcal

Dx:
Probable cirrhosis secondary to chronic hepatitis C

Tx plan:
YAZ 1 tab po
Allegra, 60 mg po qd
CT scan of liver and biopsy
Endoscopy
Test stool for occult blood
Daily I/O
Spironolactone 25 mg qid
Propranolol 40 mg bid
Soft, high-kcal, high-protein diet—small, frequent meals
Multivitamin/mineral supplement
Bed rest

UH UNIVERSITY HOSPITAL

NAME: Teresa Wilcox DOB: 3/5
AGE: 26 SEX: F
PHYSICIAN: P. Horowitz, MD

CHEMISTRY

DAY: Admit
DATE:
TIME:
LOCATION:

	NORMAL		UNITS
Albumin	3.5–5	2.1 L	g/dL
Total protein	6–8	5.4 L	g/dL
Prealbumin	16–35	15 L	mg/dL
Transferrin	250–380 (women) 215–365 (men)	187	mg/dL
Sodium	136–145	136	mEq/L
Potassium	3.5–5.5	4.8	mEq/L
Chloride	95–105	102	mEq/L
PO_4	2.3–4.7	3.6	mg/dL
Magnesium	1.8–3	2.1	mg/dL
Osmolality	285–295	293	mmol/kg/H_2O
Total CO_2	23–30	28	mEq/L
Glucose	70–110	115 H	mg/dL
BUN	8–18	16	mg/dL
Creatinine	0.6–1.2	1.2	mg/dL
Uric acid	2.8–8.8 (women) 4.0–9.0 (men)	5.9	mg/dL
Calcium	9–11	9.3	mg/dL
Bilirubin	≤0.3	3.7 H	mg/dL
Ammonia (NH_3)	9–33	33	μmol/L
ALT	4–36	62 H	U/L
AST	0–35	230 H	U/L
Alk phos	30–120	275 H	U/L
CPK	30–135 (women) 55–170 (men)	138 H	U/L
LDH	208–378	658	U/L
CHOL	120–199	199	mg/dL
HDL-C	>55 (women) >45 (men)	50 L	mg/dL
VLDL	7–32	64	mg/dL
LDL	<130	125	mg/dL
LDL/HDL ratio	<3.22 (women) <3.55 (men)		
Apo A	101–199 (women) 94–178 (men)		mg/dL
Apo B	60–126 (women) 63–133 (men)		mg/dL
TG	35–135 (women) 40–160 (men)	256 H	mg/dL
T_4	4–12		mcg/dL
T_3	75–98		mcg/dL
HbA_{1C}	3.9–5.2	4.9	%

UH *UNIVERSITY HOSPITAL*

NAME: T Wilcox DOB: 3/5
AGE: 26 SEX: F
PHYSICIAN: P Horowitz, MD

HEMATOLOGY

DAY:		Admit	
DATE:			
TIME:			
LOCATION:			
	NORMAL		UNITS
WBC	4.8–11.8	4.8	$\times 10^3/mm^3$
RBC	4.2–5.4 (women) 4.5–6.2 (men)	4.1 L	$\times 10^6/mm^3$
HGB	12–15 (women) 14–17 (men)	10.9 L	g/dL
HCT	37–47 (women) 40–54 (men)	35.9 L	%
MCV	80–96	102 H	μm^3
RETIC	0.8–2.8		%
MCH	26–32	29	pg
MCHC	31.5–36	35.4	g/dL
RDW	11.6–16.5	12.4	%
Plt Ct	140–440	342	$\times 10^3/mm^3$
Diff TYPE			
ESR	0–25 (women) 0–15 (men)		mm/hr
% GRANS	34.6–79.2	54.2	%
% LYM	19.6–52.7	20.6	%
SEGS	50–62	51	%
BANDS	3–6	4.2	%
LYMPHS	24–44	30	%
MONOS	4–8	4.2	%
EOS	0.5–4	2.8	%
Ferritin	20–120 (women) 20–300 (men)	18 L	mg/mL
ZPP	30–80		μmol/mol
Vitamin B_{12}	24.4–100	100	ng/dL
Folate	5–25	25	μg/dL
Total T cells	812–2,318		mm^3
T-helper cells	589–1,505		mm^3
T-suppressor cells	325–997		mm^3
PT	11–16	18.5 H	sec

UH UNIVERSITY HOSPITAL

NAME: Teresa Wilcox DOB: 3/5
AGE: 26 SEX: F
PHYSICIAN: P. Horowitz, MD

URINALYSIS

	NORMAL		UNITS
DAY:		Admit	
DATE:			
TIME:			
LOCATION:			
Coll meth		Random specimen	
Color		Dark	
Appear		Slightly hazy	
Sp grv	1.003–1.030	1.025	
pH	5–7	5.9	
Prot	NEG	1+	mg/dL
Glu	NEG	NEG	mg/dL
Ket	NEG	NEG	
Occ bld	NEG	NEG	
Ubil	NEG	1+	
Nit	NEG	NEG	
Urobil	<1.1	1.8	EU/dL
Leu bst	NEG	NEG	
Prot chk	NEG	NEG	
WBCs	0–5	3.8	/HPF
RBCs	0–5	2.7	/HPF
EPIs	0		/LPF
Bact	0		
Mucus	0		
Crys	0		
Casts	0		/LPF
Yeast	0		

Case Questions

I. Understanding the Disease and Pathophysiology

1. The liver is an extremely complex organ that has a particularly important role in nutrient metabolism. Identify three functions of the liver for each of the following:
   a. Carbohydrate metabolism

   b. Protein metabolism

   c. Lipid metabolism

   d. Vitamin and mineral metabolism

2. The CT scan and liver biopsy confirm the diagnosis of cirrhosis. What is cirrhosis?

3. The most common cause of cirrhosis is alcohol ingestion. What are additional causes of cirrhosis? What is the cause of this patient's cirrhosis?

4. Explain the physiological changes that occur as a result of cirrhosis.

5. List the signs and symptoms of cirrhosis, and relate each of these to the physiological changes discussed in question 4.

6. After reading this patient's history and physical, identify her signs and symptoms that are consistent with the diagnosis.

7. Hypoglycemia is a symptom that cirrhotic patients may experience. What is the physiological basis for this? Is this a potential problem? Explain.

8. What are the current medical treatments for cirrhosis?

9. What is hepatic encephalopathy? Identify the stages of encephalopathy and outline the major theories regarding the etiology of this condition.

10. Protein-energy malnutrition is commonly associated with cirrhosis. What are the potential causes of malnutrition in cirrhosis? Explain each cause.

II. Understanding the Nutrition Therapy

11. Outline the nutrition therapy for the following stages of cirrhosis with the rationale for each:

Diagnosis	Sodium	Potassium	Protein	Micronutrients	Fluid	Other Modifications
Stable cirrhosis						
Cirrhosis w/acute encephalopathy						
Cirrhosis w/ascites and esophageal varices						

III. Nutrition Assessment

12. Measurements used to assess nutritional status may be affected by the disease process and not necessarily be reflective of nutritional status. Are there any components of nutrition assessment that would be affected by cirrhosis? Explain.

A. Evaluation of Weight/Body Composition

13. Dr. Horowitz notes Ms. Wilcox has lost 10 lbs since her last exam. Assess and interpret Ms. Wilcox's weight.

14. Identify any nutrition problems using the correct diagnostic term.

B. Calculation of Nutrient Requirements

15. Calculate the patient's energy and protein needs.

16. What guidelines did you use and why?

C. Intake Domain

17. Evaluate the patient's usual nutritional intake.

18. Her appetite and intake have been significantly reduced for the past several days. Describe the factors that may have contributed to this change in her ability to eat.

19. Why was a soft, 4-g Na, high-kcalorie diet ordered? Should there be any other modifications?

20. This patient takes multiple dietary supplements. Identify the possible rationale for each and identify any that may pose a risk for someone with cirrhosis.

D. Clinical Domain

21. Examine the patient's chemistry values. Which labs support the diagnosis of cirrhosis? Explain their connection to the diagnosis.

22. Examine the patient's hematology values. Which are abnormal, and why?

23. Does she have any physical symptoms consistent with your findings?

24. What signs and/or symptoms would you monitor to determine further liver decompensation?

25. Dr. Horowitz prescribes two medications to assist with the patient's symptoms. What is the rationale for these medications, and what are the pertinent nutritional implications of each?

Rationale for Rx	Nutritional Implications
Spironolactone	
Propranolol	

26. If the patient's condition worsens (e.g., acute varices, bleeding, progression to hepatic encephalopathy), the following medications could be used. Describe each drug classification and mechanism.

Drug	Classification	Mechanism
Vasopressin		
Lactulose		
Neomycin		
Ferrous sulfate		
Bisacodyl		
Docusate		
Diphenhydramine		

E. Behavioral–Environmental Domain

27. What is the recommendation regarding alcohol intake when cirrhosis is caused by the hepatitis C virus?

IV. Nutrition Diagnosis

28. Select two high-priority nutrition problems and complete the PES statement for each.

V. Nutrition Intervention

29. Ms. Wilcox is discharged on a soft, 4-g Na diet with a 2-L fluid restriction. Do you agree with this decision?

30. Ms. Wilcox asks if she can use a salt substitute at home. What would you tell her?

31. What suggestions might you make to assist with compliance for the fluid restriction?

VI. Evaluation and Monitoring

32. When you see Ms. Wilcox 1 month later, her weight is now 140 lbs. She is wearing flip-flops because she says her shoes do not fit. What condition is she most probably experiencing? How could you confirm this?

33. Her diet history is as follows:
Breakfast: 1 slice toast with 2 tbsp peanut butter, 1 c skim milk; Lunch: 2 oz potato chips, grilled cheese sandwich (1 oz American cheese with 2 slices of whole-wheat bread; grilled with 1 tbsp margarine), 1 c skim milk; Supper: 8 barbeque chicken wings, french fries—1 c, 2 c lemonade.
What changes might you make to her nutrition therapy? Identify foods that should be eliminated and make suggestions for substitutions.

34. Over the next 6 months, Terri's cirrhosis worsens. She is evaluated and found to be a good candidate for a liver transplant. She is placed on a transplant list and, 20 weeks later, receives a transplant. After the liver transplant, what diet and nutritional recommendations will the patient need before discharge? For the long term?

	Immediate Posttransplant (First 2 Months)	Long-Term Posttransplant
Kcal		
Protein		
Fat		
CHO		
Sodium		
Fluid		
Calcium		
Vitamins		

Bibliography

Abou-Assi SG, Mihas AA, Gavis EA, Gilles HS, Haselbush A, Levy JR, Habib A, Heuman DM. Safety of an immune-enhancing nutrition supplement in cirrhotic patients with history of encephalopathy. *J Parenter Enteral Nutr.* 2006;30(2):91–96.

American Academy of Family Physicians. Cirrhosis and chronic liver failure: What you should know. *Am Fam Physician.* 2006;74(5):781.

American Dietetic Association. *Nutrition Care Manual.* Available at: http://www.nutritioncaremanual.org.

Buchman AL. Total parenteral nutrition: Challenges and practice in the cirrhotic patient. *Transplant Proc.* 2006;38(6):1659–1663.

Campillo B, Richardet JP, Bories PN. Validation of body mass index for the diagnosis of malnutrition in patients with liver cirrhosis. *Gastroenterol Clin Biol.* 2006;30(10):1137–1143.

Carvalho L, Parise ER. Evaluation of nutritional status of nonhospitalized patients with liver cirrhosis. *Arq Gastroenterol.* 2006;43(4):269–274.

Cave M, Deaciuc I, Mendez C, Song Z, Joshi-Barve S, Barve S, McClain C. Nonalcoholic fatty liver disease: Predisposing factors and the role of nutrition. *J Nutr Biochem.* 2007;18(3):184–195.

Escott-Stump S. *Nutrition and Diagnosis-Related Care.* 6th ed. Baltimore, MD: Williams & Wilkins; 2007.

Federico A, Trappoliere M, Loguercio C. Treatment of patients with non-alcoholic fatty liver disease: Current views and perspectives. *Dig Liver Dis.* 2006;38(11):789–801.

Figueiredo FA, Perez RM, Freitas MM, Kondo M. Comparison of three methods of nutritional assessment in liver cirrhosis: Subjective global assessment, traditional nutritional parameters, and body composition analysis. *J Gastroenterol.* 2006;41(5):476–482.

Foster S, Tyler V. *Tyler's Honest Herbal.* Binghamton, NY: Haworth Press; 2000.

Fukagawa NK. Sparing of methionine requirements: Evaluation of human data takes sulfur amino acids beyond protein. *J Nutr.* 2006;136(6 Suppl):1676S–1681S.

Grattagliano I, Portincasa P, Palmieri VO, Palasciano G. Managing nonalcoholic fatty liver disease: Recommendations for family physicians. *Can Fam Physician.* 2007;53(5):857–863.

Gunsar F, Raimondo ML, Jones S, Terreni N, Wong C, Patch D, Sabin C, Burroughs AK. Nutritional status and prognosis in cirrhotic patients. *Aliment Pharmacol Ther.* 2006;24(4):563–572.

Heidelbaugh JJ, Sherbondy M. Cirrhosis and chronic liver failure: Part II. Complications and treatment. *Am Fam Physician.* 2006;74(5):767–776.

Heyman JK, Whitfield CJ, Brock KE, McCaughan GW, Donaghy AJ. Dietary protein intakes in patients with hepatic encephalopathy and cirrhosis: Current practice in NSW and ACT. *Med J Aust.* 2006;185(10):542–543.

Kalaitzakis E, Simrén M, Olsson R, Henfridsson P, Hugosson I, Bengtsson M, Björnsson E. Gastrointestinal symptoms in patients with liver cirrhosis: Associations with nutritional status and health-related quality of life. *Scand J Gastroenterol.* 2006;41(12):1464–1472.

Matsumoto H, Fakui Y. Pharmokinetics of ethanol: A review of methodology. *Addition Biology.* 2002;7:5–14.

Mattfeldt-Beman M. Diseases of the hepatobiliary: Liver, gallbladder, exocrine pancreas. In: Nelms M, Sucher K, Long S. *Nutrition Therapy and Pathophysiology.* Belmont, CA: Thomson/Brooks-Cole; 2007:509–547.

Moore KP, Aithal GP. Guidelines on the management of ascites in cirrhosis. *Gut.* 2006;55(Suppl 6):vi1–12.

Morgan MY, Madden AM, Soulsby CT, Morris RW. Derivation and validation of a new global method for assessing nutritional status in patients with cirrhosis. *Hepatology.* 2006;44(4):823–835.

Moriwaki H. Nutritional assessment in liver cirrhosis. *J Gastroenterol.* 2006;41(5):511–512.

Müller MJ. Malnutrition and hypermetabolism in patients with liver cirrhosis. *Am J Clin Nutr.* 2007;85(5):1167–1168.

Nakaya Y, Okita K, Suzuki K, Moriwaki H, Kato A, Miwa Y, Shiraishi K, Okuda H, Onji M, Kanazawa H, Tsubouchi H, Kato S, Kaito M, Watanabe A, Habu D, Ito S, Ishikawa T, Kawamura N, Arakawa Y; Hepatic Nutritional Therapy (HNT) Study Group. BCAA-enriched snack improves nutritional state of cirrhosis. *Nutrition.* 2007;23(2):113–120.

Norman K, Kirchner H, Lochs H, Pirlich M. Malnutrition affects quality of life in gastroenterology patients. *World J Gastroenterol.* 2006;12(21):3380–3385.

Panagaria N, Varma K, Nijhawan S, Mathur A, Rai RR. Comparison of nutritional status between patients with alcoholic and non-alcoholic liver cirrhosis. *Trop Gastroenterol.* 2006;27(2):75–79.

Peng S, Plank LD, McCall JL, Gillanders LK, McIlroy K, Gane EJ. Body composition, muscle function, and energy expenditure in patients with liver cirrhosis: A comprehensive study. *Am J Clin Nutr.* 2007;85(5):1257–1266.

Plauth M, Cabré E, Riggio O, Assis-Camilo M, Pirlich M, Kondrup J; DGEM (German Society for Nutritional Medicine), Ferenci P, Holm E, Vom Dahl S, Müller MJ, Nolte W; ESPEN (European Society for Parenteral and Enteral Nutrition). ESPEN Guidelines on Enteral Nutrition: Liver disease. *Clin Nutr.* 2006;25(2):285–294.

Sargent S. Pathophysiology and management of hepatic encephalopathy. *Br J Nurs.* 2007;16(6):335–339.
Schouten J, Michielsen PP. Treatment of cirrhotic ascites. *Acta Gastroenterol Belg.* 2007;70(2):217–222.
Tendler D, Lin S, Yancy WS Jr, Mavropoulos J, Sylvestre P, Rockey DC, Westman EC. The effect of a low-carbohydrate, ketogenic diet on nonalcoholic fatty liver disease: A pilot study. *Dig Dis Sci.* 2007;52(2):589–593.
Wagnerberger S, Schäfer C, Schwarz E, Bode C, Parlesak A. Is nutrient intake a gender-specific cause for enhanced susceptibility to alcohol-induced liver disease in women? *Alcohol.* 2008;43(1):9–14.

Internet Resources

Centers for Disease Control, National Center for Infectious Diseases: Viral Hepatitis C Fact Sheet. http://www.cdc.gov/ncidod/diseases/hepatitis/c/fact.htm
Hepatitis Central: Cirrhosis. http://hepatitis-central.com/hcv/liver/causes.html
Hepatitis Foundation International. http://www.hepfi.org
MedlinePlus Health Information: Drugs, Supplements, and Herbal Information. http://www.nlm.nih.gov/medlineplus/druginformation.html
Merck Manuals Online. http://www.merck.com/mmhe/index.html
National Institute of Diabetes, Digestive and Kidney Diseases. http://digestive.niddk.nih.gov/ddiseases/pubs/cirrhosis/
Office of Dietary Supplements. http://ods.od.nih.gov/

Unit Five

NUTRITION THERAPY FOR NEUROLOGICAL AND PSYCHIATRIC DISORDERS

The first case in this unit approaches neurological conditions through one of the most common diagnoses: stroke. According to the National Institute of Neurological Diseases and Stroke (2005), stroke is the third leading cause of death in the United States. The health consequences resulting from stroke are significant and, as in many neurological conditions, may involve an impairment in the ability to obtain food. Symptoms, such as impaired vision or ambulation, may result in the inability to shop or prepare adequate meals. Depending on the severity of the stroke, symptoms may interfere with chewing, swallowing, or feeding oneself. These problems are often not easily identified or easily solved. Each situation is highly individualized and requires a comprehensive nutrition assessment. Throughout the course of the disease and rehabilitation, nutrition therapy plays a crucial role in the maintenance of nutritional status and quality of life.

The prevalence of Alzheimer's disease (AD) continues to increase as the U.S. population ages, and is the seventh leading cause of death in the United States (available from: http://www.cdc.gov/nchs/fastats/deaths.htm; accessed January 7, 2008). Consequences of this devastating disease interfere with all phases of obtaining, eating, and enjoying food. This case examines these consequences, and also allows for discussion about nutrition interventions for a terminal illness.

Case 18

Stroke

Objectives

After completing this case, the student will be able to:

1. Apply working knowledge of stroke pathophysiology to the nutrition care process.
2. Analyze nutrition assessment data to establish baseline nutritional status.
3. Identify and explain the role of nutrition support in recovery and rehabilitation from stroke.
4. Assess and identify nutritional risks in dysphagia.
5. Establish the nutrition diagnosis and compose a PES statement.
6. Create strategies to maximize calorie and protein intake.
7. Analyze current recommendations for nutritional supplementation and determine appropriate nutrition interventions.

Mrs. Ruth Noland, a 77-year-old woman, is transported to the emergency room of University Hospital with the symptoms of slurred speech, numbness on the left side of her face, and weakness of her left arm and leg.

Name: Ruth Noland
DOB: 10/6 (age 77)
Physician: Stephen Young, MD

ADMISSION DATABASE

BED #	DATE:	TIME:	TRIAGE STATUS (ER ONLY):
1	8/12	1035	☒ Red ☐ Yellow ☐ Green ☐ White

Initial Vital Signs

TEMP:	RESP:	SAO$_2$:	
98.8	19	92%	
HT:	WT (lb):	B/P:	PULSE:
5′2″	165	138/88	91
LAST TETANUS		LAST ATE	LAST DRANK
		this morning	this morning

PRIMARY PERSON TO CONTACT:
Name: Robert Noland
Home #: 555-421-8456
Work #: N/A

ORIENTATION TO UNIT: ☒ Call light ☒ Television/telephone ☒ Bathroom ☒ Visiting ☒ Smoking ☒ Meals ☒ Patient rights/responsibilities

CHIEF COMPLAINT/HX OF PRESENT ILLNESS

"My wife woke up this morning with everything pretty normal but in the middle of the morning, she became dizzy, and then she couldn't talk or move one side of her body."

ALLERGIES: Meds, Food, IVP Dye, Seafood: Type of Reaction

NKA

PREVIOUS HOSPITALIZATIONS/SURGERIES

2 children over 30 years ago

Hysterectomy 10 years previous

PERSONAL ARTICLES: (Check if retained/describe)
☐ Contacts ☐ R ☐ L ☒ Dentures ☒ Upper ☒ Lower
☒ Jewelry: wedding band
☐ Other:

VALUABLES ENVELOPE:
☒ Valuables instructions

INFORMATION OBTAINED FROM:
☒ Patient ☐ Previous record
☒ Family ☐ Responsible party

Signature *Robert Noland*

Home Medications (including OTC) **Codes: A = Sent home** **B = Sent to pharmacy** **C = Not brought in**

Medication	Dose	Frequency	Time of Last Dose	Code	Patient Understanding of Drug
captopril	25 mg	twice daily	this AM	C	husband–yes
lovastatin	20 mg	once daily	this AM	C	husband–yes

Do you take all medications as prescribed? ☒ Yes ☐ No

PATIENT/FAMILY HISTORY

☐ Cold in past two weeks
☐ Hay fever
☐ Emphysema/lung problems
☐ TB disease/positive TB skin test
☐ Cancer
☐ Stroke/past paralysis
☐ Heart attack
☐ Angina/chest pain
☒ Heart problems Patient
☒ High blood pressure Patient
☒ Arthritis Patient
☐ Claustrophobia
☐ Circulation problems
☐ Easy bleeding/bruising/anemia
☐ Sickle cell disease
☐ Liver disease/jaundice
☐ Thyroid disease
☐ Diabetes
☐ Kidney/urinary problems
☐ Gastric/abdominal pain/heartburn
☐ Hearing problems
☐ Glaucoma/eye problems
☐ Back pain
☐ Seizures
☐ Other

RISK SCREENING

Have you had a blood transfusion? ☐ Yes ☒ No
Do you smoke? ☐ Yes ☒ No
If yes, how many pack(s)?
Does anyone in your household smoke? ☐ Yes ☒ No
Do you drink alcohol? ☐ Yes ☒ No
If yes, how often? ________ How much?
When was your last drink? ______/______/______
Do you take any recreational drugs? ☐ Yes ☒ No
If yes, type:______ Route:
Frequency:______ Date last used:______/______/______

FOR WOMEN Ages 12–52

Is there any chance you could be pregnant? ☐ Yes ☒ No
If yes, expected date (EDC):
Gravida/Para:

ALL WOMEN

Date of last Pap smear: 6 months ago
Do you perform regular breast self-exams? ☒ Yes ☐ No

ALL MEN

Do you perform regular testicular exams? ☐ Yes ☐ No

Additional comments:

x *Nessa Nordeen, RN*
Signature/Title

Client name: Ruth Noland
DOB: 10/6
Age: 77
Sex: Female
Education: High school diploma
Occupation: Retired hairdresser
Hours of work: N/A
Household members: Lives with husband age 82. Has been married for 56 years.
Ethnic background: European American
Religious affiliation: Protestant
Referring physician: S. Young, MD

Chief complaint:
"My wife woke up this morning with everything pretty normal, but in the middle of the morning, she became dizzy, and then she couldn't talk or move one side of her body."

Patient history:
Onset of disease: N/A
Type of Tx: R/O stroke
PMH: Hypertension × 10 years; Hyperlipidemia × 2 years
Meds: Captopril 25 mg twice daily; lovastatin 20 mg once daily
Smoker: No
Family Hx: Noncontributory

Physical exam:
General appearance: Elderly female who is unable to speak; unable to move right side
Vitals: Temp 98.6°F, BP 138/88 mm Hg, HR 91 bpm, RR 18 bpm
Heart: Regular rate and rhythm, no gallops or rubs, point of maximal impulse at the fifth intercostal space in the midclavicular line
HEENT:
  Head: Normocephalic
  Eyes: Wears glasses for myopia
  Ears: Tympanic membranes normal
  Nose: WNL
  Throat: Slightly dry mucous membranes w/out exudates or lesions
Genitalia: Normal w/out lesions
Neurologic: New-onset weakness of the right side involving right arm and leg. Face and arm weakness is disproportionate to leg weakness, and sensation is impaired on the contralateral side. Dysarthria with tongue deviation. Cranial nerves III, V, VII, and XII impaired.
Motor function tone and strength diminished. Plantar reflex decreased on right side. Blink reflex intact.
Extremities: Reduced strength, bilaterally
Skin: Normal without lesions
Chest/lungs: Respirations normal; no crackles, rhonchi, wheezes, or rubs noted
Peripheral vascular: Bilateral, 3+ pedal pulses

Abdomen: Normal bowel sounds. No hepatomegaly, splenomegaly, masses, inguinal lymph nodes, or abdominal bruits.

Nutrition Hx:

General: Mr. Noland states that his wife has a good appetite. She has not followed any special diet except for trying to avoid fried foods, and she has stopped adding salt at the table. She made these changes several years ago.

24-hour recall:

According to Mr. Noland, his wife ate the following:

Breakfast: Orange juice—1 c, raisin bran—1 c with 6 oz 2% milk, 1 banana, 8 oz coffee with 2 tbsp 2% milk with sweetener

Lunch: Chicken tortellini soup—2 c (cheese tortellini cooked in chicken broth), saltine crackers—about 8, canned pears—2 halves, 6 oz iced tea with sweetener

Dinner: Baked chicken (with skin)—4–6 oz breast, baked potato—1 medium—with 2 tbsp margarine, steamed broccoli—approx. 1 c with 1 tsp margarine, canned peaches in juice—6–8 slices, 6 oz iced tea with sweetener

Snack: 3 c popcorn or 1 c ice cream—strawberry Breyer's, 12 oz Coke

Food allergies/intolerances/aversions: None
Previous nutrition therapy? No
Food purchase/preparation: Mrs. Noland and spouse
Vit/min intake: Multivitamin/mineral supplement daily, 500 mg calcium 3 × daily
Current diet order: NPO

Dx:

R/O ischemic stroke

Tx plan:

Mrs. Noland was started on the acute stroke protocol which included:

- Administer 0.6 mg/kg intravenous rtPA over 1 hour with 10% of total dose given as an initial intravenous bolus over 1 minute. Total dose of 67.5 mg.
- Vital signs: q 15 minutes × 2 hours; then q 30 minutes × 6 hours; then q 1 hour × 16 hours.
- Neuro checks: Level of consciousness and extremity weakness (use NIHSS scoring): q 30 minutes × 6 hours, then q 1 hour × 16 hours.
- IV: 0.9 NS at 75 cc/hr
- O_2 at 2 L/minute via nasal cannula (if needed to keep O_2 sats ≥ 95%)
- Continuous cardiac monitoring
- Strict Intake/Output records
- Diet: NPO except medications for 24 hours
- Noncontrast CT scan
- Labs: Chem 16, coagulation times, CBC.
- Medications: Acetaminophen 650 mg p.o. PRN for pain q 4 to 6 hours.

No heparin, warfarin, or aspirin for 24 hours. After 24 hours: CT to exclude intracranial hemorrhage before any anticoagulants.

Hospital Course:

Noncontrast CT confirmed that Mrs. Noland had suffered a lacunar ischemic stroke—NIH Stroke Scale score of 14. After stabilizing, a bedside swallowing assessment was ordered followed by an endoscopy with modified barium swallow. The speech-language pathologist and dietitian were requested to determine staged dysphagia diet.

UH UNIVERSITY HOSPITAL

NAME: Ruth Noland DOB: 10/6
AGE: 77 SEX: F
PHYSICIAN: Stephen Young, MD

CHEMISTRY

DAY: Admission
DATE: 8/12
TIME: 1042
LOCATION:

	NORMAL	Admission	UNITS
Albumin	3.5-5	4.2	g/dL
Total protein	6-8	6.8	g/dL
Prealbumin	16-35	22	mg/dL
Transferrin	250-380 (women) 215-365 (men)	182	mg/dL
Sodium	136-145	137	mEq/L
Potassium	3.5-5.5	3.8	mEq/L
Chloride	95-105	101	mEq/L
PO_4	2.3-4.7	4.2	mg/dL
Magnesium	1.8-3	2.1	mg/dL
Osmolality	285-295	290	mmol/kg/H_2O
Total CO_2	23-30	27	mEq/L
Glucose	70-110	82	mg/dL
BUN	8-18	11	mg/dL
Creatinine	0.6-1.2	0.9	mg/dL
Uric acid	2.8-8.8 (women) 4.0-9.0 (men)	3.2	mg/dL
Calcium	9-11	9.2	mg/dL
Bilirubin	≤0.3	0.1	mg/dL
Ammonia (NH_3)	9-33	15	μmol/L
ALT	4-36	21	U/L
AST	0-35	33	U/L
Alk phos	30-120	179	U/L
CPK	30-135 (women) 55-170 (men)	113	U/L
LDH	208-378	241	U/L
CHOL	120-199	210	mg/dL
HDL-C	>55 (women) >45 (men)	40	mg/dL
VLDL	7-32		mg/dL
LDL	<130	155	mg/dL
LDL/HDL ratio	<3.22 (women) <3.55 (men)	1.3	
Apo A	101-199 (women) 94-178 (men)		mg/dL
Apo B	60-126 (women) 63-133 (men)		mg/dL
TG	35-135 (women) 40-160 (men)	198	mg/dL
T_4	4-12		mcg/dL
T_3	75-98		mcg/dL
HbA_{1C}	3.9-5.2	4.9	%

UH UNIVERSITY HOSPITAL

NAME: Ruth Noland DOB: 10/6
AGE: 77 SEX: F
PHYSICIAN: Stephen Young, MD

HEMATOLOGY

DAY:		8/12	
DATE:		1042	
TIME:			
LOCATION:			
	NORMAL		UNITS
WBC	4.8–11.8	10	$\times 10^3/mm^3$
RBC	4.2–5.4 (women)		
	4.5–6.2 (men)	4.5	$\times 10^6/mm^3$
HGB	12–15 (women)	12.7	g/dL
	14–17 (men)		
HCT	37–47 (women)	38	%
	40–54 (men)		
MCV	80–96	82	μm^3
RETIC	0.8–2.8	0.9	%
MCH	26–32	27	pg
MCHC	31.5–36	33.2	g/dL
RDW	11.6–16.5	12.1	%
Plt Ct	140–440	154	$\times 10^3/mm^3$
Diff TYPE			
ESR	0–25 (women)		mm/hr
	0–15 (men)		
% GRANS	34.6–79.2		%
% LYM	19.6–52.7		%
SEGS	50–62		%
BANDS	3–6		%
LYMPHS	24–44		%
MONOS	4–8		%
EOS	0.5–4		%
Ferritin	20–120 (women)		mg/mL
	20–300 (men)		
ZPP	30–80		μmol/mol
Vitamin B_{12}	24.4–100		ng/dL
Folate	5–25		μg/dL
Total T cells	812–2,318		mm^3
T-helper cells	589–1,505		mm^3
T-suppressor cells	325–997		mm^3
PT	11–16		sec

Case Questions

I. Understanding the Disease and Pathophysiology

1. Define *stroke*. Describe the differences between ischemic and hemorrhagic stroke.

2. What does Mrs. Noland's score for the NIH Stroke Scale indicate?

3. What are the factors that place an individual at risk for stroke?

4. What specific signs and symptoms that are noted with Mrs. Noland's exam and history are consistent with her diagnosis?

5. What is rtPA? Why was it administered?

6. Which symptoms that you identified in question 4 may place Mrs. Noland at nutritional risk? Explain your rationale.

II. Understanding the Nutrition Therapy

7. Define *dysphagia*.

8. What is the primary nutrition implication of dysphagia?

9. Describe the four phases of swallowing:
   a. Oral preparation

   b. Oral transit

   c. Pharyngeal

   d. Esophageal

10. It is determined that Mrs. Noland's dysphagia is centered in the esophageal transit phase, and she has reduced esophageal peristalsis. Which dysphagia diet level is appropriate to try with Mrs. Noland?

11. Describe a bedside swallowing assessment. What are the background and training requirements of a speech-language pathologist?

12. Describe a modified barium swallow or fiberoptic endoscopic evaluation of swallowing.

13. What is the National Dysphagia Diet? Describe the major differences among the four levels of the diet.

14. Thickening agents and specialty food products are often used to provide the texture changes needed for the dysphagia diet. Describe one of these products and how they may be incorporated into the diet.

III. Nutrition Assessment

A. Evaluation of Weight/Body Composition

15. Mrs. Noland's usual body weight is approximately 165 lbs. Calculate and interpret her BMI.

B. Calculation of Nutrient Requirements

16. Estimate Mrs. Noland's energy and protein requirements. Should weight loss or weight gain be included in this estimation? What is your rationale?

C. Intake Domain

17. Using Mrs. Noland's usual dietary intake, calculate the total number of kilocalories she consumed as well as the energy distribution of kilocalories for protein, carbohydrate, and fat.

18. Compare this to the recommended intake for an individual with hyperlipidemia and hypertension. Do these recommendations apply to Mrs. Noland at the present?

19. Estimate Mrs. Noland's fluid needs using the following methods: weight; age and weight; and energy needs.

20. Which method of fluid estimation appears most reasonable for Mrs. Noland? Explain.

21. From the information gathered within the intake domain, list possible nutrition problems using the diagnostic term.

D. Clinical Domain

22. Review Mrs. Noland's labs upon admission. Identify any that are abnormal. Using the following table, describe their clinical significance for Mrs. Noland.

Chemistry	Normal Value	Mrs. Noland's Value	Reason for Abnormality	Nutritional Implications

23. From the information gathered within the clinical domain, list possible nutrition problems using the diagnostic term.

IV. Nutrition Diagnosis

24. Select two high-priority nutrition problems and complete the PES statement for each.

V. Nutrition Intervention

25. For each of the PES statements that you have written, establish an ideal goal (based on the signs and symptoms) and an appropriate intervention (based on the etiology).

VI. Nutrition Monitoring and Evaluation

26. To maintain or attain normal nutritional status while reducing danger of aspiration and choking, texture (of foods) and/or viscosity (of fluids) are personalized for a patient with dysphagia. In the following table, define each term used to describe characteristics of foods and give an example.

Term	Definition	Example
Consistency		
Texture		
Viscosity		

27. Using Mrs. Noland's 24-hour recall, make suggestions for consistency changes or food substitutions (if needed) to Mrs. Noland and her family.

Orange juice ______________________________

Raisin bran ______________________________

2% milk ______________________________

Banana ______________________________

Coffee ______________________________

Sweetener ______________________________

Chicken tortellini soup ______________________________

Saltine crackers ______________________________

Canned pears ______________________________

Iced tea ______________________________

Baked chicken ______________________________

Baked potato ______________________________

Steamed broccoli ______________________________

Margarine ______________________________

Canned peaches ______________________________

Popcorn ______________________________

Coca-Cola ______________________________

Ice cream ______________________________

28. Describe Mrs. Noland's potential nutritional problems upon discharge. What recommendations could you make to her husband to prevent each problem you identified? How would you monitor her progress?

29. Would Mrs. Noland be an appropriate candidate for a stroke rehabilitation program? Why or why not?

Bibliography

Adams HP Jr, del Zoppo G, Alberts MJ, Bhatt DL, Brass L, Furlan A, Grubb RL, Higashida RT, Jauch EC, Kidwell C, Lyden PD, Morgenstern LB, Qureshi AI, Rosenwasser RH, Scott PA, Wijdicks EF; American Heart Association; American Stroke Association Stroke Council, Clinical Cardiology Council, Cardiovascular Radiology and Intervention Council, and the Atherosclerotic Peripheral Vascular Disease and Quality of Care Outcomes in Research Interdisciplinary Working Groups. Guidelines for the early management of adults with ischemic stroke: A guideline from the American Heart Association/American Stroke Association Stroke Council, Clinical Cardiology Council, Cardiovascular Radiology and Intervention Council, and the Atherosclerotic Peripheral Vascular Disease and Quality of Care Outcomes in Research Interdisciplinary Working Groups: The American Academy of Neurology affirms the value of this guideline as an educational tool for neurologists. *Stroke.* 2007;38:1655–1711.

Bath PM, Bath FJ, Smithard DG. Interventions for dysphagia in acute stroke. *Cochrane Database Syst Rev.* 2000;(2):CD000323.

Brody RA, Touger-Decker R, VonHagen S, Maillet JO. Role of registered dietitians in dysphagia screening. *J Am Diet Assoc.* 2000;100(9):1029–1037.

Brynningsen PK, Damsgaard EM, Husted SE. Improved nutritional status in elderly patients 6 months after stroke. *J Nutr Health Aging.* 2007;11(1):75–79.

Buettner C, Phillips RS, Davis RB, Gardiner P, Mittleman MA. Use of dietary supplements among United States adults with coronary artery disease and atherosclerotic risks. *Am J Cardiol.* 2007;99(5):661–666.

Bushnell CD, Johnston DCC, Goldstein LB. Retrospective assessment of initial stroke severity. Comparison of the NIH Stroke Scale and the Canadian Neurological Scale. *Stroke.* 2001;32:656.

Dennis M, Lewis S, Cranswick G, Forbes J; FOOD Trial Collaboration. FOOD: A multicentre randomised trial evaluating feeding policies in patients admitted to hospital with a recent stroke. *Health Technol Assess.* 2006;10(2):iii–iv,ix–x,1–120.

Dennis MS, Lewis SC, Warlow C; FOOD Trial Collaboration. Effect of timing and method of enteral tube feeding for dysphagic stroke patients (FOOD): A multicentre randomised controlled trial. *Lancet.* 2005;365(9461):764–772.

Ickenstein GW, Stein J, Ambrosi D, Goldstein R, Horn M, Bogdahn U. Predictors of survival after severe dysphagic stroke. *J Neurol.* 2005;252(12):1510–1516.

Ignarro LJ, Balestrieri ML, Napoli C. Nutrition, physical activity, and cardiovascular disease: an update. *Cardiovasc Res.* 2007;73(2):326–340.

Koretz RL, Avenell A, Lipman TO, Braunschweig CL, Milne AC. Does enteral nutrition affect clinical outcome? A systematic review of the randomized trials. *Am J Gastroenterol.* 2007;102(2):412–429.

Maasland L, Koudstaal PJ, Habbema JD, Dippel DW. Knowledge and understanding of disease process, risk factors and treatment modalities in patients with a recent TIA or minor ischemic stroke. *Cerebrovasc Dis.* 2007;23(5–6):435–440.

Milne AC, Avenell A, Potter J. Meta-analysis: Protein and energy supplementation in older people. *Ann Intern Med.* 2006;144(1):37–48.

Sacco RL. Newer risk factors for stroke. *Neurology.* 2001;57(5 Suppl 2):S31-S34.

Thom T, Haase N, Rosamond W, Howard VJ, Rumsfeld JR, Manolio T, Zheng ZJ, Flegal K, O'Donnell C, Kittner S, Lloyd-Jones D, Goff DC Jr, Hong Y, Adams R, Friday G, Furie K, Gorelick P, Kissela B, Marler J, Meigs J, Roger V, Sidney S, Sorlie P, Steinberger J, Wasserthiel-Smoller S, Wilson M, Wolf P; American Heart Association Statistics Committee and Stroke Statistics Subcommittee. Heart disease and stroke statistics—2006 update: A report from the American Heart Association Statistics Committee and Stroke Statistics Subcommittee. *Circulation.* 2006;113(6):e85–e151. Available at: http://circ.ahajournals.org/cgi/content/full/113/6/e85. Accessed August 12, 2007.

Internet Resources

American Speech Hearing Language Association. http://www.asha.org

American Stroke Association: A Division of American Heart Association. http://www.strokeassociation.org

Evaluation of Stages of Swallowing. http://www.linkstudio.info/images/portfolio/medani/Swallow.swf

National Institute of Neurological Disorders and Stroke: NINDS Stroke Information Page. http://www.ninds.nih.gov/disorders/stroke/stroke.htm

National Institutes of Health: NIH Stroke Scale and Scoring. http://www.strokecenter.org/trials/scales/nihss.html

U.S. Agency for Health Care Policy and Research: Stroke Scales Overview. http://www.strokecenter.org/trials/scales/scales-overview.htm

Case 19

Alzheimer's Disease

Objectives

After completing this case, the student will be able to:

1. Apply knowledge of the consequences of Alzheimer's disease to identify and explain common nutritional problems associated with this condition.
2. Evaluate current recommendations for nutritional supplementation in wound healing.
3. Identify potential drug–nutrient interactions.
4. Analyze nutrition assessment data to evaluate nutritional status and identify specific nutrition problems.
5. Determine nutrition diagnoses and write appropriate PES statements.
6. Develop a nutrition care plan with appropriate measurable goals, interventions, and strategies for monitoring and evaluation that addresses the nutrition diagnoses of this case.

Ralph McCormick is admitted to the acute care setting for treatment of a nonhealing wound. He currently resides in an Alzheimer's unit at a local veteran's long-term care facility.

ADMISSION DATABASE

Name: Ralph McCormick
DOB: 10/19 (age 89)
Physician: B. Byrd, MD

BED #	DATE:	TIME:	TRIAGE STATUS (ER ONLY):
4	8/12	0730	☐ Red ☐ Yellow ☐ Green ☐ White

Initial Vital Signs

TEMP:	RESP:	SAO$_2$:	
100.3	32	96%	
HT:	WT (lb):	B/P:	PULSE:
5′11″	138	94/68	85
LAST TETANUS		LAST ATE	LAST DRANK
3 years ago		last PM	last PM

PRIMARY PERSON TO CONTACT:
Name: Gerald McCormick
Home #: 555-223-4231
Work #: 555-221-9243

ORIENTATION TO UNIT: ☐ Call light ☐ Television/telephone ☐ Bathroom ☐ Visiting ☐ Smoking ☐ Meals ☐ Patient rights/responsibilities

CHIEF COMPLAINT/HX OF PRESENT ILLNESS

Multiple abrasions and nonhealing wound on right hip. Admitted through ER from regional Veteran's Long-Term Care Facility/ Alzheimer's Unit.

PERSONAL ARTICLES: (Check if retained/describe)
☐ Contacts ☐ R ☐ L ☒ Dentures ☒ Upper ☒ Lower
☐ Jewelry:
☒ Other: glasses

ALLERGIES: Meds, Food, IVP Dye, Seafood: Type of Reaction

penicillin

VALUABLES ENVELOPE:
☐ Valuables instructions

PREVIOUS HOSPITALIZATIONS/SURGERIES

s/p CABG 18 years previous; s/p R hip replacement 8 years previous

INFORMATION OBTAINED FROM:
☐ Patient ☐ Previous record
☒ Family ☒ Responsible party

Signature *Louise Weber, CNA*
(transfer aide from Veteran's Home)

Home Medications (including OTC) Codes: A=Sent home B=Sent to pharmacy C=Not brought in

Medication	Dose	Frequency	Time of Last Dose	Code	Patient Understanding of Drug
furosemide	80 mg	daily	0730	C	N/A
atenolol	25 mg	daily	0730	C	N/A
lisinopril	20 mg	daily	0730	C	N/A
Zocor	40 mg	daily	0730	C	N/A
haloperidol	0.5 mg	AM and PM	0730	C	N/A
warfarin	5 mg	daily	0730	C	N/A
donepezil	10 mg	PM	1400	C	N/A

Do you take all medications as prescribed? ☒ Yes ☐ No If no, why?

PATIENT/FAMILY HISTORY

☐ Cold in past two weeks
☐ Hay fever
☐ Emphysema/lung problems
☐ TB disease/positive TB skin test
☐ Cancer
☐ Stroke/past paralysis
☒ Heart attack Patient
☒ Angina/chest pain Patient
☒ Heart problems Patient
☒ High blood pressure Patient
☐ Arthritis
☐ Claustrophobia
☐ Circulation problems
☒ Easy bleeding/bruising/anemia Patient
☐ Sickle cell disease
☐ Liver disease/jaundice
☐ Thyroid disease
☐ Diabetes
☐ Kidney/urinary problems
☐ Gastric/abdominal pain/heartburn
☒ Hearing problems Patient
☐ Glaucoma/eye problems
☐ Back pain
☐ Seizures
☒ Other Alzheimer's vs. vascular dementia-Patient

RISK SCREENING

Have you had a blood transfusion? ☐ Yes ☒ No
Do you smoke? ☐ Yes ☒ No
If yes, how many pack(s)?
Does anyone in your household smoke? ☐ Yes ☒ No
Do you drink alcohol? ☐ Yes ☒ No
If yes, how often?_______ How much?
When was your last drink? ______/______/______
Do you take any recreational drugs? ☐ Yes ☐ No
If yes, type:_______ Route:
Frequency:_______ Date last used: ______/______/______

FOR WOMEN Ages 12–52

Is there any chance you could be pregnant? ☐ Yes ☐ No
If yes, expected date (EDC):
Gravida/Para:

ALL WOMEN

Date of last Pap smear:
Do you perform regular breast self-exams? ☐ Yes ☐ No

ALL MEN

Do you perform regular testicular exams? ☐ Yes ☒ No

Additional comments:

× *S. Beattie, RN*
Signature/Title

Client name: Ralph McCormick
DOB: 10/19
Age: 89
Sex: Male
Education: HS Diploma
Occupation: Retired from AT&T as telephone technician
Hours of work: N/A
Household members: Divorced—ex-spouse deceased; one adult child who lives in the area; has been a resident of the Veteran's Long-Term Care Facility for past 3 years
Ethnic background: Caucasian
Religious affiliation: None
Referring physician: B. Byrd, MD

Chief complaint:
Transfer from Veteran's Long-Term Care Facility. Patient was in a combative episode with roommate. He fell and hit his hip on the corner of a bed. He is admitted for evaluation of this nonhealing wound.

Patient history:
Onset of disease: Four years ago, patient history indicates that patient was having difficulty taking care of his life-long home and immediate medical needs. Reported some forgetfulness but could manage with assistance. His son moved him to an assisted living facility nearby to son's home, and he was able to live there for approximately 1 year. His son reports that 3 years ago, his "Alzheimer's" became acutely worse. His needs at the assisted care facility had significantly increased, and even with a nurse's aide coming in twice a day in addition to the assisted care, the patient's safety was questioned. Patient began wandering away from the facility and became combative. He was admitted to an Alzheimer's unit at a local nursing home, and then approximately 3 weeks later, was transferred to the Veteran's Home when an opening was available.
PMH: s/p MI × 2 at ages 45 and 62; s/p 4 vessel CABG at age 62. R hip replacement 5 years ago. HTN × 44 years.

Meds:

furosemide	80 mg	daily
atenolol	25 mg	daily
lisinopril	20 mg	daily
Zocor	40 mg	daily
haloperidol	0.5 mg	AM and PM
warfarin	5 mg	daily
donepezil	10 mg	PM

Smoker: Yes but quit over 20 years ago
Family Hx: What? Cardiac disease; Alzheimer's *Who?* Father, uncles, brother—all died before age 50 of MI. Mother had Alzheimer's or some type of dementia.

Physical exam:
General appearance: Frail, thin, elderly gentleman who is obviously confused and agitated.
Vitals: Temp: 100.3°F, BP 94/68 mm Hg, HR 85 bpm, RR 32 bpm (increases with minimal activity).
Heart: PMI sustained and displaced laterally; Normal S_1; S_2; $+S_3$ at apex.
HEENT: Pupils are small and react to light sluggishly; Ocular fundus is pale; negative thyromegaly and adenopathy. + JVD—increased 4 cm above sternal angle at 45°.
Genitalia: Normal
Rectal: Not performed
Neurologic: Disoriented to time, place, and person.
Extremities: Cool to touch, pale with bruising; 2+ radial pulses, 1+ dorsalis pedis, and 1+ posterior tibial pulses bilaterally; DRT 2+ and symmetrical; strength 2/5 throughout.
Skin: Transparent with decreased turgor, pale, cool with multiple ecchymosis;
Open, draining purulent wound approximately 2 cm × 2 cm × 8 cm located on right posterior thigh.
Chest/lungs: Clear to auscultation and percussion with no rubs.
Abdomen: Nontender with bowel sounds present.

Nutrition Hx:
Usual dietary intake: (prior to current illness)
The inpatient RD at the hospital found the following information after discussion with the dietary manager employed at the Veteran's Home. As a resident in a long-term care facility, the patient was prescribed a regular diet with specific modifications made for the Alzheimer's unit. These include finger foods and access to snacks for a minimum of three times daily. Mr. McCormick did not have a good appetite and had difficulty attending to the task of eating. He required assistance with all meals. His best meal of the day was in the morning with cornflakes, banana, and a high-calorie, high-protein shake. Intake after that was highly variable. Patient's son indicates that his dad's weight had been about 170 lbs until 4 years ago. He lost quite a bit of weight during the transition to assisted living and then to a long-term care facility. Weight has been stable for the past 6 months, but overall, he has lost over 40 lbs over the past 4 years, with most of that in the first year.
Food allergies/intolerances/aversions: None
Previous nutrition therapy? N/A
Food purchase/preparation: Long-term care facility
Vit/min intake: Multivitamin

Dx:
Nonhealing wound/cellulitis; Alzheimer's disease; Hx of CAD, HTN

Tx plan/Hospital course:
Surgical consult indicates that patient has a Stage III full-thickness nonpressure wound (laceration) with purulent drainage and foul odor. Cultures for wound exudate are pending. Patient was started on 1.5 g of Ampicillin-Sulbactam IV every 6 hours. Patient is scheduled for initial wound debridement with a consult for wound management and nutrition consult.

UH UNIVERSITY HOSPITAL

NAME: Ralph McCormick DOB: 10/19
AGE: 89 SEX: M
PHYSICIAN: B. Byrd, MD

CHEMISTRY

	NORMAL		UNITS
DAY:		Admission	
DATE:		8/12	
TIME:		0830	
LOCATION:			
Albumin	3.5-5	2.9 L	g/dL
Total protein	6-8	5.5 L	g/dL
Prealbumin	16-35	14 L	mg/dL
Transferrin	250-380 (women) 215-365 (men)	165 L	mg/dL
Sodium	136-145	136	mEq/L
Potassium	3.5-5.5	3.5	mEq/L
Chloride	95-105	96	mEq/L
PO_4	2.3-4.7	2.5	mg/dL
Magnesium	1.8-3	1.9	mg/dL
Osmolality	285-295	291	mmol/kg/H_2O
Total CO_2	23-30	27	mEq/L
Glucose	70-110	82	mg/dL
BUN	8-18	22 H	mg/dL
Creatinine	0.6-1.2	1.3 H	mg/dL
Uric acid	2.8-8.8 (women) 4.0-9.0 (men)	5.1	mg/dL
Calcium	9-11	9	mg/dL
Bilirubin	≤ 0.3	0.1	mg/dL
Ammonia (NH_3)	9-33	24	µmol/L
ALT	4-36	25	U/L
AST	0-35	21	U/L
Alk phos	30-120	145	U/L
CPK	30-135 (women) 55-170 (men)	5	U/L
LDH	208-378	257	U/L
CHOL	120-199	155	mg/dL
HDL-C	> 55 (women) > 45 (men)	33	mg/dL
VLDL	7-32		mg/dL
LDL	< 130	121	mg/dL
LDL/HDL ratio	< 3.22 (women) < 3.55 (men)	3.6	
Apo A	101-199 (women) 94-178 (men)		mg/dL
Apo B	60-126 (women) 63-133 (men)		mg/dL
TG	35-135 (women) 40-160 (men)	153	mg/dL
T_4	4-12		mcg/dL
T_3	75-98		mcg/dL
HbA_{1C}	3.9-5.2	4.9	%

UH UNIVERSITY HOSPITAL

NAME: Ralph McCormick DOB: 10/19
AGE: 89 SEX: M
PHYSICIAN: B. Byrd, MD

HEMATOLOGY************************************

DAY: Admission
DATE: 8/12
TIME: 0830
LOCATION:

	NORMAL		UNITS
WBC	4.8-11.8	16.0 H	$\times 10^3/mm^3$
RBC	4.2-5.4 (women)	5.1	$\times 10^6/mm^3$
	4.5-6.2 (men)		$\times 10^6/mm^3$
HGB	12-15 (women)	13.5 L	g/dL
	14-17 (men)		
HCT	37-47 (women)	39%	%
	40-54 (men)		
MCV	80-96	77 L	μm^3
RETIC	0.8-2.8	2.7	%
MCH	26-32	24 L	pg
MCHC	31.5-36	30 L	g/dL
RDW	11.6-16.5	17.8 H	%
Plt Ct	140-440	145	$\times 10^3/mm^3$
Diff TYPE			
ESR	0-25 (women)	15	mm/hr
	0-15 (men)		
% GRANS	34.6-79.2	75	%
% LYM	19.6-52.7	10 L	%
SEGS	50-62	50	%
BANDS	3-6	5	%
LYMPHS	24-44	10 L	%
MONOS	4-8	5	%
EOS	0.5-4	1	%
Ferritin	20-120 (women)	18 L	mg/mL
	20-300 (men)		
ZPP	30-80		$\mu mol/mol$
Vitamin B_{12}	24.4-100		ng/dL
Folate	5-25		$\mu g/dL$
Total T cells	812-2,318		mm^3
T-helper cells	589-1,505		mm^3
T-suppressor cells	325-997		mm^3
PT	11-16	15	sec

Case Questions

I. Understanding the Disease and Pathophysiology

1. Define *dementia.* Define *Alzheimer's disease (AD).* How do they differ?

2. What is the current theory regarding the etiology of AD?

3. Based on Mr. McCormick's medical record, does he present with any risk factors for the development of AD?

4. Based on Mr. McCormick's PMH, what are his other concurrent diagnoses? Could any of these contribute to his symptoms?

5. How is AD diagnosed?

6. What are the current medical interventions available for the management of AD? What are the goals of these interventions?

7. Mr. McCormick has a Stage III full-thickness nonpressure wound. What does that mean?

8. Describe the normal stages of wound healing.

II. Understanding the Nutrition Therapy

9. Name a minimum of three factors that support wound healing. Name a minimum of three factors that may impair wound healing. Identify the most probable factors that may have contributed to Mr. McCormick's poor wound healing.

10. Describe the potential roles of zinc, vitamin C, vitamin E, and arginine in wound healing.

III. Nutrition Assessment

A. Evaluation of Weight/Body Composition

11. Assess this patient's available anthropometric data. Calculate UBW, percent UBW, and BMI. Which of these is the most pertinent in identifying the patient's nutrition risk? Why?

12. Discuss the progressive weight loss that Mr. McCormick has experienced. Why is this a concern? What factors may have contributed to this weight loss?

B. Calculation of Nutrient Requirements

13. Calculate energy and protein requirements for Mr. McCormick. Identify the formula/calculation method you used and explain the rationale for using it.

14. How would you determine the levels of micronutrients that Mr. McCormick needs? List those that you believe will need to be supplemented.

C. Intake Domain

15. From the information gathered within the intake domain, list possible nutrition problems using the diagnostic term.

D. Clinical Domain

16. Identify all medications that Mr. McCormick is prescribed. Describe the basic function of each.

17. Using his admission chemistry and hematology values, which biochemical measures are abnormal?

a. Which values can be used to further assess his nutritional status? Explain.

b. Which laboratory measures (see lab report, pages 229–230) are related to his infection and wound?

c. Which laboratory measures (see lab report, pages 229–230) are related to any of Mr. McCormick's concurrent diagnoses? Explain.

18. From the information gathered within the clinical domain, list possible nutrition problems using the diagnostic term.

19. Do you think this patient is malnourished? If so, why? Will this impact the success of interventions for wound healing?

E. Behavioral–Environmental Domain

20. Identify issues from Mr. McCormick's primary diagnosis that could potentially interfere with his ability to consume an adequate diet.

21. Are residents in a long-term care facility at higher nutritional risk than elders living independently? Why or why not?

22. From the information gathered within the behavioral–environmental domain, list possible nutrition problems using the diagnostic term.

IV. Nutrition Diagnosis

23. Select two high-priority nutrition problems and complete the PES statement for each.

V. Nutrition Intervention

24. For each of the PES statements that you have written, establish an ideal goal (based on the signs and symptoms) and an appropriate intervention (based on the etiology).

25. What specific dosage recommendations would you make about supplementing zinc, vitamin C, vitamin E, and arginine to promote wound healing for Mr. McCormick?

26. Mr. McCormick drinks high-calorie, high-protein milkshakes at the Veteran's Home. What products might you recommend for him during his hospitalization?

27. What specific interventions might you recommend for a patient with Alzheimer's that could improve his oral intake during a hospitalization?

VI. Nutrition Monitoring and Evaluation

28. What measures would be appropriate to use to measure adequacy of oral intake during Mr. McCormick's hospitalization?

29. If Mr. McCormick's intake is inadequate, is he a candidate for enteral feeding? Outline the pros and cons for recommending nutrition support for this patient. What are the ethical considerations?

Bibliography

Baden LR, Eisenstein BI. Impact of antibiotic resistance on the treatment of Gram-negative sepsis. *Curr Infect Dis Rep.* 2000;2(5):409–416.

Bourdel-Marchasson I, Barateau M, Rondeau V, Dequae-Merchadou L, Salles-Montaudon N, Emeriau JP, Manciet G, Dartigues JF. A multi-center trial of the effects of oral nutritional supplementation in critically ill older inpatients. GAGE Group. Groupe Aquitain Gériatrique d'Evaluation. *Nutrition.* 2000;16:1–5.

Cullum N, McInnes E, Bell-Syer SEM, Legood R. Support surfaces for pressure ulcer prevention. *Cochrane Database of Systematic Reviews.* 2004;3:CD001735.

Houwing R, Rozendaal M, Wouters-Wesseling W, Beulens JWJ, Buskens E, Haalboom J. A randomized, double-blind assessment of the effect of nutritional supplementation on the prevention of pressure ulcers in hip-fracture patients. *Clin Nutr.* 2003;22(4):401–405.

Langer G, Schloemer G, Knerr A, Kuss O, Behrens J. Nutritional interventions for preventing and treating pressure ulcers. *Cochrane Database of Systematic Reviews.* 2003;(4):CD003216.

Lansdown A, Mirastschijski U, Stubbs N, Scanlon E, Ågren M. Zinc in wound healing: Theoretical, experimental, and clinical aspects. *Wound Repair Regen.* 2007;15(1):2–16.

Nelms MN. Assessment of nutrition status and risk. In: Nelms M, Sucher K, Long S. *Nutrition Therapy and Pathophysiology.* Belmont, CA: Thomson/Brooks-Cole; 2007:101–135.

Nelms MN. Cellular and physiological response to injury. In: Nelms M, Sucher K, Long S. *Nutrition Therapy and Pathophysiology.* Belmont, CA: Thomson/Brooks-Cole; 2007:219–236.

Nelms MN. Diseases of the neurological system. In: Nelms M, Sucher K, Long S. *Nutrition Therapy and Pathophysiology.* Belmont, CA: Thomson/Brooks-Cole; 2007:687–714.

Rypkema G, Adang E, Dicke H, Naber T, de Swart B, Disselhorst L, Goluke-Willemse G, Olde Rikkert M. Cost-effectiveness of an interdisciplinary intervention in geriatric inpatients to prevent malnutrition. *J Nutr Health Aging.* 2004;8(2):122–127.

Sorensen JL, Jorgensen B, Gottrup F. Surgical treatment of pressure ulcers. *Am J Surg.* 2004;188(1A Suppl):42–51.

Sullivan DH, Johnson LE, Bopp MM, Roberson PK. Prognostic significance of monthly weight fluctuations among older nursing home residents. *J Gerontol Ser A: Biol Sci Med Sci.* 2004;59:M633–M639.

Thomas DR. Improving the outcome of pressure ulcers with nutritional intervention: A review of the evidence. *Nutrition.* 2001;17:121–25.

Thompson C, Fuhrman MP. Nutrients and wound healing: Still searching for the magic bullet. *Nutr Clin Pract.* 2005;20:331–347.

Whitney J, Phillips L, Aslam R, Barbul A, Gottrup F, Gould L, Robson M, Rodeheaver G, Thomas D, Stotts N. Guidelines for the treatment of pressure ulcers. *Wound Repair Regen.* 2006;14(6):663–679.

Internet Resources

Alzheimer Research Forum. http://www.alzforum.org

National Alzheimer's Association. http://www.alz.org

National Institute of Neurological Disorders and Stroke: NINDS Alzheimer's Disease Information Page. http://www.ninds.nih.gov/disorders/alzheimersdisease/alzheimersdisease.htm

Mayo Clinic: Alzheimer's disease. http://www.mayoclinic.com/health/alzheimers-disease/DS00161

Unit Six

NUTRITION THERAPY FOR PULMONARY DISORDERS

The two cases in this section portray the interrelationship between nutrition and the respiratory system. In a healthy individual, the respiratory system receives oxygen for cellular metabolism and expires waste products—primarily carbon dioxide. Fuels—carbohydrate, protein, and lipid—are metabolized, using oxygen and producing carbon dioxide. The type of fuel an individual receives can affect physiological conditions and interfere with normal respiratory function.

Nutritional status and pulmonary function are interdependent. Malnutrition can evolve from pulmonary disorders and can contribute to declining pulmonary status. The incidence of malnutrition is common for people with COPD, ranging anywhere from 25 percent to 50 percent. In respiratory disease, maintaining nutritional status improves muscle strength needed for breathing, decreases risk of infection, facilitates weaning from mechanical ventilation, and improves ability for physical activity.

The American Thoracic Society defines chronic obstructive pulmonary disease (COPD) as a disease process of chronic airway obstruction caused by chronic bronchitis, emphysema, or both. These conditions place a significant burden on the health care systems in the United States, with an estimated cost of over $30 billion each year. Prevalence is increasing as well, with approximately 16 million people affected in the United States alone. COPD is the fourth leading cause of death and is the only diagnosis in the United States with an increasing death rate.

In Cases 20 and 21, nutritional assessment and evaluation demonstrate the effects of COPD on nutritional status. As patients are started on nutrition support, you will examine the impact of nutrition on declining respiratory status.

Case 20

Chronic Obstructive Pulmonary Disease

Objectives

After completing this case, the student will be able to:

1. Apply knowledge of the pathophysiology of chronic obstructive pulmonary disease in order to identify and explain common nutritional problems associated with this disease.
2. Identify the effects of malnutrition on pulmonary status.
3. Identify the effects of nutrient metabolism on pulmonary function.
4. Analyze nutrition assessment data to evaluate nutritional status and identify specific nutrition problems.
5. Determine nutrition diagnoses and write appropriate PES statements.
6. Determine interventions to increase an individual's intake of energy and protein.

Stella Bernhardt is initially diagnosed with Stage 1 COPD (emphysema). She is now admitted with increasing shortness of breath and possible upper respiratory infection.

Name: Stella Bernhardt
DOB: 10/23 (age 62)
Physician: D. Bradshaw, MD

ADMISSION DATABASE

BED #	DATE:	TIME:	TRIAGE STATUS (ER ONLY):
2	1/25	1340	☐ Red ☐ Yellow ☐ Green ☐ White

Initial Vital Signs

TEMP:	RESP:	SAO$_2$:	
98.8	22		
HT:	**WT (lb):**	**B/P:**	**PULSE:**
5'3"	119	130/88	92
LAST TETANUS		**LAST ATE**	**LAST DRANK**
unknown		today noon	today noon

PRIMARY PERSON TO CONTACT:
Name: Pete Bernhardt
Home #: 555-339-6543
Work #: N/A

ORIENTATION TO UNIT: ☒ Call light ☒ Television/telephone ☒ Bathroom ☒ Visiting ☒ Smoking ☒ Meals ☒ Patient rights/responsibilities

CHIEF COMPLAINT/HX OF PRESENT ILLNESS

"I'm hardly able to do anything for myself right now. Even taking a bath or getting dressed makes me so short of breath. My husband had to help me out of the shower this morning."

ALLERGIES: Meds, Food, IVP Dye, Seafood: Type of Reaction

NKA

PREVIOUS HOSPITALIZATIONS/SURGERIES

4 live births

Last hospitalization–January 1 year ago for pneumonia

PERSONAL ARTICLES: (Check if retained/describe)
☐ Contacts ☐ R ☐ L ☐ Dentures ☐ Upper ☐ Lower
☒ Jewelry: wedding band
☒ Other: glasses

VALUABLES ENVELOPE:
☐ Valuables instructions

INFORMATION OBTAINED FROM:
☒ Patient ☐ Previous record
☒ Family ☐ Responsible party

Signature *Stella Bernhardt*

Home Medications (including OTC) **Codes: A=Sent home** **B=Sent to pharmacy** **C=Not brought in**

Medication	Dose	Frequency	Time of Last Dose	Code	Patient Understanding of Drug
Combivent	2 puffs	4 times daily	today noon	A	yes

Do you take all medications as prescribed? ☒ Yes ☐ No If no, why?

PATIENT/FAMILY HISTORY

☒ Cold in past two weeks Patient
☐ Hay fever
☒ Emphysema/lung problems Patient
☐ TB disease/positive TB skin test
☒ Cancer Mother
☐ Stroke/past paralysis
☐ Heart attack
☐ Angina/chest pain
☐ Heart problems
☐ High blood pressure
☐ Arthritis
☐ Claustrophobia
☐ Circulation problems
☐ Easy bleeding/bruising/anemia
☐ Sickle cell disease
☐ Liver disease/jaundice
☐ Thyroid disease
☐ Diabetes
☐ Kidney/urinary problems
☐ Gastric/abdominal pain/heartburn
☐ Hearing problems
☐ Glaucoma/eye problems
☐ Back pain
☐ Seizures
☐ Other

RISK SCREENING

Have you had a blood transfusion? ☒ Yes ☐ No
Do you smoke? ☐ Yes ☒ No Quit last year
If yes, how many pack(s)? previously 1/day for 46 years
Does anyone in your household smoke? ☐ Yes ☒ No
Do you drink alcohol? ☐ Yes ☒ No
If yes, how often? How much?
When was your last drink?
Do you take any recreational drugs? ☐ Yes ☒ No
If yes, type:_______ Route:
Frequency:_______ Date last used:______/______/______

FOR WOMEN Ages 12–52

Is there any chance you could be pregnant? ☐ Yes ☒ No
If yes, expected date (EDC):
Gravida/Para: 6/4

ALL WOMEN

Date of last Pap smear: 9/10 this past year
Do you perform regular breast self-exams? ☒ Yes ☐ No

ALL MEN

Do you perform regular testicular exams? ☐ Yes ☐ No

Additional comments:

x *Betty Larson, RN*
Signature/Title

Client name: Stella Bernhardt
DOB: 10/23
Age: 62
Sex: Female
Education: Some college. *What grade/level?* Patient completed 2 years of college.
Occupation: Retired office manager for independent insurance agency
Hours of work: N/A
Household members: Husband age 68. PMH of CAD.
Ethnic background: Caucasian
Religious affiliation: Methodist
Referring physician: Debra Bradshaw, MD (pulmonology)

Chief complaint:
"I'm hardly able to do anything for myself right now. Even taking a bath or getting dressed makes me short of breath. My husband had to help me out of the shower this morning. I feel that I am gasping for air. I am coughing up a lot of phlegm that is a dark brownish-green. I am always short of breath, but I can tell when things change. I was at a church meeting with a lot of people—I might have caught something there. My husband says that I am confused in the morning. I know it is hard for me to get going in the morning. Do you think my confusion is related to my COPD?"

Patient history:
Onset of disease: Initially diagnosed with Stage 1 COPD (emphysema) 5 years ago. Medical records at last admission indicate pulmonary function tests: baseline $FEV^1 = 0.7$ L, $FVC = 1.5$ L, FEV^1/FVC 46%.
Type of Tx: Combivent (metered-dose inhaler)—2 inhalations four times daily (each inhalation delivers 18 mcg ipratropium bromide; 130 mcg albuterol sulfate)
PMH: No occupational exposures; bronchitis and upper respiratory infections during winter months for most of adult life. Four live births; 2 miscarriages.
Meds: Combivent (see *Type of Tx*)
Smoker: Yes. 46 years, 1 PPD history—has quit for past 1 year.
Family Hx: What? CA *Who?* Mother, 2 aunts died from lung cancer

Physical exam:
General appearance: 62-year-old female in no acute distress
Vitals: Temp: 98.8°F, BP 130/88 mm Hg, HR 92 bpm, RR 22 bpm
Heart: Regular rate and rhythm; mild jugular distension noted
HEENT
Eyes: PERRLA, no hemorrhages
Ears: Slight redness
Nose: Clear
Throat: Clear
Genitalia: Deferred
Neurologic: Alert, oriented; cranial nerves intact
Extremities: 1+ bilateral pitting edema; no cyanosis or clubbing.
Skin: Warm, dry

Chest/lungs: Decreased breath sounds, percussion hyperresonant; prolonged expiration with wheezing; rhonchi throughout; using accessory muscles at rest
Abdomen: Liver, spleen palpable; nondistended, nontender, normal bowel sounds

Nutrition Hx:
General: Patient states that her appetite is poor: "I fill up so quickly—after just a few bites." Relates that meal preparation is difficult: "By the time I fix a meal, I am too tired to eat it." In the previous 2 days, she states that she has eaten very little. Increased coughing has made it very hard to eat: "I don't think food tastes as good, either. Everything has a bitter taste." Highest adult weight was 145–150 lbs (5 years ago). States that her family constantly tells her how thin she has gotten: "I haven't weighed myself for a while, but I know my clothes are bigger." Dentures are present but fit loosely.

Usual dietary intake:

AM:	Coffee, juice or fruit, dry cereal with small amount of milk
Lunch:	Large meal of the day—meat; vegetables; rice, potato, or pasta, but patient admits she eats only very small amounts
Dinner/evening meal:	Eats very light in evening—usually soup, scrambled eggs, or sandwich. Drinks Pepsi throughout day (usually 3 12-oz cans)

24-hour recall: ½ c coffee with nondairy creamer, few sips of orange juice, ½ c oatmeal with 1 tsp sugar, ¾ c chicken noodle soup, 2 saltine crackers, ½ c coffee with nondairy creamer; sips of Pepsi throughout day and evening—estimated amount 32 oz
Food allergies/intolerances/aversions (specify): Avoids milk: "People say it will increase mucus production."
Previous nutrition therapy? No
Food purchase/preparation: Self; "My daughters come and help sometimes."
Vit/min intake: None
Anthropometric Data: Ht. 5'3", Wt. 119 lbs, UBW 145–150 lbs, last recorded weight: 139 lbs 1 year ago MAC 19.05 cm, TSF 15 mm

Dx:
Acute exacerbation of COPD, increasing dyspnea, hypercapnia, r/o pneumonia

Tx plan:
O_2 1 L/minute via nasal cannula with humidity—keep O_2 saturation 90%–91%
IVF D5 ½ NS with 20 mEq KCL @ 75 cc/hr
Solumedrol 10 mg/kg q 6 hr
Ancef 500 mg q 6 hr
Ipratropium bromide via nebulizer 2.5 mg q 30 minutes × 3 treatments then q 2 hr Albuterol sulfate via nebulizer 4 mg q 30 minutes × 3 doses then 2.5 mg q 4 hr
ABGs q 6 hours.
CXR—EPA/LAT.
Sputum cultures and Gram stain

Hospital course:

Mrs. Bernhardt was diagnosed with acute exacerbation of COPD secondary to bacterial pneumonia. This was confirmed by CXR and sputum culture. She responded well to aggressive medical treatment for her emphysema, although her physician does feel her underlying condition has progressed. She will be discharged on home O_2 therapy for the first time and referred to an outpatient pulmonary rehabilitation program. Her discharge medications will be the same (Combivent), but she will complete an oral course of corticosteroids and an additional 10-day course of Keflex. Dr. Bradshaw ordered a nutrition consult in-house with recommendations for nutritional follow-up through the pulmonary rehabilitation program.

UH UNIVERSITY HOSPITAL

NAME: Stella Bernhardt
DOB: 10/23
AGE: 62
SEX: F
PHYSICIAN: D. Bradshaw, MD

CHEMISTRY

DAY: Admit
DATE: 1/25
TIME:
LOCATION:

	NORMAL		UNITS
Albumin	3.5-5	3.3 L	g/dL
Total protein	6-8	5.8 L	g/dL
Prealbumin	16-35	16	mg/dL
Transferrin	250-380 (women) 215-365 (men)	298	mg/dL
Sodium	136-145	137	mEq/L
Potassium	3.5-5.5	3.7	mEq/L
Chloride	95-105	101	mEq/L
PO_4	2.3-4.7	3.1	mg/dL
Magnesium	1.8-3	1.8	mg/dL
Osmolality	285-295	293	mmol/kg/H_2O
Total CO_2	23-30	32 H	mEq/L
Glucose	70-110	92	mg/dL
BUN	8-18	9	mg/dL
Creatinine	0.6-1.2	0.9	mg/dL
Uric acid	2.8-8.8 (women) 4.0-9.0 (men)	3.4	mg/dL
Calcium	9-11	9.1	mg/dL
Bilirubin	≤0.3	0.1	mg/dL
Ammonia (NH_3)	9-33	25	μmol/L
ALT	4-36	8	U/L
AST	0-35	22	U/L
Alk phos	30-120	112	U/L
CPK	30-135 (women) 55-170 (men)	22	U/L
LDH	208-378	313	U/L
CHOL	120-199	145	mg/dL
HDL-C	>55 (women) >45 (men)	61	mg/dL
VLDL	7-32		mg/dL
LDL	<130	98	mg/dL
LDL/HDL ratio	<3.22 (women) <3.55 (men)		
Apo A	101-199 (women) 94-178 (men)		mg/dL
Apo B	60-126 (women) 63-133 (men)		mg/dL
TG	35-135 (women) 40-160 (men)	155	mg/dL
T_4	4-12		mcg/dL
T_3	75-98		mcg/dL
HbA_{1C}	3.9-5.2		%

UH UNIVERSITY HOSPITAL

NAME: Stella Bernhardt DOB: 10/23
AGE: 62 SEX: F
PHYSICIAN: D. Bradshaw, MD

HEMATOLOGY

DAY:		Admit	
DATE:		1/25	
TIME:			
LOCATION:			
	NORMAL		UNITS
WBC	4.8–11.8	15.0 H	$\times 10^3$/mm^3
RBC	4.2–5.4 (women) 4.5–6.2 (men)	4 L	$\times 10^6$/mm^3
HGB	12–15 (women) 14–17 (men)	11.5 L	g/dL
HCT	37–47 (women) 40–54 (men)	35 L	%
MCV	80–96		μm^3
RETIC	0.8–2.8		%
MCH	26–32		pg
MCHC	31.5–36		g/dL
RDW	11.6–16.5		%
Plt Ct	140–440		$\times 10^3$/mm^3
Diff TYPE			
ESR	0–25 (women) 0–15 (men)		mm/hr
% GRANS	34.6–79.2		%
% LYM	19.6–52.7		%
SEGS	50–62	83 H	%
BANDS	3–6	5	%
LYMPHS	24–44	10 L	%
MONOS	4–8	3	%
EOS	0.5–4	1	%
Ferritin	20–120 (women) 20–300 (men)		mg/mL
ZPP	30–80		μmol/mol
Vitamin B_{12}	24.4–100		ng/dL
Folate	5–25		μg/dL
Total T cells	812–2,318		mm^3
T-helper cells	589–1,505		mm^3
T-suppressor cells	325–997		mm^3
PT	11–16		sec

UH UNIVERSITY HOSPITAL

NAME: Stella Bernhardt DOB: 10/23
AGE: 62 SEX: F
PHYSICIAN: D. Bradshaw, MD

********************************ARTERIAL BLOOD GASES (ABGs)********************************

	NORMAL			UNITS
DAY:		Admit	3	
DATE:		1/25	1/27	
TIME:				
LOCATION:				
pH	7.35-7.45	7.29 L	7.4	
pCO_2	35-45	50.9 H	40.1	mm Hg
SO_2	≥95	92 L	90.2 L	%
CO_2 content	23-30	31 H	29.8	mmol/L
O_2 content	15-22			%
pO_2	≥80			mm Hg
Base excess	>3		6.0	mEq/L
Base deficit	<3	3.6 H		mEq/L
HCO_3^-	24-28	24.7	28	mEq/L
HGB	12-16 (women) 13.5-17.5 (men)	11.5 L		g/dL
HCT	37-47 (women) 40-54 (men)	35 L		%
COHb	<2			%
$[Na^+]$	135-148	136		mmol/L
$[K^+]$	3.5-5	3.7		mEq/L

Case Questions

I. Understanding the Disease and Pathophysiology

1. Mrs. Bernhardt was diagnosed with Stage 1 emphysema/COPD 5 years ago. What criteria are used to classify this staging?

2. COPD includes two distinct diagnoses. Outline the similarities and differences between emphysema and chronic bronchitis.

3. What risk factors does Mrs. Bernhardt have for this disease?

4. a. Identify the symptoms described in the MD's history and physical that are consistent with Mrs. Bernhardt's diagnosis, then describe the pathophysiology that may be responsible for each symptom.

Symptom	Etiology
Shortness of breath (dyspnea)	
Early morning confusion (hypercapnia)	
Increased production of brownish-green sputum	
Fatigue	
Early satiety	
Anorexia	
Dysgeusia	

b. Now identify at least four features of the physician's physical examination that are consistent with her admitting diagnosis. Describe the pathophysiology that might be responsible for each physical finding.

Physical Finding	Physiological Change/Etiology

5. Mrs. Bernhardt's medical record indicates previous pulmonary function tests as follows: baseline FEV[1] = 0.7 L, FVC = 1.5 L, FEV[1]/FVC 46%. Define FEV, FVC, and FEV/FVC, and indicate how they are used in the diagnosis of COPD. How can these measurements be used in treating COPD?

6. Look at Mrs. Bernhardt's arterial blood gas report from the day she was admitted.

a. Why would arterial blood gases (ABGs) be drawn for this patient?

b. Define each of the following and interpret Mrs. Bernhardt's values:

pH:

$PaCO_2$:

SaO_2:

HCO_3^-:

c. Mrs. Bernhardt was placed on oxygen therapy. What lab values tell you that the therapy is working?

7. Mrs. Bernhardt has quit smoking. Shouldn't her condition now improve? Explain.

8. What is a respiratory quotient? How is this figure related to nutritional intake and respiratory status?

II. Understanding the Nutrition Therapy

9. What are the most common nutritional concerns for someone with COPD? Why is the patient diagnosed with COPD at higher risk for malnutrition?

10. Is there a specific nutrition therapy prescribed for these patients?

III. Nutrition Assessment

A. Evaluation of Weight/Body Composition

11. Calculate Mrs. Bernhardt's UBW, percent UBW, and BMI. Do any of these values indicate that she is at nutritional risk? How would her 1+ bilateral pitting edema affect the evaluation of her weight?

12. Calculate arm muscle area using the anthropometric data for mid-arm muscle circumference (MAC) and triceps skinfold (TSF). How would these data be interpreted?

B. Calculation of Nutrient Requirements

13. Calculate Mrs. Bernhardt's energy and protein requirements. What activity and stress factors would you use? What is your rationale?

C. Intake Domain

14. Using Mrs. Bernhardt's nutrition history and 24-hour recall as a reference, do you feel she has an adequate oral intake? Explain, and provide any evidence.

15. From the information gathered within the intake domain, list possible nutrition problems using the diagnostic term.

D. Clinical Domain

16. Evaluate Mrs. Bernhardt's laboratory values. Identify those that are abnormal. Which of these may be used to assess her nutritional status?

17. Why may Mrs. Bernhardt be at risk for anemia? Do her laboratory values indicate that this may be a problem?

18. From the information gathered within the clinical domain, list possible nutrition problems using the diagnostic term.

E. Behavioral–Environmental Domain

19. What factors can you identify from her nutrition interview that probably contribute to her difficulty in eating?

20. From the information gathered within the behavioral–environmental domain, list possible nutrition problems using the diagnostic term.

IV. Nutrition Diagnosis

21. Select two high-priority nutrition problems and complete the PES statement for each.

V. Nutrition Intervention

22. What is the current recommendation on the appropriate mix of calories from carbohydrate, protein, and fat for this patient?

23. For each of the PES statements that you have written, establish an ideal goal (based on the signs and symptoms) and an appropriate intervention (based on the etiology).

24. What goals might you set for Mrs. Bernhardt as she is discharged and beginning pulmonary rehabilitation?

VI. Nutrition Monitoring and Evaluation

25. You are now seeing Mrs. Bernhardt at her second visit to pulmonary rehabilitation. She provides you with the following information from her food record. Her weight is now 116 lbs. She explains that adjustment to her medications and oxygen at home has been difficult, so she hasn't felt like eating very much. When you talk with her, you find she is hungriest in the morning, and often by evening, she is too tired to eat. She is having no specific intolerances. She does tell you that she hasn't consumed any milk products because she thought they would cause more sputum to be produced.

Food Diary
Monday

Breakfast: Coffee—1 c with 2 tbsp nondairy creamer, orange—½ c, 1 poached egg, ½ slice toast

Lunch: ½ tuna salad sandwich (3 tbsp tuna salad on 1 slice wheat bread), coffee—1 c with 2 tbsp nondairy creamer

Supper: Cream of tomato soup—1 c, ½ slice toast, ½ banana, Pepsi—approx 36 oz

Tuesday

Breakfast: Coffee—1 c with 2 tbsp nondairy creamer, orange juice—½ c, ½ c oatmeal with 2 tbsp brown sugar

Lunch: 1 chicken leg from Kentucky Fried Chicken, ½ c mashed potatoes with 2 tbsp gravy, coffee—1 c with 2 tbsp nondairy creamer

Supper: Cheese—2 oz, 8 saltine crackers, 1 can V8 juice (6 oz), Pepsi—approx 36 oz

a. Is she meeting her calorie and protein goals?

b. What would you tell her regarding the use of supplements and/or milk and sputum production?

c. Using the information from her food diary as a teaching tool, identify three interventions that you would propose for Mrs. Bernhardt to increase her calorie and protein intake.

Bibliography

Benedict M. Pulmonary rehab: Opportunity is knocking for RDs. *Health Care Food Nutr Focus.* 2006;23:6–7.

Bergman EA, Buergel NS. Diseases of the respiratory system. In: Nelms M, Sucher K, Long S. *Nutrition Therapy and Pathophysiology.* Belmont, CA: Thomson/Brooks-Cole; 2007:715–749.

Decramer M, De Benedetto F, Del Ponte A, Marinari S. Systemic effects of COPD. *Respir Med.* 2005;99(Suppl B):S3–S10.

Donahoe M. Nutritional aspects of lung disease. *Respir Care Clin N Am.* 1998;4:85–112.

Gronberg AM, Slinde F, Engstrom CP, Hulthen L, Larsson S. Dietary problems in patients with severe chronic obstructive pulmonary disease. *J Hum Nutr Diet.* 2005;18(6):445–452.

Hallin R, Koivisto-Hursti UK, Lindberg E, Janson C. Nutritional status, dietary energy intake and the risk of exacerbations in patients with chronic obstructive pulmonary disease (COPD). *Respir Med.* 2006;100(3):561–567.

Koehler F, Doehner W, Hoernig S, Witt C, Anker SD, John M. Anorexia in chronic obstructive pulmonary disease: Association to cachexia and hormonal derangement. *Int J Cardiol.* 2007;119:83–89.

Lerario MC, Sachs A, Lazaretti-Castro M, Saraiva LG, Jardim JR. Body composition in patients with chronic obstructive pulmonary disease: Which method to use in clinical practice? *Br J Nutr.* 2006;96:86–92.

Nici L, Donner C, Wouters E, Zuwallack R, Ambrosino N, Bourbeau J, Carone M, Celli B, Engelen M, Fahy B, Garvey C, Goldstein R, Gosselink R, Lareau S, MacIntyre N, Maltais F, Morgan M, O'Donnell D, Prefault C, Reardon J, Rochester C, Schols A, Singh S, Troosters T; ATS/ERS Pulmonary Rehabilitation Writing Committee. American Thoracic Society/European Respiratory Society statement on pulmonary rehabilitation. *Am J Respir Crit Care Med.* 2006;173(12):1390–1413.

Odencrants S, Ehnfors M, Grobe SJ. Living with chronic obstructive pulmonary disease (COPD): Part II. RNs' experience of nursing care for patients with COPD and impaired nutritional status. *Scand J Caring Sci.* 2007;21(1):56–63.

Odencrants S, Ehnfors M, Grobe SJ. Living with chronic obstructive pulmonary disease: Part I. Struggling with meal-related situations: Experiences among persons with COPD. *Scand J Caring Sci.* 2005;19(3):230–239.

Petty TL. Spirometry made simple. *Adv Manage Respir Care.* 1999;8:37–41.

Rochester DF, Esay SA. Malnutrition and the respiratory system. *Chest.* 1984;85;411–415. Available at: http://www.chestjournal.org/cgi/reprint/85/3/411.pdf. Accessed July 15, 2007.

Sergi G, Coin A, Marin S, Vianello A, Manzan A, Peruzza S, Inelmen EM, Busetto L, Mulone S, Enzi G. Body composition and resting energy expenditure in elderly male patients with chronic obstructive pulmonary disease. *Respir Med.* 2006;100(11):1918–1924.

Slinde F, Gronberg A, Engstrom CP, Rossander-Hulthen L, Larsson S. Body composition by bioelectrical impedance predicts mortality in chronic obstructive pulmonary disease patients. *Respir Med.* 2005;99(8):1004–1009.

Velloso M, Jardim JR. Study of energy expenditure during activities of daily living using and not using body position recommended by energy conservation techniques in patients with COPD. *Chest.* 2006;130(1):126–132.

Vermeeren MA, Creutzberg EC, Schols AM, Postma DS, Pieters WR, Roldaan AC, Wouters EF; on behalf of the COSMIC Study Group. Prevalence of nutritional depletion in a large out-patient population of patients with COPD. *Respir Med.* 2006;100(8):1349–1355.

Vermeeren MA, Wouters EF, Geraerts-Keeris AJ, Schols AM. Nutritional support in patients with chronic obstructive pulmonary disease during hospitalization for an acute exacerbation: A randomized controlled feasibility trial. *Clin Nutr.* 2004;23(5):1184–1192.

Internet Resources

American Lung Association: Chronic Obstructive Pulmonary Disease (COPD) Fact Sheet. http://www.lungusa.org/site/pp.asp?c=dvLUK9O0E&b=35020

COPD International Organization: COPD: What Does It Mean. http://www.copd-international.com/

National Heart Lung Blood Institute: Learn More Breathe Better: It Has a Name: COPD Chronic Obstructive Pulmonary Disease. http://www.nhlbi.nih.gov/health/public/lung/copd/index.htm

U.S. Department of Agriculture: Nutrient Data Laboratory. http://www.ars.usda.gov/ba/bhnrc/ndl

Case 21

COPD with Respiratory Failure

Objectives

After completing this case, the student will be able to:

1. Define the pathophysiology of chronic obstructive pulmonary disease and its relationship to acute respiratory failure.
2. Identify the role of nutrition in mechanical ventilation.
3. Determine the metabolic implications of acute respiratory failure.
4. Interpret biochemical indices for assessment of respiratory function.
5. Interpret biochemical indices for assessment of nutritional status.
6. Plan, interpret, and evaluate nutrition support for mechanically ventilated patients.

Daishi Hayato, a 65-year-old male, is brought to the University Hospital emergency room by his wife when he experiences severe shortness of breath. The patient has a long-standing history of COPD secondary to tobacco use.

IVERSITY HOSPITAL

ADMISSION DATABASE

Name: Daishi Hayato
DOB: 7/14 (age 65)
Physician: M. McFarland, MD

BED #	DATE:	TIME:	TRIAGE STATUS (ER ONLY):
1	3/26	1700	☐ Red ☐ Yellow ☐ Green ☐ White

Initial Vital Signs

TEMP:	RESP:	SAO$_2$:	
98	36		
HT: 5'4"	WT (lb): 122	B/P: 110/80	PULSE: 118
LAST TETANUS: unknown		LAST ATE: this AM	LAST DRANK: about 3 hours ago

PRIMARY PERSON TO CONTACT:
Name: Mrs. Mei Hayato
Home #: 555-456-3422
Work #: N/A

ORIENTATION TO UNIT: ☒ Call light ☒ Television/telephone ☒ Bathroom ☒ Visiting ☒ Smoking ☒ Meals ☒ Patient rights/responsibilities

CHIEF COMPLAINT/HX OF PRESENT ILLNESS

"My husband has had emphysema for many years. He was working in the yard today and got really short of breath. I called our doctor, and she said to go straight to the emergency room."

ALLERGIES: Meds, Food, IVP Dye, Seafood: Type of Reaction

penicillin

PREVIOUS HOSPITALIZATIONS/SURGERIES

cholecystectomy–20 years ago
dental extraction–5 years ago

PERSONAL ARTICLES: (Check if retained/describe)
☐ Contacts ☐ R ☐ L ☒ Dentures ☒ Upper ☒ Lower
☐ Jewelry:
☒ Other: eyeglasses

VALUABLES ENVELOPE:
☐ Valuables instructions

INFORMATION OBTAINED FROM:
☐ Patient ☐ Previous record
☒ Family ☐ Responsible party

Signature *M. Hayato*

Home Medications (including OTC) **Codes: A=Sent home** **B=Sent to pharmacy** **C=Not brought in**

Medication	Dose	Frequency	Time of Last Dose	Code	Patient Understanding of Drug
Combivent	2 inhalations	4 times daily	1200	A	yes
Lasix	40 mg	daily	0700	A	yes
oxygen via nasal cannula		during sleep			

Do you take all medications as prescribed? ☒ Yes ☐ No If no, why?

PATIENT/FAMILY HISTORY

☐ Cold in past two weeks
☐ Hay fever
☒ Emphysema/lung problems Patient
☐ TB disease/positive TB skin test
☒ Cancer Father
☐ Stroke/past paralysis
☐ Heart attack
☐ Angina/chest pain
☐ Heart problems
☐ High blood pressure
☐ Arthritis
☐ Claustrophobia
☒ Circulation problems Patient
☐ Easy bleeding/bruising/anemia
☐ Sickle cell disease
☐ Liver disease/jaundice
☐ Thyroid disease
☐ Diabetes
☐ Kidney/urinary problems
☐ Gastric/abdominal pain/heartburn
☐ Hearing problems
☐ Glaucoma/eye problems
☐ Back pain
☐ Seizures
☐ Other

RISK SCREENING

Have you had a blood transfusion? ☐ Yes ☒ No
Do you smoke? ☒ Yes ☐ No
If yes, how many pack(s)? 2/day for 50 years
Does anyone in your household smoke? ☐ Yes ☒ No
Do you drink alcohol? ☒ Yes ☐ No
If yes, how often? 1-2 × week How much? 1-2 drinks
When was your last drink? last week
Do you take any recreational drugs? ☐ Yes ☒ No
If yes, type:_______ Route:
Frequency:_______ Date last used:______/______/______

FOR WOMEN Ages 12–52

Is there any chance you could be pregnant? ☐ Yes ☐ No
If yes, expected date (EDC):
Gravida/Para:

ALL WOMEN

Date of last Pap smear:
Do you perform regular breast self-exams? ☐ Yes ☐ No

ALL MEN

Do you perform regular testicular exams? ☐ Yes ☒ No

Additional comments:

× *Carolyn Masterson, RN*
Signature/Title

Client name: Daishi Hayato
DOB: 7/14
Age: 65
Sex: Male
Education: Bachelor's degree
Occupation: Retired manager of local grocery chain
Hours of work: N/A
Household members: Wife age 62, well; four adult children not living in the area
Ethnic background: Asian American
Religious affiliation: Methodist
Referring physician: Marie McFarland, MD (pulmonary)

Chief complaint:
"My husband has had emphysema for many years. He was working in the yard today and got really short of breath. I called our doctor, and she said to go straight to the emergency room."

Patient history:
Onset of disease: The patient has a long-standing history of COPD secondary to chronic tobacco use, 2 PPD for 50 years. He was in his usual state of health today with marked limitation of his exercise capacity due to dyspnea on exertion. He also notes two-pillow orthopnea, swelling in both lower extremities. Today, while performing some yard work, he noted the sudden onset of marked dyspnea. His wife brought him to the emergency room right away. There, a chest radiograph showed a tension pneumothorax involving the left lung. Patient also states that he gets cramping in his right calf when he walks.
PMH: Had cholecystectomy 20 years ago. Total dental extraction 5 years ago. Patient describes intermittent claudication. Claims to be allergic to penicillin. Diagnosed with emphysema more than 10 years ago. Has been treated successfully with Combivent (metered dose inhaler)—2 inhalations qid (each inhalation delivers 18 mcg ipratropium bromide; 130 mcg albuterol sulfate).
Meds: Combivent, Lasix, O_2 2 L/hour via nasal cannula at night
Smoker: Yes, 2 PPD for 50 years
Family Hx: What? Lung cancer *Who?* Father

Physical exam:
General appearance: Acutely dyspneic Asian American male in acute respiratory distress
Vitals: Temp 97.6°F, BP 110/80 mm Hg, HR 118 bpm, RR 36 bpm
Heart: Normal heart sounds; no murmurs or gallops
HEENT: Within normal limits; funduscopic exam reveals AV nicking
Eyes: Pupil reflex normal
Ears: Slight neurosensory deficit acoustically
Nose: Unremarkable
Throat: Jugular veins appear distended. Trachea is shifted to the right. Carotids are full, symmetrical, and without bruits.
Genitalia: Unremarkable
Rectal: Prostate normal; stool hematest negative
Neurologic: DTR full and symmetric; alert and oriented × 3

Extremities: Cyanosis, 1+ pitting edema
Skin: Warm, dry to touch
Chest/lungs: Hyperresonance to percussion over the left chest anteriorly and posteriorly. Harsh inspiratory breath sounds are noted over the right chest with absent sounds on the left. Using accessory muscles at rest.
Abdomen: Old surgical scar RUQ. No organomegaly or masses. BS reduced.
Circulation: R femoral bruit present. Right PT and DP pulses were absent.

Nutrition Hx:
General: Wife relates general appetite is only fair. Usually, breakfast is the largest meal. His appetite has been decreased for past several weeks. She states that his highest weight was 135 lbs, but feels he weighs much less than that now.

Usual dietary intake:
AM: Egg, hot cereal, bread or muffin, hot tea (with milk and sugar)
Lunch: Soup, sandwich, hot tea (with milk and sugar)
Dinner: Small amount of meat, rice, 2–3 kinds of vegetables, hot tea (with milk and sugar)

24-hour recall: 2 scrambled eggs, few bites of Cream of Wheat, sips of hot tea, bite of toast; ate nothing rest of day—sips of hot tea
Food allergies/intolerances/aversions: NKA
Previous nutrition therapy? No
Food purchase/preparation: Wife
Vit/min intake: None
Anthropometric data: Ht 5′4″, Wt 122 lbs, UBW 135 lbs

Dx:
Acute respiratory distress, COPD, peripheral vascular disease with intermittent claudication

Tx plan:
ABG, pulse oximetry, CBC, chemistry panel, UA
Chest X-ray, ECG, Proventil 0.15 in 1.5 cc NS q 30 min × 3 followed by Proventil 0.3 cc in 3 cc normal saline q 2 hr per HHN (hand-held nebulizer)
Spirogram post nebulizer Tx
IVF D5½ NS at TKO Solumedrol 10-40 mg q 4–6 hr; high dose = 30 mg/kg q 4–6 hr (2 days max)
NPO

Hospital course:
In the emergency room, a chest tube was inserted into the left thorax with drainage under suction. Subsequently, the oropharynx was cleared. A resuscitation bag and mask was used to ventilate the patient with high-flow oxygen. Endotracheal intubation was then carried out, using the laryngoscope so that the trachea could be directly visualized. The patient was then ventilated with the help of a volume-cycled ventilator. Ventilation is 15 breaths/min with an FiO_2 of 100%, a positive end-expiratory pressure of 6, and a tidal volume of 700 mL. Daily chest radiographs and ABGs were used

each AM to guide settings on the ventilator. A nutrition consult was completed on day 2 of admission, and enteral feedings were initiated. Due to high residuals, the patient was started on ProcalAmine. Enteral feedings were restarted on day 4. Respiratory status actually became worse on day 5 but improved thereafter. ProcalAmine was discontinued on day 5, and enteral feeding continued until day 8. The patient was weaned from the ventilator on day 8, and discharged to home on day 11.

UH UNIVERSITY HOSPITAL

NAME: Daishi Hayato DOB: 7/14
AGE: 65 SEX: M
PHYSICIAN: M. McFarland, MD

CHEMISTRY

	NORMAL			UNITS
DAY:		3/26	3/29	
DATE:		1	4	
TIME:				
LOCATION:				
Albumin	3.5–5	3.6	3.5	g/dL
Total protein	6–8	6.1	6.2	g/dL
Prealbumin	16–35	26	17	mg/dL
Transferrin	250–380 (women) 215–365 (men)	170	173	mg/dL
Sodium	136–145	138	137	mEq/L
Potassium	3.5–5.5	3.9	3.5	mEq/L
Chloride	95–105	101	104	mEq/L
PO_4	2.3–4.7	3.2	2.6	mg/dL
Magnesium	1.8–3	1.9	1.8	mg/dL
Osmolality	285–295	293	285	mmol/kg/H_2O
Total CO_2	23–30	29	30	mEq/L
Glucose	70–110	108	110	mg/dL
BUN	8–18	11	15	mg/dL
Creatinine	0.6–1.2	0.7	0.9	mg/dL
Uric acid	2.8–8.8 (women) 4.0–9.0 (men)	3.9		mg/dL
Calcium	9–11	9.1		mg/dL
Bilirubin	≤0.3	0.8		mg/dL
Ammonia (NH_3)	9–33	9		µmol/L
ALT	4–36	15		U/L
AST	0–35	22		U/L
Alk phos	30–120	114		U/L
CPK	30–135 (women) 55–170 (men)	152		U/L
LDH	208–378	412		U/L
CHOL	120–199	155		mg/dL
HDL-C	>55 (women) >45 (men)	32 L		mg/dL
VLDL	7–32			mg/dL
LDL	<130	142 H		mg/dL
LDL/HDL ratio	<3.22 (women) <3.55 (men)	4.4 H		
Apo A	101–199 (women) 94–178 (men)			mg/dL
Apo B	60–126 (women) 63–133 (men)			mg/dL
TG	35–135 (women) 40–160 (men)	155		mg/dL
T_4	4–12			mcg/dL
T_3	75–98			mcg/dL
HbA_{1C}	3.9–5.2			%

UH UNIVERSITY HOSPITAL

NAME: Daishi Hayato DOB: 7/14
AGE: 65 SEX: M
PHYSICIAN: M. McFarland, MD

**************************************HEMATOLOGY**************************************

DAY:		3/26	
DATE:		1	
TIME:			
LOCATION:			
	NORMAL		UNITS
WBC	4.8–11.8	5.6	$\times 10^3/mm^3$
RBC	4.2–5.4 (women)	4.7	$\times 10^6/mm^3$
	4.5–6.2 (men)		
HGB	12–15 (women)	13.2	g/dL
	14–17 (men)		
HCT	37–47 (women)	39	%
	40–54 (men)		
MCV	80–96		μm^3
RETIC	0.8–2.8		%
MCH	26–32		pg
MCHC	31.5–36		g/dL
RDW	11.6–16.5		%
Plt Ct	140–440		$\times 10^3/mm^3$
Diff TYPE			
ESR	0–25 (women)		mm/hr
	0–15 (men)		
% GRANS	34.6–79.2	52.3	%
% LYM	19.6–52.7	48.5	%
SEGS	50–62	83 H	%
BANDS	3–6	5	%
LYMPHS	24–44	10 L	%
MONOS	4–8	3 L	%
EOS	0.5–4		%
Ferritin	20–120 (women)		mg/mL
	20–300 (men)		
ZPP	30–80		μmol/mol
Vitamin B_{12}	24.4–100		ng/dL
Folate	5–25		μg/dL
Total T cells	812–2,318		mm^3
T-helper cells	589–1,505		mm^3
T-suppressor cells	325–997		mm^3
PT	11–16		sec

UH *UNIVERSITY HOSPITAL*

NAME: Daishi Hayato DOB: 7/14
AGE: 65 SEX: M
PHYSICIAN: M. McFarland, MD

*********************************ARTERIAL BLOOD GASES (ABGs)*********************************

DAY:		1	2	3	5	
DATE:						
TIME:						
LOCATION:						
	NORMAL					UNITS
pH	7.35–7.45	7.2 L	7.30 L	7.36	7.22 L	
pCO_2	35–45	65 H	59 H	50 H	66 H	mm Hg
SO_2	≥ 95					%
CO_2 content	23–30	35	30	29	36	mmol/L
O_2 content	15–22					%
pO_2	≥ 80	56 L	58 L	60 L	57 L	mm Hg
Base excess	> 3					mEq/L
Base deficit	< 3					mEq/L
HCO_3^-	24–28	38 H	33 H	32 H	37 H	mEq/L
HGB	12–16 (women) 13.5–17.5 (men)					g/dL
HCT	37–47 (women) 40–54 (men)					%
COHb	< 2					%
$[Na^+]$	135–148					mmol/L
$[K^+]$	3.5–5					mEq/L

Name: Daishi Hayato
Physician: M. McFarland, MD

PATIENT CARE SUMMARY SHEET

Date: 3 Room: Wt Yesterday: 121 Today: 122

Temp °F	NIGHTS								DAYS								EVENINGS							
	00	01	02	03	04	05	06	07	08	09	10	11	12	13	14	15	16	17	18	19	20	21	22	23
105																								
104																								
103																								
102																								
101																								
100																								
99																								
98																								
97v																								
96																								
Pulse	80																							
Respiration																								
BP	110/80																							
Blood Glucose																								
Appetite/Assist																								
INTAKE	NPO																							
Oral																								
IV	50	→	→	→	→	→	→	→	50	→	→	→	→	→	→	→	100	100	100	100	100	100	100	100
TF Formula/Flush	25	25	25	25	25/50	25	25	25	25	25	25/50	25	25	25	25	25	**							
Shift Total	650								650								800							
OUTPUT																								
Void	125				200				275				300				200	175	150		240			
Cath.																								
Emesis																								
BM	200													100										
Drains																								
Shift Total	525								675								765							
Gain	+125																+35							
Loss									50															
Signatures	Mary Rogers, RT								Patricia Elkins, RN								Frannie Lowe, RN							

** tube-feeding held due to high residuals

Case Questions

I. Understanding the Disease and Pathophysiology

1. Mr. Hayato was diagnosed with emphysema more than 10 years ago. Define *emphysema* and its underlying pathophysiology.

2. Define the following terms found in the history and physical for Mr. Hayato.

a. *Dyspnea:*

b. *Orthopnea:*

c. *Pneumothorax:*

d. *Endotracheal intubation:*

e. *Cyanosis:*

3. Identify features of the physician's physical examination that are consistent with his admitting diagnosis. Describe the pathophysiology that might be responsible for each physical finding.

II. Understanding the Nutrition Therapy

4. What is the relationship between nutritional status and respiratory function? Define *respiratory quotient (RQ)*. What dietary factors affect RQ?

5. Do nutrition support and nutritional status play a role in the ability to be weaned from a respiratory ventilator? Explain.

III. Nutrition Assessment

A. Evaluation of Weight/Body Composition

6. Evaluate Mr. Hayato's admitting anthropometric data for nutritional assessment.

B. Calculation of Nutrient Requirements

7. Determine Mr. Hayato's energy and protein requirements using the Harris-Benedict equation, the Ireton-Jones equation, and the COPD predictive equations. Compare them. As Mr. Hayato's clinician, which would you set as your goal for meeting his energy needs?

8. Determine Mr. Hayato's fluid requirements.

C. Intake Domain

9. From the information gathered within the intake domain, list possible nutrition problems using the diagnostic term.

D. Clinical Domain

10. Evaluate Mr. Hayato's biochemical indices for nutritional assessment on day 1.

11. From the information gathered within the clinical domain, list possible nutrition problems using the diagnostic term.

IV. Nutrition Diagnosis

12. Select two high-priority nutrition problems and complete the PES statement for each.

V. Nutrition Intervention

13. Mr. Hayato was started on Isosource @ 25 cc/hr continuously over 24 hours.
    a. At this current rate, how many kcalories and grams of protein should he receive per day?

    b. Calculate his nutrition prescription utilizing this enteral formula. Include goal rate, free water requirements, and the appropriate progression of the rate.

14. What type of formula is Isosource? What is the percentage of kilocalories from carbohydrate, protein, and lipid? Should the patient have been started on a disease-specific formula? Support your responses. What is the rationale for pulmonary formulas?

VI. Nutrition Monitoring and Evaluation

15. Examine the patient care summary sheet. How much enteral feeding did the patient receive?

16. You read in the physician's orders that the patient experienced high gastric residuals and the enteral feeding was discontinued. What does this mean, and what is the potential cause of the problem?

17. Dr. McFarland elected to begin peripheral parenteral nutrition using a formula called ProcalAmine. She began the PPN @ 100 cc/hr and discontinued Mr. Hayato's regular IV of D5 ½ NS at TKO. What is ProcalAmine, and how much nutrition does this provide?

18. Was this adequate to meet the patient's nutritional needs? Explain.

19. Do you feel it was a good idea to begin peripheral parenteral nutrition (PPN)? What are the pros and cons? What are the limitations of using this form of nutrition support? Were other nutrition support options available for the health care team?

20. On day 4, the enteral feeding was restarted at 25 cc/hr and then increased to 50 cc/hr after 12 hours. You document that the ProcalAmine @ 100 cc/hr was also continued. What would have been the total energy intake for Mr. Hayato?

21. Examine the values documented for arterial blood gases (ABGs).
    a. On the day Mr. Hayato was intubated, his ABGs were as follows: pH 7.2, pCO_2 65, CO_2 35, pO_2 56, and HCO_3^- 38. What can you determine from each of these values?

    b. On day 3 while Mr. Hayato was on the ventilator, his ABGs were as follows: pH 7.36, pCO_2 50, CO_2 29, pO_2 60, and HCO_3^- 32. What can you determine from each of these values?

c. On day 5, after restarting enteral feeding and continuing on ProcalAmine, his ABGs were as follows: pH 7.22, pCO_2 66, pO_2 57, CO_2 36, and HCO_3^- 37. In addition, indirect calorimetry indicated a RQ of 0.95 and measured energy intake to be 1,350 kcal. How does the patient's measured energy intake compare to your previous calculations? What does the RQ indicate?

22. As Mr. Hayato is prepared for discharge, what nutritional goals might you set with him and his wife to improve his overall nutritional status?

Bibliography

Akrabawi SS, Mobarhan S, Stoltz RR, Ferguson PW. Gastric emptying, pulmonary function, gas exchange, and respiratory quotient after feeding a moderate versus high fat enteral formula meal in chronic obstructive pulmonary disease patients. *Nutrition.* 1996;12(4):260–265.

Bergman EA, Buergel NS. Diseases of the respiratory system. In: Nelms M, Sucher K, Long S. *Nutrition Therapy and Pathophysiology.* Belmont, CA: Thomson/Brooks-Cole; 2007:715–749.

Charney P. Enteral nutrition: Indications, options, and formulations. In: Gottschlich M, ed. *The Science and Practice of Nutrition Support.* Dubuque, IA: Kendall/Hunt; 2001:141–166.

Confalonieri M, Rossi A. Burden of chronic obstructive pulmonary disease. *Lancet.* 2000;356(Suppl):S56–S65.

Demling RH, De Santi L. Effect of a catabolic state with involuntary weight loss on acute and chronic respiratory disease. Available at: http://www.medscape.com/viewprogram/1816_pnt. Accessed July 17, 2007.

Engelen M, Wouters E, Deutz N, Menheere P, Schols A. Factors contributing to alterations in skeletal muscle and plasma amino acid profiles in patients with chronic obstructive pulmonary disease. *Am J Clin Nutr.* 2000;72(6):1480–1487.

Hogg JH, Klapholz A, Reid-Hector J. Pulmonary disease. In: Gottschlich M, ed. *The Science and Practice of Nutrition Support.* Dubuque, IA: Kendall/Hunt; 2001:491–516.

Ireton-Jones CS, Borman KR, Turner WW Jr. Nutrition considerations in the management of ventilator dependent patients. *Nutr Clin Prac.* 1993;8(2):60–64.

Nelms MN. Assessment of nutrition status and risk. In: Nelms M, Sucher K, Long S. *Nutrition Therapy and Pathophysiology.* Belmont, CA: Thomson/Brooks-Cole; 2007:101–133.

Nelms MN. Metabolic stress. In: Nelms M, Sucher K, Long S. *Nutrition Therapy and Pathophysiology.* Belmont, CA: Thomson/Brooks-Cole; 2007:785–804.

Novartis. Technical features Isosource standard. Available at: http://www.novartisnutrition.com/us/productDetail?id=3. Accessed April 15, 2008.

RxList. Procalamine. Available at: http://www.rxlist.com/cgi/generic/procalamine_ids.htm. Accessed July 17, 2007.

Schols AMWJ. Nutrition in chronic obstructive pulmonary disease. *Curr Opin in Pulm Med.* 2000;6(2):110–115.

Skipper A, Nelms MN. Methods of nutrition support. In: Nelms M, Sucher K, Long S. *Nutrition Therapy and Pathophysiology.* Belmont, CA: Thomson/Brooks-Cole; 2007:149–179.

Vermeeren MA, Creutzberg EC, Schols AM, Postma DS, Pieters WR, Roldaan AC, Wouters EF; on behalf of the COSMIC Study Group. Prevalence of nutritional depletion in a large out-patient population of patients with COPD. *Respir Med.* 2006;100(8):1349–1355.

Vermeeren MA, Wouters EF, Geraerts-Keeris AJ, Schols AM. Nutritional support in patients with chronic obstructive pulmonary disease during hospitalization for an acute exacerbation: A randomized controlled feasibility trial. *Clin Nutr.* 2004;23(5):1184–1192.

Internet Resources

American Lung Association: Chronic Obstructive Pulmonary Disease (COPD) Fact Sheet. http://www.lungusa.org/site/pp.asp?c=dvLUK9O0E&b=35020

COPD International Organization: COPD: What Does It Mean. http://www.copd-international.com/

National Heart Lung Blood Institute: Learn More Breathe Better: It Has a Name: COPD Chronic Obstructive Pulmonary Disease. http://www.nhlbi.nih.gov/health/public/lung/copd/index.htm

Unit Seven

NUTRITION THERAPY FOR ENDOCRINE DISORDERS

The most common of all endocrine disorders is diabetes mellitus. Diabetes mellitus is actually a group of diseases characterized by hyperglycemia resulting from cessation of insulin production or impairment in insulin secretion and/or insulin action. There are four major categories of diabetes mellitus: type 1, type 2, gestational, and diabetes secondary to other diseases. The first three cases in this section focus on this chronic disease in three of the most common forms—type 1, type 2, and gestational diabetes.

Diabetes mellitus is a chronic disease that has no cure and is the sixth leading cause of death in the United States. New diagnoses of diabetes have tripled in the last 20 years. Almost 16 million people in the United States have diabetes, and many more are undiagnosed (available from: http://www.cdc.gov/diabetes/statistics/incidence; accessed April 17, 2008).

Diabetes affects men and women equally, but minorities (especially American Indians and Alaska Natives) are almost twice as likely as non-Hispanic whites to develop diabetes in their lifetime. In addition, diabetes is one of the most costly health problems in the United States. In 2002, health care and other direct medical costs, as well as indirect costs (such as loss of productivity), were approximately $132 billion. Each year, more than 200,000 people die as a result of diabetes and its complications. For example, diabetes is the leading cause of new blindness in the United States and the leading cause of nephropathy, which leads to end-stage renal disease requiring dialysis or organ transplant for survival (available from: http://diabetes.niddk.nih.gov/dm/pubs/statistics/#14; accessed December 1, 2007).

Medical nutrition therapy is integral to total diabetes care and management. The Diabetes Control and Complications Trial (DCCT) corroborated the significance of integrating nutrition and blood glucose self-management education in achieving and maintaining target blood glucose levels. Nutrition and meal planning are among the most challenging aspects of diabetes care for the person with diabetes and the health care team. The major components of successful nutrition management are learning about nutrition therapy, altering eating habits, implementing new behaviors, participating in exercise, evaluating changes, and integrating this information into diabetes care. Observance of meal-planning principles requires people with diabetes to make demanding lifestyle changes. To be effective, the registered dietitian must be able to customize his or her approach to the personal lifestyle and diabetes management goals of the individual with diabetes. Cases 22, 23, and 24 allow you to put this guideline into practice. The final case in this section explores the nutritional concerns associated with the most common hormonal reproductive problem for women of childbearing age: polycystic ovary syndrome (PCOS). Women with PCOS are at high risk of developing diabetes, hypertension, and heart disease. Maintaining a normal weight is a crucial component of the plan of care for this disorder. Many of the medications used to treat PCOS create nutritional concerns and form an important component of this case.

Case 22

Type 1 Diabetes Mellitus

Objectives

After completing this case, the student will be able to:

1. Describe the pathophysiology of type 1 diabetes mellitus.
2. Apply knowledge of the pathophysiology of type 1 diabetes mellitus to identify and explain short- and long-term complications associated with type 1 diabetes mellitus.
3. Demonstrate understanding of the role of nutrition therapy as an adjunct to pharmacotherapy and other medical treatment for type 1 diabetes mellitus.
4. Interpret laboratory parameters for nutritional implications and significance.
5. Determine nutrition diagnoses and write appropriate PES statements.
6. Develop a nutrition care plan with appropriate measurable goals, interventions, and strategies for monitoring and evaluation that addresses the nutrition diagnoses of this case.
7. Describe the pathophysiology of diabetic ketoacidosis (DKA).
8. Integrate the pathophysiology of DKA into MNT recommendations.
9. Interpret laboratory parameters to analyze fluid and electrolyte status and acid-base balance.

Susan Cheng, a 15-year-old high school student, is admitted to the hospital complaining of thirst, hunger, problems with urination, and fatigue. There is a history of diabetes in the family.

Name: Susan Cheng
DOB: 9/25 (age 15)
Physician: P. Green, MD

ADMISSION DATABASE

BED #	DATE: 1/8	TIME: 1400	TRIAGE STATUS (ER ONLY): ☐ Red ☐ Yellow ☐ Green ☐ White

Initial Vital Signs

TEMP: 98.6	RESP: 18	SAO$_2$:	
HT: 5′2″	WT (lb): 100	B/P: 124/70	PULSE: 85
LAST TETANUS: 4 years ago		LAST ATE: this AM	LAST DRANK: 1 hour ago

PRIMARY PERSON TO CONTACT:
Name: Mai or David Cheng (parents)
Home #: 555-390-8217
Work #: 555-390-2234

ORIENTATION TO UNIT: ☒ Call light ☒ Television/telephone ☒ Bathroom ☒ Visiting ☒ Smoking ☒ Meals ☒ Patient rights/responsibilities

CHIEF COMPLAINT/HX OF PRESENT ILLNESS

excessive thirst, urination, hunger, and fatigue

PERSONAL ARTICLES: (Check if retained/describe)
☒ Contacts ☒ R ☒ L ☐ Dentures ☐ Upper ☐ Lower
☐ Jewelry:
☐ Other:

ALLERGIES: Meds, Food, IVP Dye, Seafood: Type of Reaction

none

VALUABLES ENVELOPE:
☐ Valuables instructions

PREVIOUS HOSPITALIZATIONS/SURGERIES

none

INFORMATION OBTAINED FROM:
☒ Patient ☐ Previous record
☒ Family ☐ Responsible party

Signature *Susan Cheng*

Home Medications (including OTC) **Codes: A=Sent home** **B=Sent to pharmacy** **C=Not brought in**

Medication	Dose	Frequency	Time of Last Dose	Code	Patient Understanding of Drug

Do you take all medications as prescribed? ☐ Yes ☐ No If no, why?

PATIENT/FAMILY HISTORY

☐ Cold in past two weeks
☐ Hay fever
☐ Emphysema/lung problems
☐ TB disease/positive TB skin test
☐ Cancer
☐ Stroke/past paralysis
☐ Heart attack
☐ Angina/chest pain
☐ Heart problems
☐ High blood pressure
☐ Arthritis
☐ Claustrophobia
☐ Circulation problems
☐ Easy bleeding/bruising/anemia
☐ Sickle cell disease
☐ Liver disease/jaundice
☐ Thyroid disease
☒ Diabetes Maternal grandmother
☐ Kidney/urinary problems
☐ Gastric/abdominal pain/heartburn
☐ Hearing problems
☐ Glaucoma/eye problems
☐ Back pain
☐ Seizures
☐ Other

RISK SCREENING

Have you had a blood transfusion? ☐ Yes ☒ No
Do you smoke? ☐ Yes ☒ No
If yes, how many pack(s)?
Does anyone in your household smoke? ☐ Yes ☒ No
Do you drink alcohol? ☐ Yes ☒ No
If yes, how often?_______ How much?
When was your last drink?
Do you take any recreational drugs? ☐ Yes ☒ No
If yes, type:_______ Route:
Frequency:_______ Date last used:______/______/______

FOR WOMEN Ages 12–52

Is there any chance you could be pregnant? ☐ Yes ☒ No
If yes, expected date (EDC):
Gravida/Para:

ALL WOMEN

Date of last Pap smear: 1 year ago
Do you perform regular breast self-exams? ☒ Yes ☐ No

ALL MEN

Do you perform regular testicular exams? ☐ Yes ☐ No

Additional comments:

× *Francis Miller, RN*
Signature/Title

Client name: Susan Cheng
DOB: 9/25
Age: 15
Sex: Female
Education: Less than high school. *What grade/level?* 9th grade, HS student
Occupation: Student
Hours of work: N/A
Household members: Mother age 40, father age 42, sister age 16, brother age 9—all in excellent health
Ethnic background: Asian American
Religious affiliation: Protestant
Referring physician: Pryce Green, MD (endocrinology)

Chief complaint:
"I've been so thirsty and hungry. I haven't slept through the night for 2 weeks. I have to get up several times a night to go to the bathroom. It's a real pain. I've also noticed that my clothes are getting loose. My mom and dad think I must be losing weight."

Patient history:
Onset of disease: Susan is a 15-year-old female who lives with her parents, brother, and sister. She is in the 9th grade and a member of the girls' volleyball team. She has had an uneventful medical history with no significant illness until the past several weeks. Her parents brought her to the office with c/o polydipsia, polyuria, polyphagia, weight loss, and fatigue. Blood was drawn in the ER to measure blood glucose and glycated hemoglobin levels.
PMH: Normal adolescence
Meds: None PTA
Smoker: No
Family Hx: What? DM *Who?* Maternal grandmother

Physical exam:
General appearance: Tired-appearing adolescent female
Vitals: Temp 98.6°F, BP 124/70 mm Hg, HR 85 bpm, RR 18 bpm
Heart: Regular rate and rhythm, heart sounds normal
HEENT: Noncontributory
Genitalia: Normal adolescent female
Neurologic: Alert and oriented
Extremities: Noncontributory
Skin: Smooth, warm, and dry; excellent turgor; no edema
Chest/lungs: Lungs are clear
Peripheral vascular: Pulse 4+ bilaterally, warm, no edema
Abdomen: Nontender, no guarding

Nutrition Hx:
General: Mother describes appetite as good. Meals are somewhat irregular due to Susan's volleyball practice/game schedule. She is a starter on the girls' volleyball team, practices four evenings per week, and participates in approximately two games per week, some of which are away games. Susan eats lunch in the school cafeteria.

Usual dietary intake:

AM: 1½ c dry cereal, usually sugar-coated, 1 c 2% milk, 1 c orange juice (unsweetened), hot chocolate in the winter (made from 1 packet mix)

Lunch: 6-in pepperoni pizza, 1 c mixed salad w/Thousand Island dressing (¼ c), 1 can Coke (regular), Snickers candy bar

Snack: Peanut butter and jelly sandwich (2 slices white bread, 1 tbsp grape jelly, 2–3 tbsp crunchy peanut butter), 1 12-oz can Coke

PM: Spaghetti w/meat sauce (about 2 c noodles and ½ c sauce w/ 1 oz ground beef), steamed broccoli—3 stalks (will eat with salt and butter, but prefers cheese sauce), 16 oz 2% milk

HS snack: 2 c ice cream (different flavors) or popcorn with melted butter and salt (about 6 c popcorn with ¼ c melted butter) with 12 oz Coke

24-hour recall: N/A
Food allergies/intolerances/aversions: NKA
Previous nutrition therapy? No
Food purchase/preparation: Parents
Vit/min intake: None
Current diet order: 2,400 kcal (300 g CHO, 55–65 g protein, 80 g lipid)

Dx:
Type 1 diabetes mellitus

Tx plan:
Achieve glycemic control
Evaluate serum lipid levels
Monitor blood glucose levels
Initiate self-management training for patient and parents on insulin administration, nutrition prescription, meal planning, signs/symptoms and Tx of hypo-/hyperglycemia, monitoring instructions (SBGM, urine ketones, and use of record system), exercise
Baseline visual examination
Contraception education

UH UNIVERSITY HOSPITAL

NAME: Susan Cheng DOB: 9/25
AGE: 15 SEX: F
PHYSICIAN: P. Green, MD

***************************************CHEMISTRY**

DAY:		Admit	d/c	
DATE:				
TIME:				
LOCATION:				
	NORMAL			UNITS
Albumin	3.5-5	4.2	4.5	g/dL
Total protein	6-8	7.5	7.6	g/dL
Prealbumin	16-35	40	39	mg/dL
Transferrin	250-380 (women) 215-365 (men)			mg/dL
Sodium	136-145	140	138	mEq/L
Potassium	3.5-5.5	4.5	4.1	mEq/L
Chloride	95-105	98	99	mEq/L
PO_4	2.3-4.7	3.7	3.8	mg/dL
Magnesium	1.8-3	2.1	1.9	mg/dL
Osmolality	285-295	304 H	297 H	mmol/kg/H_2O
Total CO_2	23-30			mEq/L
Glucose	70-110	250 H	120 H	mg/dL
BUN	8-18	20 H	18	mg/dL
Creatinine	0.6-1.2	0.9	0.8	mg/dL
Uric acid	2.8-8.8 (women) 4.0-9.0 (men)			mg/dL
Calcium	9-11	9.5	9.7	mg/dL
Bilirubin	≤0.3			mg/dL
Ammonia (NH_3)	9-33			µmol/L
ALT	4-36			U/L
AST	0-35			U/L
Alk phos	30-120			U/L
CPK	30-135 (women) 55-170 (men)			U/L
LDH	208-378			U/L
CHOL	120-199	169	170	mg/dL
HDL-C	>55 (women) >45 (men)			mg/dL
VLDL	7-32			mg/dL
LDL	<130	109		mg/dL
LDL/HDL ratio	<3.22 (women) <3.55 (men)			
Apo A	101-199 (women) 94-178 (men)			mg/dL
Apo B	60-126 (women) 63-133 (men)			mg/dL
TG	35-135 (women) 40-160 (men)			mg/dL
T_4	4-12			mcg/dL
T_3	75-98			mcg/dL
HbA_{1C}	3.9-5.2	7.95 H		%

Case Questions

I. Understanding the Disease and Pathophysiology

1. Define *insulin.* Describe its major functions within normal metabolism.

2. What are the current opinions regarding the etiology of type 1 diabetes mellitus (DM)?

3. What genes have been identified that indicate susceptibility to type 1 diabetes mellitus?

4. After examining Susan's medical history, can you identify any risk factors for type 1 DM?

5. What are the established diagnostic criteria for type 1 DM? How can the physicians distinguish between type 1 and type 2 DM?

6. Describe the metabolic events that led to Susan's symptoms (polyuria, polydipsia, polyphagia, weight loss, and fatigue) and integrate these with the pathophysiology of the disease.

7. List the microvascular and neurologic complications associated with type 1 diabetes.

8. When Susan's blood glucose level is tested at 2 AM, she is hypoglycemic. In addition, her plasma ketones are elevated. When she is tested early in the morning before breakfast, she is hyperglycemic. Describe the dawn phenomenon. Is Susan likely to be experiencing this? How might this be prevented?

9. What precipitating factors may lead to the complication of diabetic ketoacidosis? List these factors and describe the metabolic events that result in the signs and symptoms associated with DKA.

II. Nutrition Assessment

A. Evaluation of Weight/Body Composition

10. Determine Susan's stature for age and weight for age percentiles.

11. Interpret these values using the appropriate growth chart.

B. Calculation of Nutrient Requirements

12. Estimate Susan's daily energy and protein needs. Be sure to consider Susan's age.

13. What would the clinician monitor in order to determine whether or not the prescribed energy level is adequate?

C. Intake Domain

14. Using a computer dietary analysis program or food composition table, calculate the kcalories, protein, fat (saturated, polyunsaturated, and monounsaturated), CHO, fiber, and cholesterol content of Susan's typical diet.

15. What dietary assessment tools can Susan use to coordinate her eating patterns with her insulin and physical activity?

16. Dietitians must obtain and use information from all components of a nutrition assessment to develop appropriate interventions and goals that are achievable for the patient. This assessment is ongoing and continuously modified and updated throughout the nutrition therapy process. For each of the following components of an initial nutrition assessment, list at least three assessments you would perform for each component:

Component	**Assessments You Would Perform**
Clinical data	
Nutrition history	
Weight history	
Physical activity history	
Monitoring	
Psychosocial/economic	
Knowledge and skills level	
Expectations and readiness to change	

D. Clinical Domain

17. Does Susan have any laboratory results that support her diagnosis?

18. Why did Dr. Green order a lipid profile?

19. Evaluate Susan's laboratory values:

Chemistry	Normal Value	Susan's Value	Reason for Abnormality	Nutritional Implications

20. Compare the pharmacological differences in insulins:

Type of Insulin	Brand Name	Onset of Action	Peak of Action	Duration of Action
lispro				
aspart				
glulisine				
NPH				
glargine				
detemir				
70/30 premix				
50/50 premix				
60/40 premix				

21. Once Susan's blood glucose levels were under control, Dr. Green prescribed the following insulin regimen: 24 units of glargine in PM with the other 24 units as lispro divided between meals and snacks. How did Dr. Green arrive at this dosage?

E. Behavioral–Environmental Domain

22. Identify at least three specific potential nutrition problems within this domain that will need to be addressed for Susan and her family.

23. Just before Susan is discharged, her mother asks you, "My friend who owns a health food store told me that Susan should use stevia instead of artificial sweeteners or sugar. What do you think?" What will you tell Susan and her mother?

F. Nutrition Diagnosis

24. Select two high-priority nutrition problems and complete the PES statement for each.

III. Nutrition Intervention

25. For each of the PES statements that you have written, establish an ideal goal (based on the signs and symptoms) and an appropriate intervention (based on the etiology).

26. Does the current diet order meet Susan's overall nutritional needs? If yes, explain why it is appropriate. If no, what would you recommend? Justify your answer.

IV. Nutrition Monitoring and Evaluation

27. Susan is discharged Friday morning. She and her family have received information on insulin administration, SMBG, urine ketones, recordkeeping, exercise, signs, symptoms, and Tx of hypo-/hyperglycemia, meal planning (CHO counting), and contraception. Susan and her parents verbalize understanding of the instructions and have no further questions at this time. They are instructed to return in 2 weeks for appointments with the outpatient dietitian and CDE. When you come in to work Monday morning, you see that Susan was admitted through the ER Saturday night with a BG of 50 mg/dL. You see her when you make rounds and review her chart. During an interview, Susan tells you she was invited to a party Saturday night after her discharge on Friday. She tested her blood glucose before going to the party, and it measured 95 mg/dL. She took 2 units of insulin and knew she needed to have a snack that contained approximately 15 grams of CHO, so she drank one beer when she arrived at the party. She remembers getting lightheaded and then woke up in the ER. What happened to Susan physiologically?

28. What kind of educational information will you give her before this discharge? Keep in mind that she is underage for legal consumption of alcohol.

Bibliography

American Association for Clinical Chemistry. Diabetes-related autoantibodies (last reviewed 1/30/06). *Lab Tests Online.* Available at: http://www.labtestsonline.org/understanding/analytes/diabetes_auto/sample.html. Accessed April 16, 2008.

American College of Physicians. Diabetic testing and monitoring. Available at: http://www.acponline.org/mle/diabetic_test.htm. Accessed January 7, 2008.

American Diabetes Association. Diagnosis and classification of diabetes mellitus. *Diabetes Care.* 2006;29(Suppl 1):S43–S48.

American Diabetes Association. Hyperglycemic crisis in diabetes. *Diabetes Care.* 2004;27(Suppl 1):S94–S102.

American Diabetes Association. National Diabetes Fact Sheet, 2005. Available at: http://www.diabetes.org/diabetes-statistics.jsp. Accessed January 7, 2008.

American Diabetes Association. Nutrition principles and recommendations in diabetes. *Diabetes Care.* 2004;27(Suppl 1):S36.

American Diabetes Association. Physical activity/exercise and diabetes. *Diabetes Care.* 2004;27(Suppl 1):S58–S62.

American Diabetes Association. Standards of medical care in diabetes—2006. *Diabetes Care.* 2006;29(Suppl 1):S4.

American Diabetes Association. *The Diabetes Ready-Reference Guide for Health Care Professionals.* Alexandria, VA: American Diabetes Association; 2000.

American Diabetes Association. Translation of the diabetes nutrition recommendations for health care institutions. *Diabetes Care.* 2002;35(Suppl 1):S61–S63.

American Diabetes Association and American Dietetic Association. *Choose Your Foods: Exchange Lists for Diabetics.* Alexandria, VA: American Diabetes Association; 2007.

American Dietetic Association. *Nutrition Care Manual.* Available at: http://nutritioncaremanual.org. Accessed January 7, 2008.

Anderson EJ, Delahanty L, Richardson M, Castle G, Cercone S, Lyon R, Mueller D, Snetselaar L. Nutrition interventions for intensive therapy in the diabetes control and complications trial. *J Am Diet Assoc.* 1993;93:768–772.

Baynes K, Betteridge DJ. Diabetes mellitus, lipoprotein disorders and other metabolic diseases. In Axford J, O'Callaghan C, eds. *Medicine.* 2nd Ed. Oxford, UK: Blackwell Science; 2004.

Centers for Disease Control and Prevention. National diabetes fact sheet: General information and national estimates on diabetes in the United States, 2005. Atlanta, GA: U.S. Department of Health and Human Services. Centers for Disease Control and Prevention; 2005.

Cryer PE, Davis SN, Shamoon H. Hypoglycemia in diabetes. *Diabetes Care.* 2003;26(6):1902–1912.

DCCT Research Group. Nutrition interventions for intensive therapy in the Diabetes Control and Complications Trial. *J Am Diet Assoc.* 1993;93:768–772.

The effect of intensive treatment of diabetes on the development and progression of long-term complications in insulin-dependent diabetes mellitus. Diabetes Control and Complications Trial Research Group. *N Engl J Med.* 1993;329(14):977–986.

Franz MJ, Bantle JP, Beebe CA, Brunzell JD, Chiasson J, Garg A, Holzmeister LA, Hoogwerf B, Mayer E, Mooradian AD, Purnell JQ, Wheeler M. Evidence-based nutrition principles and recommendations for the treatment and prevention of diabetes and related complications. *Diabetes Care.* 2002;25(1):148–198.

Gillespie S, Kulkarni K, Daly A. Using carbohydrate counting in diabetes clinical practice. *J Am Diet Assoc.* 1998;98(8):897–899.

Hayes C. Physical activity and exercise. In: Ross TA, Boucher JL, O'Connell BS, eds. *American Dietetic Association Guide to Diabetes: Medical Nutrition Therapy and Education.* Chicago, IL: American Dietetic Association; 2005:71–80.

Hollowell JG, Staehling NW, Flanders WD, Hannon WH, Gunter EW, Spencer CE, Braverman LE. Serum TSH, T_4, and thyroid antibodies in the United States population (1988 to 1994): National Health and Nutrition Examination Survey (NHANES III). *J Clin Endocrinol Metab.* 2002;87(2):489–499.

Holzmeister LA, Geil P. Evidence-based nutrition care and recommendations. In: Ross TA, Boucher JL, O'Connell BS, eds. *American Dietetic Association Guide to Diabetes: Medical Nutrition Therapy and Education.* Chicago, IL: American Dietetic Association; 2005:61–70.

Kulkarni K. Pattern management. In: Ross TA, Boucher JL, O'Connell BS, eds. *American Dietetic Association Guide to Diabetes: Medical Nutrition Therapy and Education.* Chicago. IL: American Dietetic Association; 2005:116–127.

Mulcahy K, Lumber T. *The Diabetes Ready Reference for Health Professionals.* 2nd ed. Alexandria, VA: American Diabetes Association; 2004.

National Center for Chronic Disease Prevention and Health Promotion. Incidence of Diabetes in the Population Aged 18-79 Years. Available at: http://www.cdc.gov/diabetes/statistics/incidence. Accessed April 15, 2008.

Notkins AL, Lernmark A. Autoimmune type 1 diabetes: Resolved and unresolved issues. *J Clin Invest.* 2001;108(9):1247–1252.

Nuttall FQ. Carbohydrate and dietary management of clients with insulin-requiring diabetes. *Diabetes Care.* 1993;16(7):1039–1042.
O'Connell BS. Diabetes classification, pathophysiology, and diagnosis. In: Ross TA, Boucher JL, O'Connell BS, eds. *American Dietetic Association Guide to Diabetes: Medical Nutrition Therapy and Education.* Chicago, IL: American Dietetic Association; 2005:39–48.
Pastors JG, Waslaski J, Gunderson H. Diabetes meal-planning strategies. In: Ross TA, Boucher JL, O'Connell BS, eds. *American Dietetic Association Guide to Diabetes: Medical Nutrition Therapy and Education.* Chicago, IL: American Dietetic Association; 2005:201–217.
Rystrom JK. Insulin therapy. In: Ross TA, Boucher JL, O'Connell BS, eds. *American Dietetic Association Guide to Diabetes: Medical Nutrition Therapy and Education.* Chicago, IL: American Dietetic Association; 2005:89–105.
Umpierrez GE, Murphy MB, Kitabchi AE. Diabetic ketoacidosis and hyperglycemic hyperosmolar syndrome. *Diabetes Spectrum.* 2002;15(1):28–36.
Vinik AI, Maser RE, Mitchell BD, Freeman R. Diabetic autonomic neuropathy. *Diabetes Care.* 2003;26(5):1553–1579.
Wheeler ML. Long-term complications. In: Ross TA, Boucher JL, O'Connell BS, eds. *American Dietetic Association Guide to Diabetes: Medical Nutrition Therapy and Education.* Chicago, IL: American Dietetic Association; 2005:139–145.

Internet Resources

American Association of Diabetes Educators. http://www.diabeteseducator.org/
American Diabetes Association. http://www.diabetes.org
American Dietetic Association. http://www.eatright.org
Centers for Disease Control: Diabetes Public Health Resource. http://www.cdc.gov/diabetes/
Diabetes.com. http://www.diabetes.com/
eMedicineHealth: Diabetes. http://www.emedicinehealth.com/diabetes/article_em.htm
MedicineNet.com: Diabetes Mellitus. http://www.medicinenet.com/diabetes_mellitus/article.htm
MedlinePlus: Diabetes. http://www.nlm.nih.gov/medlineplus/diabetes.html
National Diabetes Information Clearinghouse: Diabetes Overview. http://diabetes.niddk.nih.gov/dm/pubs/overview/index.htm
WebMD: Diabetes Health Center. http://diabetes.webmd.com/

Case 23

Type 2 Diabetes Mellitus

Objectives

After completing this case, the student will be able to:

1. Describe the pathophysiology of type 2 diabetes mellitus.
2. Apply knowledge of the pathophysiology of type 2 diabetes mellitus in order to identify and explain short- and long-term complications associated with this disease.
3. Demonstrate understanding of the role of nutrition therapy as an adjunct to pharmacotherapy and other medical treatments for type 2 diabetes mellitus.
4. Interpret laboratory parameters for nutritional implications and significance.
5. Determine nutrition diagnoses and write appropriate PES statements.
6. Develop a nutrition care plan with appropriate measurable goals, interventions, and strategies for monitoring and evaluation that addresses the nutrition diagnoses of this case.
7. Describe the pathophysiology of hyperglycemic hyperosmolar nonketotic syndrome (HHNS).
8. Integrate the pathophysiology of HHNS into MNT recommendations.
9. Demonstrate working knowledge of fluid and electrolyte requirements.

Eileen Douglas is a 71-year-old woman who was admitted for surgical debridement of a nonhealing foot wound. On admission, Mrs. Douglas was found to be hyperglycemic, and a diagnosis of type 2 diabetes mellitus was determined.

Name: Eileen Douglas
DOB: 7/27 (age 71)
Physician: R. Case, MD; D. Shyne, MD

ADMISSION DATABASE

BED #	DATE:	TIME:	TRIAGE STATUS (ER ONLY):
1	6/8	1523	☐ Red ☐ Yellow ☐ Green ☐ White

Initial Vital Signs

TEMP:	RESP:	SAO$_2$:	
99.2	12		
HT: 5′0″	WT (lb): 155	B/P: 150/97	PULSE: 75
LAST TETANUS		LAST ATE	LAST DRANK

PRIMARY PERSON TO CONTACT:
Name: Connie Locher
Home #: 555-8217
Work #: 555-7512

ORIENTATION TO UNIT: ☒ Call light ☒ Television/telephone ☒ Bathroom ☒ Visiting ☒ Smoking ☒ Meals ☒ Patient rights/responsibilities

CHIEF COMPLAINT/HX OF PRESENT ILLNESS

unhealed ulcer on leg

PERSONAL ARTICLES: (Check if retained/describe)
☐ Contacts ☐ R ☐ L ☒ Dentures ☒ Upper ☒ Lower
☐ Jewelry:
☐ Other:

ALLERGIES: Meds, Food, IVP Dye, Seafood: Type of Reaction

none

VALUABLES ENVELOPE:
☐ Valuables instructions

PREVIOUS HOSPITALIZATIONS/SURGERIES

INFORMATION OBTAINED FROM:
☒ Patient ☐ Previous record
☐ Family ☐ Responsible party

Signature *Eileen Douglas*

Home Medications (including OTC) **Codes: A=Sent home** **B=Sent to pharmacy** **C=Not brought in**

Medication	Dose	Frequency	Time of Last Dose	Code	Patient Understanding of Drug
Capoten (captopril)	50 mg	bid	yesterday	C	

Do you take all medications as prescribed? ☒ Yes ☐ No If no, why?

PATIENT/FAMILY HISTORY

☐ Cold in past two weeks
☐ Hay fever
☐ Emphysema/lung problems
☐ TB disease/positive TB skin test
☐ Cancer
☐ Stroke/past paralysis
☐ Heart attack
☐ Angina/chest pain
☐ Heart problems
☒ High blood pressure Patient
☐ Arthritis
☐ Claustrophobia
☐ Circulation problems
☐ Easy bleeding/bruising/anemia
☐ Sickle cell disease
☐ Liver disease/jaundice
☐ Thyroid disease
☒ Diabetes Sibling
☐ Kidney/urinary problems
☐ Gastric/abdominal pain/heartburn
☐ Hearing problems
☐ Glaucoma/eye problems
☐ Back pain
☐ Seizures
☐ Other

RISK SCREENING

Have you had a blood transfusion? ☐ Yes ☒ No
Do you smoke? ☐ Yes ☒ No
If yes, how many pack(s)?
Does anyone in your household smoke? ☐ Yes ☒ No
Do you drink alcohol? ☐ Yes ☒ No
If yes, how often?_______ How much?
When was your last drink?
Do you take any recreational drugs? ☐ Yes ☒ No
If yes, type:_______ Route:
Frequency:_______ Date last used:______/______/______

FOR WOMEN Ages 12–52

Is there any chance you could be pregnant? ☐ Yes ☐ No
If yes, expected date (EDC):
Gravida/Para:

ALL WOMEN

Date of last Pap smear: over a year ago
Do you perform regular breast self-exams? ☒ Yes ☐ No

ALL MEN

Do you perform regular testicular exams? ☐ Yes ☐ No

Additional comments:

x *Ruth Long, RN*
Signature/Title

Client name: Eileen Douglas
DOB: 7/27
Age: 71
Sex: Female
Education: Less than high school *What grade/level?* 10th grade
Occupation: Homemaker
Hours of work: N/A
Household members: Sister age 80, Dx with type 2 DM 10 years ago. Mrs. Douglas cares for her sister.
Ethnic background: African American
Religious affiliation: Protestant
Referring physicians: Richard Case, MD (internal medicine); Dennis Shyne, MD (general surgery)

Chief complaint:
"This cut on my foot happened over 2 months ago and has not healed. And I don't think I see as well. Maybe I need my eyes checked again. I have been having trouble reading the newspaper for the past few months."

Patient history:
Onset of disease: Mrs. Douglas is a 71-year-old widow who lives with her 80-year-old sister, whom she cares for. They live in a two-bedroom, low-income housing apartment. In addition to the unhealed wound and blurry vision, Mrs. Douglas complains of frequent bladder infections, which are documented in her clinic chart, and a slight tingling and numbness in her feet. On admission to the hospital, her blood glucose measured 325 mg/dL. Surgical debridement of wound is indicated, along with normalization of blood glucose and alleviation of blurred vision.
Type of Tx: Surgical debridement of wound, sliding scale insulin, 1,200-kcal diet, DM self-management education
PMH: HTN
Meds: Capoten (captopril), 50 mg PO bid
Smoker: No
Family Hx: What? DM *Who?* Sister, for 10 years

Physical exam:
General appearance: Overweight elderly African American female
Vitals: Temp 99.2°F, BP 150/97 mm Hg, HR 75 bpm, RR 12 bpm
Heart: Regular rate and rhythm, no gallops or rubs, point of maximal impulse at the fifth intercostal space in the midclavicular line
HEENT:
Head: Normocephalic
Eyes: Wears glasses for myopia, mild retinopathy
Ears: Tympanic membranes normal
Nose: Dry mucous membranes w/out lesions
Throat: Slightly dry mucous membranes w/out exudates or lesions
Genitalia: Normal w/out lesions
Neurologic: Alert and oriented. Cranial nerves II–XII grossly intact, strength 5/5 throughout, sensation to light touch intact in hands, mildly diminished in feet, normal gait, normal reflexes

Extremities: Normal muscular tone for age, normal ROM, nontender
Skin: Warm and dry, 2–3-cm ulcer on lateral left foot
Chest/lungs: Respirations normal; no crackles, rhonchi, wheezes, or rubs noted
Peripheral vascular: Pulse 2-bilaterally, cool, mild edema
Abdomen: Audible bowel sounds, soft and nontender, w/out masses or organomegaly

Nutrition Hx:
General: Because her sister "has sugar," Mrs. Douglas does not purchase cakes, candy, and other desserts. In fact, Mrs. Douglas reports that she and her sister try to avoid "all starchy foods" because that's what they were told to do when her sister received a printed diet sheet from her MD (10 years ago). Once a month, though, she and her sister have cake and ice cream at the Senior Center birthday party.

Usual dietary intake:
AM: One egg—fried in bacon fat, 2 strips of bacon or sausage, 1 cup coffee—black, ½ c orange juice—unsweetened
Lunch: Lunch meat sandwich: 2 slices enriched white bread, 1 slice (1 oz) bologna, 1 slice (1 oz) American cheese, mustard; 1 glass (8 oz) iced tea—unsweetened
PM: 1 c turnip greens seasoned with (1 oz) fatback, salt, and pepper (simmered on stove top for at least 3 hours); 2 small new potatoes, boiled, seasoned with salt and pepper; 2-inch square of cornbread with 1 tsp butter; 1 c beans and ham (Great Northern beans cooked with ham, approximately ¾ c beans and ¼ c or 1 oz ham); 1 c coffee—black
Snack: 2 vanilla wafers

24-hour recall: N/A
Food allergies/intolerances/aversions: N/A
Previous nutrition therapy? No
Food purchase/preparation: Self
Vit/min intake: None
Current diet order: 1,200 kcal ADA exchange diet

Dx:
Cellulitis; type 2 diabetes mellitus

Tx plan:
Debride wound
Normalize blood glucose levels
Begin self-management training on nutrition prescription, meal planning, signs/symptoms, and Tx of hypo-/hyperglycemia, SMBG, appropriate exercise, potential food–drug interaction
Initiate Lipitor 10 mg gd, continue Capoten 50 mg bid

UH UNIVERSITY HOSPITAL

NAME: Eileen Douglas　　DOB: 7/27
AGE: 71　　SEX: F
PHYSICIAN: R. Case, MD

CHEMISTRY

DAY:		Admit	d/c	
DATE:				
TIME:				
LOCATION:				
	NORMAL			UNITS
Albumin	3.5-5	4.0	4.1	g/dL
Total protein	6-8	7	7.2	g/dL
Prealbumin	16-35	23	24.5	mg/dL
Transferrin	250-380 (women) 215-365 (men)	310	305	mg/dL
Sodium	136-145	140	145	mEq/L
Potassium	3.5-5.5	4.2	4.5	mEq/L
Chloride	95-105	103	100	mEq/L
PO_4	2.3-4.7	3.6	3.2	mg/dL
Magnesium	1.8-3	2.1	1.8	mg/dL
Osmolality	285-295	315 H	314 H	mmol/kg/H_2O
Total CO_2	23-30	25	26	mEq/L
Glucose	70-110	325 H	121 H	mg/dL
BUN	8-18	26 H	26 H	mg/dL
Creatinine	0.6-1.2	1.2	1.2	mg/dL
Uric acid	2.8-8.8 (women) 4.0-9.0 (men)			mg/dL
Calcium	9-11			mg/dL
Bilirubin	$\leq$0.3			mg/dL
Ammonia (NH_3)	9-33			μmol/L
ALT	4-36			U/L
AST	0-35			U/L
Alk phos	30-120			U/L
CPK	30-135 (women) 55-170 (men)			U/L
LDH	208-378			U/L
CHOL	120-199	300 H	250 H	mg/dL
HDL-C	>55 (women) >45 (men)	35 L	37 L	mg/dL
VLDL	7-32			mg/dL
LDL	<130	140 H	138 H	mg/dL
LDL/HDL ratio	<3.22 (women) <3.55 (men)	4.0 H	3.7 H	
Apo A	101-199 (women) 94-178 (men)			mg/dL
Apo B	60-126 (women) 63-133 (men)			mg/dL
TG	35-135 (women) 40-160 (men)	400 H	300 H	mg/dL
T_4	4-12			mcg/dL
T_3	75-98			mcg/dL
HbA_{1C}	3.9-5.2	8.5 H		%

UH UNIVERSITY HOSPITAL

NAME: Eileen Douglas DOB: 7/27
AGE: 71 SEX: F
PHYSICIAN: R. Case, MD

HEMATOLOGY*

	NORMAL			UNITS
DAY:		1	5	
DATE:				
TIME:				
LOCATION:				
WBC	4.8–11.8			× $10^3/mm^3$
RBC	4.2–5.4 (women) 4.5–6.2 (men)			× $10^6/mm^3$
HGB	12–15 (women) 14–17 (men)	9.9 L	10.1 L	g/dL
HCT	37–47 (women) 40–54 (men)	30.4 L	27.7 L	%
MCV	80–96			μm^3
RETIC	0.8–2.8			%
MCH	26–32			pg
MCHC	31.5–36			g/dL
RDW	11.6–16.5			%
Plt Ct	140–440			× $10^3/mm^3$
Diff TYPE				
ESR	0–25 (women) 0–15 (men)			mm/hr
% GRANS	34.6–79.2			%
% LYM	19.6–52.7			%
SEGS	50–62			%
BANDS	3–6			%
LYMPHS	24–44			%
MONOS	4–8			%
EOS	0.5–4			%
Ferritin	20–120 (women) 20–300 (men)			mg/mL
ZPP	30–80			µmol/mol
Vitamin B_{12}	24.4–100			ng/dL
Folate	5–25			µg/dL
Total T cells	812–2,318			mm^3
T-helper cells	589–1,505			mm^3
T-suppressor cells	325–997			mm^3
PT	11–16			sec

Case Questions

I. Understanding the Disease and Pathophysiology

1. What is the difference between type 1 diabetes mellitus and type 2 diabetes mellitus?

2. How would you clinically distinguish between type 1 and type 2 diabetes mellitus?

3. What are the risk factors for development of type 2 diabetes mellitus? What risk factors does Mrs. Douglas present with?

4. What are the common complications associated with diabetes mellitus? Describe the pathophysiology associated with these complications, specifically addressing the role of chronic hyperglycemia.

5. Does Mrs. Douglas present with any complications of diabetes mellitus? If yes, which ones?

6. Identify at least four features of the physician's physical examination as well as her presenting signs and symptoms that are consistent with her admitting diagnosis. Describe the pathophysiology that might be responsible for each physical finding.

Physical Finding	Physiological Change/Etiology

7. Prior to admission, Mrs. Douglas had not been diagnosed with diabetes mellitus. How could she present with complications?

8. Briefly describe hyperglycemic hyperosmolar nonketotic syndrome (HHNS). How is this syndrome different from ketoacidosis?

9. What are the symptoms of HHNS?

10. What factors may lead to HHNS? Is Mrs. Douglas at risk?

11. What is the immediate aim of treatment for HHNS? If HHNS is not treated, how would you expect the condition of HHNS to progress?

II. Nutrition Assessment

A. Evaluation of Weight/Body Composition

12. Calculate Mrs. Douglas's body mass index (BMI).

13. What are the health implications for a BMI in this range?

B. Calculation of Nutrient Requirements

14. Calculate Mrs. Douglas's energy needs using the Mifflin-St. Jeor equation. Should Mrs. Douglas's weight be adjusted for obesity?

15. Calculate Mrs. Douglas's protein needs.

16. Is the diet order of 1,200 kcal appropriate?

17. If yes, explain why it is appropriate. If no, what would you recommend? Justify your answer.

C. Intake Domain

18. Does Mrs. Douglas's "usual" dietary intake meet the USDA Food Guide/MyPyramid guidelines? Is she deficient in any food groups? If so, which ones?

19. Using a computer dietary analysis program or food composition table, calculate the kcalories, protein, fat, CHO, fiber, cholesterol, and Na content of Mrs. Douglas's diet.

20. How would you compare Mrs. Douglas's "usual" dietary intake to her current nutritional needs?

D. Clinical Domain

21. Compare the patient's laboratory values that were out of range on admission with normal values. How would you interpret this patient's labs? Make sure explanations are pertinent to *this* situation.

Parameter	Normal Value	Patient's Value	Reason for Abnormality	Nutritional Implications
Glucose (mg/dL)				
HbA_{1c} (%)				
Cholesterol (mg/dL)				
LDL-cholesterol (mg/dL)				
HDL-cholesterol (mg/dL)				
Triglycerides (mg/dL)				

22. Identify two lab values that should be monitored regularly.

23. Why wasn't HbA_{1c} measured at discharge?

24. Why is regular insulin used to correct hyperglycemia in patients with HHNS?

25. When HHNS is treated, the initial target serum glucose level is typically set at the 250 mg/dL range instead of a normal blood glucose level. Why?

26. Compare the pharmacologic differences among the oral hypoglycemic agents.

Class	Brand Name(s) (& Generic Name)	Mechanism of Action	Efficacy	↑ or ↓ Effect on Plasma Insulin Levels	Effect on Body Weight	Effect on Plasma Lipids	Side Effects & Contraindications	Adult Daily Maintenance Dose (mg)	Number of Daily Doses
α–Glucosidase inhibitors									
Biguanides									
Meglitinides									
Sulfonylureas First generation									
Second generation									
Thiainedioneszolid									

27. Avandia is often used to help control blood glucose levels. Describe the (medication) action of Avandia.

28. The goal for healthy elderly patients with diabetes should be near-normal, fasting plasma glucose levels without hypoglycemia. Although acceptable glucose control must be carefully individualized, the elderly tend to be predisposed to hypoglycemia. List five factors that predispose elderly patients to hypoglycemia.

E. Behavioral–Environmental Domain

29. Identify at least three factors that may interfere with Mrs. Douglas's compliance and success with her diabetes treatment.

III. Nutrition Diagnosis

30. Select two high-priority nutrition problems and complete the PES statement for each.

IV. Nutrition Intervention

31. What was the most important nutritional concern when the patient was originally admitted to the hospital (time of Dx)?

32. What additional information does the dietitian need to collect before he or she can mutually develop clinical and behavioral outcomes with the patient and health care team?

33. For each of the PES statements that you have written, establish an ideal goal (based on the signs and symptoms) and an appropriate intervention (based on the etiology).

V. Monitoring

34. Mrs. Douglas was d/c with instructions for a non–kilocaloric-restricted, low-fat ($\leq$ 30% total kcal), high-CHO ($\geq$ 50% total kcal) diet, in combination with a walking program, and a prescription for captopril to control her HTN. Glucose levels were well controlled for 6 months, but she became unable to afford the necessary supplies to check her BG or urine acetone levels. After 6 months, she was readmitted with a BG of 905 mg/dL, a slight temperature, BP of 68/100 mm Hg, tachycardia, and shallow, tachypneic breathing (Kussmal respirations). She was Dx with pneumonia, dehydration, and hyperglycemic hyperosmolar nonketotic syndrome (HHNS). What is the MNT for patients with HHNS?

Bibliography

Ahmann AJ, Riddle MC. Current oral agents for type 2 diabetes: Many options, but which to choose when? *Postgrad Med.* 2002;111(5):32–34,37–40,43–46.

American College of Physicians. Diabetic testing and monitoring. Available at: http://www.acponline.org/mle/diabetic_test.htm. Accessed January 4, 2008.

American Diabetes Association. Diagnosis and classification of diabetes mellitus. *Diabetes Care.* 2006;29(Suppl 1):S43–S48.

American Diabetes Association. Hyperglycemic crisis in diabetes. *Diabetes Care.* 2004;27(Suppl 1):S94–S102.

American Diabetes Association. Implications of the United Kingdom prospective diabetes study. *Diabetes Care.* 2002;25(Suppl 1):S28.

American Diabetes Association. National Diabetes Fact Sheet, 2005. Available at: http://diabetes.org/diabetes-statistics.jsp. Accessed January 5, 2008.

American Diabetes Association. Nutrition principles and recommendations in diabetes. *Diabetes Care.* 2004;27(Suppl 1):S36.

American Diabetes Association. Standards of medical care in diabetes—2006. *Diabetes Care.* 2006;29(Suppl 1):S4.

American Diabetes Association. *The Diabetes Ready-Reference Guide for Health Care Professionals.* Alexandria, VA: American Diabetes Association; 2000.

American Diabetes Association. Translation of the diabetes nutrition recommendations for health care institutions. *Diabetes Care.* 2002;35(Suppl 1):S61–S63.

American Diabetes Association and American Dietetic Association. *Choose Your Foods: Exchange Lists for Diabetics.* Alexandria, VA: American Diabetes Association; 2007.

American Dietetic Association. *Nutrition Care Manual.* Available at: http://nutritioncaremanual.org. Accessed January 5, 2008.

Avandia. How Avandia works. Available at: http://www.avandia.com. Accessed January 5, 2008.

Aronoff S, Rosenblatt S, Braithwaite S, Egan JW, Mathisen AL, Scheider RL, The Proglitazone 001 Study Group. Pioglitazone hydrochloride monotherapy improves glycemic control in the treatment of patients with type 2 diabetes. *Diabetes Care.* 2000;23(11):1605–1611.

Centers for Disease Control and Prevention. National diabetes fact sheet: General information and national estimates on diabetes in the United States, 2005. Atlanta, GA: U.S. Department of Health and Human Services. Centers for Disease Control and Prevention; 2005.

Cryer PE, Davis SN, Shamoon H. Hypoglycemia in diabetes. *Diabetes Care.* 2003;26(6):1902–1912.

Franz MJ, Bantle JP, Beebe CA, Brunzell JD, Chiasson J, Garg A, Holzmeister LA, Hoogwerf B, Mayer E, Mooradian AD, Purnell JQ, Wheeler M. Evidence-based nutrition principles and recommendations for the treatment and prevention of diabetes and related complications. *Diabetes Care.* 2002;25(1):148–198.

Freeman J. Oral diabetes medications. In: Ross TA, Boucher JL, O'Connell BS, eds. *American Dietetic Association Guide to Diabetes: Medical Nutrition Therapy and Education.* Chicago, IL: American Dietetic Association; 2005:81–88.

Hayes C. Physical activity and exercise. In: Ross TA, Boucher JL, O'Connell BS, eds. *American Dietetic Association Guide to Diabetes: Medical Nutrition Therapy and Education.* Chicago, IL: American Dietetic Association; 2005:71–80.

Holzmeister LA, Geil P. Evidence-based nutrition care and recommendations. In: Ross TA, Boucher JL, O'Connell BS, eds. *American Dietetic Association Guide to Diabetes: Medical Nutrition Therapy and Education.* Chicago, IL: American Dietetic Association; 2005:61–70.

Inzucchi SE. Oral antihyperglycemic therapy for type 2 diabetes. *JAMA.* 2002;287(3):360–372.

Kulkarni K. Pattern management. In: Ross TA, Boucher JL, O'Connell BS, eds. *American Dietetic Association Guide to Diabetes: Medical Nutrition Therapy and Education.* Chicago, IL: American Dietetic Association; 2005:116–127.

Long S. Diseases of the endocrine system. In: Nelms M, Sucher K, Long S. *Nutrition Therapy and Pathophysiology.* Belmont, CA: Thomson/Brooks-Cole; 2007:549–608.

Luna B, Feinglos MN. Oral agents in the management of type 2 diabetes mellitus. *Am Fam Physician.* 2001;63(9):1747–1756.

Mulcahy K, Lumber T. *The Diabetes Ready Reference for Health Professionals.* 2nd ed. Alexandria, VA: American Diabetes Association; 2004.

National Center for Chronic Disease Prevention and Health Promotion. Incidence of Diabetes in the Population Aged 18-79 Years. Available at: http://www.cdc.gov/diabetes/statistics/incidence. Accessed April 15. 2008.

O'Connell BS. Diabetes classification, pathophysiology, and diagnosis. In: Ross TA, Boucher JL, O'Connell BS, eds. *American Dietetic Association Guide to Diabetes: Medical Nutrition Therapy and Education.* Chicago, IL: American Dietetic Association; 2005:39–48.

Pastors JG, Waslaski J, Gunderson H. Diabetes meal-planning strategies. In: Ross TA, Boucher JL, O'Connell BS, eds. *American Dietetic Association*

Guide to Diabetes: Medical Nutrition Therapy and Education. Chicago, IL: American Dietetic Association; 2005:201–217.

Pronsky ZM. *Powers and Moore's Food–Medication Interactions,* 15th ed. Birchrunville, PA: Food–Medication Interactions; 2008.

Stratton IM, Adler AI, Neil HA, Matthews DR, Manley SE, Cull CA, Hadden D, Turner RC, Homan RR. Association of glycaemia with macrovascular and microvascular complications of type 2 diabetes (UKPDS 35): Prospective observational study. *BMJ.* 2000;321(7258):405–412.

UK Prospective Diabetes Study (UKPDS) Group. Intensive blood-glucose control with sulphonylureas or insulin compared with conventional treatment and risk of complications in patients with type 2 diabetes (UKPDS 33). *Lancet.* 1998;352:937–853.

Umpierrez GE, Murphy MB, Kitabchi AE. Diabetic ketoacidosis and hyperglycemic hyperosmolar syndrome. *Diabetes Spectrum.* 2002;15(1):28–36.

Vinik AI, Maser RE, Mitchell BD, Freeman R. Diabetic autonomic neuropathy. *Diabetes Care.* 2003;26(5):1553–1579.

Votey SR, Peters AL. Diabetes mellitus, type 2: A review. Available at: http://www.emedicine.com/emerg/TOPIC134.HTM. Accessed April 15, 2008.

Welschen LMC, Bloemendal E, Nijpels G, Dekker JM, Heine RJ, Stalman WAB, Bouter LM. Self-monitoring of blood glucose in patients with type 2 diabetes who are not using insulin. *Diabetes Care.* 2005;28(6):1510–1517.

Wheeler ML. Long-term complications. In: Ross TA, Boucher JL, O'Connell BS, eds. *American Dietetic Association Guide to Diabetes: Medical Nutrition Therapy and Education.* Chicago, IL: American Dietetic Association; 2005:139–145.

White JR. The pharmacologic management of patients with type II diabetes mellitus in the era of new oral agents and insulin analogs. *Diabetes Spectrum.* 1996;9(4):227–234.

Internet Resources

American Association of Diabetes Educators. http://www.diabeteseducator.org/

American Diabetes Association. http://www.diabetes.org

American Dietetic Association. http://www.eatright.org

Centers for Disease Control: Diabetes Public Health Resource. http://www.cdc.gov/diabetes/

Diabetes.com. http://www.diabetes.com/

eMedicineHealth: Diabetes. http://www.emedicinehealth.com/diabetes/article_em.htm

MedicineNet.com: Diabetes Mellitus. http://www.medicinenet.com/diabetes_mellitus/article.htm

MedlinePlus: Diabetes. http://www.nlm.nih.gov/medlineplus/diabetes.html

National Diabetes Information Clearinghouse: Diabetes Overview. http://diabetes.niddk.nih.gov/dm/pubs/overview/index.htm

WebMD: Diabetes Health Center. http://diabetes.webmd.com/

Case 24

Gestational Diabetes Mellitus

Objectives

After completing this case, the student will be able to:

1. Discuss the etiology and risk factors for development of gestational diabetes mellitus (GDM).
2. Describe the diagnostic criteria for GDM.
3. Identify classes of medications used to treat GDM and determine possible drug–nutrient interactions.
4. Identify possible complications of hyperglycemia during pregnancy.
5. Apply knowledge of nutrition therapy guidelines for treatment of GDM.
6. Analyze nutrition assessment data consistent with the physiological changes in pregnancy to evaluate nutritional status and identify specific nutrition problems.
7. Determine nutrition diagnoses and write appropriate PES statements.
8. Develop a nutrition care plan with appropriate measurable goals, interventions, and strategies for monitoring and evaluation that addresses the nutrition diagnoses of this case.

A 31-year-old woman in the second trimester of pregnancy is admitted when her doctor suspects gestational diabetes mellitus.

Name: Veronica Delgado
DOB: 3/22 (age 31)
Physician: Patricia Ortez, MD

ADMISSION DATABASE

BED #	DATE:	TIME:	TRIAGE STATUS (ER ONLY):
1	5/11	0300	☐ Red ☐ Yellow ☐ Green ☐ White

Initial Vital Signs

TEMP:	RESP:	SAO$_2$:	
98.9	18		
HT: 5′3″	WT (lb): 175 (pre-preg 165)	B/P: 131/92	PULSE: 83
LAST TETANUS		LAST ATE	LAST DRANK

PRIMARY PERSON TO CONTACT:
Name: Michael Delgado
Home #: 555-233-5643
Work #: 555-235-7855

ORIENTATION TO UNIT: ☒ Call light ☒ Television/telephone ☒ Bathroom ☒ Visiting ☒ Smoking ☒ Meals ☒ Patient rights/responsibilities

CHIEF COMPLAINT/HX OF PRESENT ILLNESS

"When I was at my last prenatal visit, my doctor was worried about my blood pressure and that I was maybe diabetic—or that my blood sugar was high. I have not been feeling well at all."

ALLERGIES: Meds, Food, IVP Dye, Seafood: Type of Reaction

PREVIOUS HOSPITALIZATIONS/SURGERIES

Wisdom teeth removed at age 18
Appendectomy at age 14

PERSONAL ARTICLES: (Check if retained/describe)
☒ Contacts ☐ R ☐ L ☐ Dentures ☐ Upper ☐ Lower
☐ Jewelry:
☐ Other:

VALUABLES ENVELOPE:
☐ Valuables instructions

INFORMATION OBTAINED FROM:
☒ Patient ☐ Previous record
☒ Family ☐ Responsible party

Signature *Veronica Delgado*

Home Medications (including OTC) **Codes: A=Sent home** **B=Sent to pharmacy** **C=Not brought in**

Medication	Dose	Frequency	Time of Last Dose	Code	Patient Understanding of Drug
prenatal vitamins	1	daily	this AM	A	yes

Do you take all medications as prescribed? ☒ Yes ☐ No If no, why?

PATIENT/FAMILY HISTORY

☐ Cold in past two weeks
☐ Hay fever
☐ Emphysema/lung problems
☐ TB disease/positive TB skin test
☐ Cancer
☐ Stroke/past paralysis
☒ Heart attack Paternal grandfather
☒ Angina/chest pain Paternal grandfather
☒ Heart problems Paternal grandfather

☒ High blood pressure Paternal grandfather
☐ Arthritis
☐ Claustrophobia
☐ Circulation problems
☐ Easy bleeding/bruising/anemia
☐ Sickle cell disease
☐ Liver disease/jaundice
☐ Thyroid disease
☒ Diabetes Mother

☐ Kidney/urinary problems
☐ Gastric/abdominal pain/heartburn
☐ Hearing problems
☐ Glaucoma/eye problems
☐ Back pain
☐ Seizures
☐ Other

RISK SCREENING

Have you had a blood transfusion? ☐ Yes ☒ No
Do you smoke? ☐ Yes ☒ No
If yes, how many pack(s)?
Does anyone in your household smoke? ☐ Yes ☒ No
Do you drink alcohol? ☒ Yes ☐ No
If yes, how often? not during pregnancy How much?
When was your last drink?
Do you take any recreational drugs? ☐ Yes ☒ No
If yes, type:_______ Route:
Frequency:_______ Date last used:______/______/______

FOR WOMEN Ages 12–52

Is there any chance you could be pregnant? ☒ Yes ☐ No
If yes, expected date (EDC): 09/12
Gravida/Para: 1/0

ALL WOMEN

Date of last Pap smear: 08/20
Do you perform regular breast self-exams? ☒ Yes ☐ No

ALL MEN

Do you perform regular testicular exams? ☐ Yes ☐ No

Additional comments:

x *Farkhrosh Shariyeh RN, BSN*
Signature/Title

Client Name: Veronica Delgado
DOB: 3/22
Age: 31
Gender: Female
Education: Bachelor's degree
Occupation: Paralegal
Hours of work: 9–5:30
Household members: Husband age 32, well
Ethnic background: Hispanic
Religious affiliation: Catholic
Referring Physician: Dr. Patricia Ortez

Chief complaint:
"I am in my 22nd week of my pregnancy. This will be my first child. My doctor is a little concerned about my blood pressure and that my blood sugar is high. I haven't felt very good for the past week or so."

Patient History:
Onset of disease: Gravida 1 Para 0. In 22nd week of gestation. Result of 50-g oral glucose tolerance test (OGTT) was 172 mg/dL at 2 hours.
Type of Tx: None at present
PMH: Noncontributory—wisdom teeth removal and appendectomy only previous hospitalizations
Meds: Prenatal vitamins
Smoker: No
Family Hx: Patient's mother has type 2 diabetes mellitus currently treated with oral agent. Patient believes that her mother and her aunts also had problems with blood sugar during their pregnancies. Patient weighed 9 lbs 10 oz at birth. All of her siblings weighed greater than 9 lbs.

Physical Exam:
General appearance: 31-year-old pregnant female in no acute distress
Vitals: Temperature: 98.8°F, BP 135/90 mm Hg, HR 87 bpm, RR 16 bpm
Heart: RRR without murmurs or gallops
HEENT:
  Eyes: PERRLA, normal fundi
  Ears: Noncontributory
  Nose: Noncontributory
  Throat: Pharynx clear
Genitalia: Normal
Neurologic: Alert and oriented × 3
Extremities: No edema, pulses full, no bruits; normal strength, sensation, and DTR
Skin: Warm, dry
Chest/lungs: Lungs clear to auscultation and percussion
Abdomen: Bowel sounds present

Nutrition Hx:
General: Patient states that her appetite has been increased lately and that she has found that she is more thirsty than normal. She attributed this to normal pregnancy symptoms prior to this admission.

Recent dietary intake:

AM:	Cereal, 2% milk; toast, bagel, or doughnut; juice or fruit
Lunch:	If packs her lunch, it includes sandwich (ham or turkey) from home, raw vegetables, chips, cola. If she eats out for lunch, usually orders salad, breadsticks, and occasionally a hamburger-type sandwich
Dinner:	Only eats small amount of meat; pasta or rice; some type of bread; 1–2 vegetables or a tossed salad
Bedtime snack:	Cheese and crackers, cookies, cola, or popcorn

24-hour recall:

AM:	Cheerios—1 c; 1 c 2% milk; 1 4″ doughnut; 8 oz grapefruit juice
Lunch:	Chicken Caesar salad from Wendy's, 2 breadsticks, 16 oz milkshake
Dinner:	Grilled squash, zucchini, and carrots—2 c; 2 slices French bread; 1 c peaches and blueberries; 16 oz 2% milk
Bedtime:	4–5 Oreos with 1 c milk

Food allergies/intolerances/aversions: Mrs. Delgado states that she likes most foods but historically has not eaten much meat. When she does eat meat, it is usually fish or chicken. During her pregnancy, she has eaten red meat more often than before.
Previous nutrition therapy? No
Food purchase/preparation: Self and husband
Vit/min intake: Prenatal vitamins daily
Anthropometric data: Ht. 5′3″ Adm wt: 175 lbs; pre-pregnancy wt: 165 lbs

Dx:
R/O gestational diabetes mellitus; R/O preeclampsia

Tx Plan:
Activity: Ad lib; *Diet:* 2,000 kcal; *Lab:* CBC, Chem 16; 100-g oral glucose tolerance test (OGTT); 24-hour urine collection; monitor fetal heart tones
Vitals: Routine; SBGM ac q meal

Hospital Course:
100-g oral glucose tolerance test: fasting—126 mg/dL; 1 hour—175 mg/dL; 2 hour—170 mg/dL; 3 hour—155 mg/dL. 24-hour urine collection was positive for glycosuria but negative for protein and/or ketones. Diabetic education consult was requested. Patient is to be discharged on a combination of 3–4 U Aspart prior to each meal and 10 U Lantus given at bedtime.

UH UNIVERSITY HOSPITAL

NAME: Veronica Delgado DOB: 3/22
AGE: 31 SEX: F
PHYSICIAN: P. Ortez, MD

CHEMISTRY

DAY:		Admit	
DATE:		5/11	
TIME:			
LOCATION:			
	NORMAL		UNITS
Albumin	3.5-5	3.8	g/dL
Total protein	6-8	7.2	g/dL
Prealbumin	16-35	34	mg/dL
Transferrin	250-380 (women) 215-365 (men)	350	mg/dL
Sodium	136-145	138	mEq/L
Potassium	3.5-5.5	3.7	mEq/L
Chloride	95-105	101	mEq/L
PO_4	2.3-4.7	2.8	mg/dL
Magnesium	1.8-3	1.9	mg/dL
Osmolality	285-295	296 H	mmol/kg/H_2O
Total CO_2	23-30	27	mEq/L
Glucose	70-110	186 H	mg/dL
BUN	8-18	9	mg/dL
Creatinine	0.6-1.2	0.7	mg/dL
Uric acid	2.8-8.8 (women) 4.0-9.0 (men)	2.8	mg/dL
Calcium	9-11	9.1	mg/dL
Bilirubin	≤0.3	0.3	mg/dL
Ammonia (NH_3)	9-33	9	μmol/L
ALT	4-36	12	U/L
AST	0-35	8	U/L
Alk phos	30-120	111	U/L
CPK	30-135 (women) 55-170 (men)		U/L
LDH	208-378	210	U/L
CHOL	120-199	180	mg/dL
HDL-C	>55 (women) >45 (men)	56	mg/dL
VLDL	7-32		mg/dL
LDL	<130	125	mg/dL
LDL/HDL ratio	<3.22 (women) <3.55 (men)	2.8	
Apo A	101-199 (women) 94-178 (men)		mg/dL
Apo B	60-126 (women) 63-133 (men)		mg/dL
TG	35-135 (women) 40-160 (men)	155 H	mg/dL
T_4	4-12		mcg/dL
T_3	75-98		mcg/dL
HbA_{1C}	3.9-5.2	8.5 H	%

UH UNIVERSITY HOSPITAL

NAME: Veronica Delgado DOB: 3/22
AGE: 31 SEX: F
PHYSICIAN: P. Ortez, MD

HEMATOLOGY

DAY:		Admit	
DATE:		5/11	
TIME:			
LOCATION:			
	NORMAL		UNITS
WBC	4.8-11.8	7.2	× 10^3/mm^3
RBC	4.2-5.4 (women) 4.5-6.2 (men)	4.1 L	× 10^6/mm^3
HGB	12-15 (women) 14-17 (men)	12	g/dL
HCT	37-47 (women) 40-54 (men)	36.5 L	%
MCV	80-96	93	μm^3
RETIC	0.8-2.8	0.9	%
MCH	26-32	29	pg
MCHC	31.5-36	32	g/dL
RDW	11.6-16.5	13.2	%
Plt Ct	140-440	320	× 10^3/mm^3
Diff TYPE			
ESR	0-25 (women) 0-15 (men)	10	mm/hr
% GRANS	34.6-79.2	38.6	%
% LYM	19.6-52.7	21.4	%
SEGS	50-62	55	%
BANDS	3-6	4	%
LYMPHS	24-44	28	%
MONOS	4-8	5	%
EOS	0.5-4	1	%
Ferritin	20-120 (women) 20-300 (men)	12 L	mg/mL
ZPP	30-80		μmol/mol
Vitamin B_{12}	24.4-100		ng/dL
Folate	5-25		μg/dL
Total T cells	812-2,318		mm^3
T-helper cells	589-1,505		mm^3
T-suppressor cells	325-997		mm^3
PT	11-16		sec

UH UNIVERSITY HOSPITAL

NAME: Veronica Delgado
AGE: 31
PHYSICIAN: P. Ortez, MD

DOB: 3/22
SEX: F

***URINALYSIS**

DAY: Admit
DATE: 5/11
TIME:
LOCATION:

	NORMAL		UNITS
Coll meth		Random specimen	
Color		Straw	
Appear		Hazy	
Sp grv	1.003-1.030	1.010	
pH	5-7	5	
Prot	NEG	NEG	mg/dL
Glu	NEG	+2	mg/dL
Ket	NEG	NEG	
Occ bld	NEG	NEG	
Ubil	NEG	NEG	
Nit	NEG	NEG	
Urobil	<1.1	0	EU/dL
Leu bst	NEG	0	
Prot chk	NEG	0	
WBCs	0-5	0	/HPF
RBCs	0-5	0	/HPF
EPIs	0	0	/LPF
Bact	0	0	
Mucus	0	0	
Crys	0	0	
Casts	0	0	/LPF
Yeast	0	0	

Case Questions

I. Understanding the Disease and Pathophysiology

1. Define *gestational diabetes mellitus (GDM).*
2. What physiological changes during pregnancy contribute to the development of hyperglycemia in GDM?
3. What are the criteria for diagnosis of GDM?
4. Identify the risk factors that are associated with GDM. Examine the admission information and the physician's history and physical. Which risk factors does Mrs. Delgado have?
5. What are the risks of untreated or poorly controlled hyperglycemia in pregnancy?
6. Is it possible that Mrs. Delgado may continue to have diabetes mellitus after the delivery of her baby?
7. Dr. Ortez was also concerned about the possibility of preeclampsia for Mrs. Delgado. What is eclampsia or preeclampsia? What symptoms would Mrs. Delgado present with if she were preeclamptic? What are the risks of eclampsia?

II. Understanding the Nutrition Therapy

8. Describe the major components of nutrition therapy for GDM. Do they differ from recommendations for other diagnoses of diabetes mellitus?

III. Nutrition Assessment

A. Evaluation of Body Weight/Body Composition

9. Evaluate Mrs. Delgado's anthropometric information. Compare her weight gain at this time to the recommended rate of weight gain during pregnancy.

10. Mrs. Delgado expresses a desire to lose a "little weight" during pregnancy. What would you recommend to her?

11. Identify any nutrition problems regarding body weight and body composition using the correct diagnostic term.

B. Calculation of Nutrient Requirements

12. Estimate Mrs. Delgado's caloric and protein requirements. Be sure to consider not only her GDM, but also her needs for pregnancy.

C. Intake Domain

13. Evaluate her 24-hour recall. Calculate her total caloric intake as well as grams of carbohydrate, protein, and fat.

14. From the information gathered within the intake domain, list nutrition problems using the correct diagnostic term.

D. Clinical Domain

15. Evaluate Mrs. Delgado's laboratory parameters. Note any that are abnormal and suggest a rationale for the abnormality.

Chemistry	Normal Value	Mrs. Delgado's Value	Reason for Abnormality	Nutritional Implications

16. Evaluate the results of her urinalysis. What do they indicate?

17. Mrs. Delgado was started initially on a combination of Aspart before each meal and snack, with Lantus at bedtime. What are the initial reaction times, peak, and duration for each of these types of insulin?

	Aspart	Lantus
Classification		
Onset of action		
Peak (hours)		
Duration (hours)		

18. List any nutrition problems in the clinical domain using the correct diagnostic term.

E. Behavioral–Environmental Domain

19. Mrs. Delgado states that her mother has diabetes and that her mother avoids all sweets. Mrs. Delgado describes her concerns about having to change her eating habits in such a drastic manner. Address her concerns within the context of the principles of nutrition therapy for GDM.

20. Mrs. Delgado relates that she would like to continue her exercise class for pregnant women held at the local community center. Would you encourage this activity? How might this affect her blood glucose control? What specific information would you give to her to ensure she exercises safely?

21. List any nutrition problems in the behavioral–environmental domain using the correct diagnostic term.

IV. Nutrition Diagnosis

22. Select two high-priority nutrition problems and complete the PES statement for each.

V. Nutrition Intervention

23. For each of the PES statements that you have written, establish an ideal goal (based on the signs and symptoms) and an appropriate intervention (based on the etiology).

VI. Nutrition Monitoring and Evaluation

24. Mrs. Delgado received education regarding self-monitoring of blood glucose. What would be an appropriate goal for her postprandial glucose level?

25. Using Mrs. Delgado's diet history and 24-hour recall, design a meal plan that would incorporate the MNT prescription you outlined in question 8.

26. Write an initial nutrition assessment note that you would enter in Mrs. Delgado's medical record outlining your nutrition recommendations and suggestions for educational requirements.

Bibliography

American Diabetes Association. Gestational diabetes mellitus. *Diabetes Care.* 2004;27(Suppl 1):S88–S90.

American Diabetes Association. Nutrition principles and recommendations in diabetes. *Diabetes Care.* 2004;27(Suppl 1):S36–S46.

Case J, Willoughby D, Haley-Zitlin V, Maybee P. Preventing type 2 diabetes after gestational diabetes. *Diabetes Educ.* 2006;32(6):877–886.

Cheng YW, Caughey AB. Gestational diabetes mellitus: What is the optimal treatment modality? *J Perinatol.* 2007;27(5):257–258.

Gabbe SG. Case study: A 34 year old woman in her second pregnancy at 24 weeks gestation. *Clin Diabetes.* 1999;17:140–141.

Gindlesberger D, Schrager S, Johnson S, Neher JO. Clinical inquiries. What's the best treatment for gestational diabetes? *J Fam Pract.* 2007;56(9):757–758.

Gutierrez YM, Reader DM. Medical nutrition therapy. In: Thomas AM, Gutierrez YM, eds. *American Dietetic Association Guide to Gestational Diabetes Mellitus.* Chicago, IL: American Dietetic Association; 2005:45–64.

Henriksen T. Nutrition and pregnancy outcome. *Nutr Rev.* 2006;64(5 Pt 2):S19-S23; discussion S72-S91.

Jovanovic L. The role of continuous glucose monitoring in gestational diabetes mellitus. *Diabetes Technol Ther.* 2000;2(Suppl 1):S67–S71.

Mottola MF. The role of exercise in the prevention and treatment of gestational diabetes mellitus. *Curr Sports Med Rep.* 2007;6(6):381–386.

Reader D, Splett P, Gunderson EP. Diabetes Care and Education Dietetic Practice Group. Impact of gestational diabetes mellitus nutrition practice guidelines implemented by registered dietitians on pregnancy outcomes. *J Am Diet Assoc.* 2006;106(9):1426–1433.

Scollan-Koliopoulos M, Guadagno S, Walker EA. Gestational diabetes management: Guidelines to a healthy pregnancy. *Nurse Pract.* 2006;31(6):14–23

Zhang C, Liu S, Solomon CG, Hu FB. Dietary fiber intake, dietary glycemic load, and the risk for gestational diabetes mellitus. *Diabetes Care.* 2006;29(10):2223–2230.

Internet Resources

American Diabetes Association: Gestational Diabetes. http://www.diabetes.org/gestational-diabetes.jsp

Mayo Clinic: Gestational Diabetes. http://www.mayoclinic.com/health/gestational-diabetes/DS00316

National Diabetes Information Clearinghouse: What I Need to Know About Gestational Diabetes. http://diabetes.niddk.nih.gov/dm/pubs/gestational/

National Institute of Child Health and Family Development: Gestational Diabetes. http://www.nichd.nih.gov/health/topics/Gestational_Diabetes.cfm

Case 25

Polycystic Ovarian Syndrome (PCOS)

Objectives

After completing this case, the student will be able to:

1. Describe the pathophysiology of polycystic ovary disease (PCOS).
2. Identify and explain common nutritional problems associated with this disease.
3. Explain the role of nutrition therapy as an adjunct to the pharmacotherapy of PCOS.
4. Interpret laboratory parameters for nutritional implications and significance.
5. Analyze nutrition assessment data to evaluate nutritional status and identify specific nutrition problems.
6. Determine nutritional diagnoses and write appropriate PES statements.
7. Develop a nutrition care plan with appropriate measurable goals, interventions, and strategies for monitoring and evaluation that addresses the nutrition diagnoses of this case.

Gracie Moore is a 34-year-old graduate student working on her doctoral degree. She was diagnosed with polycystic ovarian syndrome 6 years ago.

Name: Gracie Moore
DOB: 10/15 (age 34)
Physician: G. Davis, MD

ADMISSION DATABASE

BED #	DATE: 12/28	TIME:	TRIAGE STATUS (ER ONLY): ☐ Red ☐ Yellow ☐ Green ☐ White

Initial Vital Signs

TEMP: 98.7	RESP: 12	SAO$_2$:	
HT: 5′5″	WT (lb): 180	B/P:	PULSE:
LAST TETANUS		LAST ATE	LAST DRANK

PRIMARY PERSON TO CONTACT:
Name: D. Moore
Home #: 555-555-7512
Work #: 555-555-6263

ORIENTATION TO UNIT: ☐ Call light ☐ Television/telephone ☐ Bathroom ☐ Visiting ☐ Smoking ☐ Meals ☐ Patient rights/responsibilities

CHIEF COMPLAINT/HX OF PRESENT ILLNESS

Weight gain

PERSONAL ARTICLES: (Check if retained/describe)
☐ Contacts ☐ R ☐ L ☐ Dentures ☐ Upper ☐ Lower
☐ Jewelry:
☐ Other:

ALLERGIES: Meds, Food, IVP Dye, Seafood: Type of Reaction

NKA

VALUABLES ENVELOPE:
☐ Valuables instructions

PREVIOUS HOSPITALIZATIONS/SURGERIES

None

INFORMATION OBTAINED FROM:
☒ Patient ☐ Previous record
☒ Family ☐ Responsible party

Signature *Debbie Moore*

Home Medications (including OTC) **Codes: A=Sent home** **B=Sent to pharmacy** **C=Not brought in**

Medication	Dose	Frequency	Time of Last Dose	Code	Patient Understanding of Drug
OrthoNovum 1/35	1 mg	1 q d	this AM	C	excellent
	norethindrone				
	and 0.035 mg				
	ethinyl				
	estradiol				

Do you take all medications as prescribed? ☒ Yes ☐ No If no, why?

PATIENT/FAMILY HISTORY

☐ Cold in past two weeks
☐ Hay fever
☐ Emphysema/lung problems
☐ TB disease/positive TB skin test
☐ Cancer
☐ Stroke/past paralysis
☐ Heart attack
☐ Angina/chest pain
☐ Heart problems
☒ High blood pressure Patient
☐ Arthritis
☒ Claustrophobia Patient
☐ Circulation problems
☐ Easy bleeding/bruising/anemia
☐ Sickle cell disease
☐ Liver disease/jaundice
☐ Thyroid disease
☒ Diabetes Patient
☐ Kidney/urinary problems
☐ Gastric/abdominal pain/heartburn
☐ Hearing problems
☐ Glaucoma/eye problems
☐ Back pain
☐ Seizures
☐ Other

RISK SCREENING

Have you had a blood transfusion? ☐ Yes ☒ No
Do you smoke? ☐ Yes ☒ No
If yes, how many pack(s)?
Does anyone in your household smoke? ☒ Yes ☐ No
Do you drink alcohol? ☒ Yes ☐ No
If yes, how often? weekends How much? 3-5 drinks
When was your last drink? 12/23
Do you take any recreational drugs? ☐ Yes ☒ No
If yes, type:______ Route:
Frequency:______ Date last used:______/______/______

FOR WOMEN Ages 12–52

Is there any chance you could be pregnant? ☐ Yes ☒ No
If yes, expected date (EDC):
Gravida/Para: 2/0

ALL WOMEN

Date of last Pap smear: 5/27
Do you perform regular breast self-exams? ☒ Yes ☐ No

ALL MEN

Do you perform regular testicular exams? ☐ Yes ☐ No

Additional comments:

× *J. Higginbotham, RN*
Signature/Title

Client name: Gracie Moore
DOB: 10/15
Age: 34
Sex: Female
Education: Graduate student working on doctoral degree
Occupation: Graduate teaching assistant
Hours of work: 8–5
Household members: Husband & (adopted) infant daughter
Ethnic background: European American
Religious affiliation: Unitarian
Referring physician: Dr. Geoff Davis

Chief complaint:
"I just keep gaining weight, no matter what I do! The more weight I gain, the more hair shows up on my body. And I just found out I have sleep apnea and I have to use a CPAP at night!"

Patient history:
Gracie stopped menstruating in college. She was placed on oral contraceptives to control her hormone levels. As a student at the university, Gracie had access to the University Recreation Center and was able to control her weight through regular physical activity and eating a healthy diet. After she married and entered graduate school while working full time, it became more difficult to control her weight. Through high school and college, Gracie was able to maintain her weight at 140 lbs through extracurricular activities and regular physical activity. But once she graduated with her bachelor's degree and began graduate school and working, she was often "too busy" to exercise, and she gained an average of 4 lbs each year until reaching the present weight of 180 lbs. As she gained weight, the hirsutism became worse, as did many PCOS symptoms. Her physician prescribed oral contraceptives, which helped bring her testosterone level back into normal range. Gracie and her husband have tried to conceive since they were married. Gracie conceived twice, but did not carry the pregnancies to term. They adopted an infant daughter a year ago. The stress of juggling a career, graduate school, and a family has exacerbated her symptoms, causing her to seek further medical intervention.
Onset of disease: 6 years ago
Type of Tx: OrthoNovum 1/35
PMH: Unremarkable other than that noted above
Meds: Oral contraceptives. *Allergies:* None
Smoker: No
Family Hx: What? Diabetes *Who?* Father

Physical exam:
General appearance: Overweight female who looks her stated age
Vitals: Temp 98.4°F, BP 139/85 mm Hg, HR 70 bpm, RR 20 bpm
Heart: No murmurs, clicks, or rubs
HEENT:
Head: Normocephalic, thin hair with dandruff
Eyes: EOMI, fundoscopic exam WNL; no evidence of atherosclerosis, diabetic retinopathy, or early hypertensive changes

Ears: TM normal bilaterally
Nose: WNL
Throat: Tonsils not infected, uvula midline, gag normal
Genitalia: Grossly physiologic
Neurologic: No focal localizing abnormalities; DTR symmetric bilaterally
Extremities: Pulses intact; no edema
Skin: Dry and pale; acne noted on upper back; hirsutism noted on chest, stomach, and face; skin tags around neck; acanthosis nigricans noted on breasts, neck, and thighs
Chest/lungs: Lungs clear to auscultation and percussion
Peripheral vascular: WNL
Abdomen: BS WNL. No hepatomegaly, splenomegaly, masses, inguinal lymph nodes, or abdominal bruits.
Height/Weight: 65″, 180 lbs; Waist 36″; Hips 49″

Nutrition Hx:
General: Appetite good.

24-hour recall:

Breakfast:	8 oz calcium-fortified orange juice and 6 oz black coffee
Midmorning snack:	1 c mixed (salted) nuts (including cashews, walnuts, macadamia nuts, Brazil nuts, almonds, and pecans) and 10 oz iced tea (unsweet)
Lunch:	Cheeseburger (3 oz meat, 1 oz American cheese) with mustard, pickle, and onion; and small fries from Wendy's, 18 oz Diet Coke
Dinner:	1½ c ham and beans (about 1 oz ham) and 2 corn muffins, 12 oz Diet Coke
Snack:	Skinny Cow ice cream sandwich

Food allergies/intolerances/aversions: None
Previous Nutrition Therapy? Yes *If yes, when:* 6 years ago when diagnosed *Where?* University dietitian
Food purchase/preparation: Self and spouse
Vit/min intake: Multivitamin/mineral with iron

Dx:
R/O polycystic ovary disease syndrome

Tx plan:
CBC with differential, metabolic panel, lipid panel, thyroid panel with TSH, testosterone level, 2-hour GTT; YAZ 1 tablet PO qd; Glucophage 850 mg PO qd; Aldactone 100 mg/d PO; Vaniqua apply thin layer to affected and adjacent involved area q 8 h, do not wash treated area for at least 4 h after application; nutrition consultation

UH *UNIVERSITY HOSPITAL*

NAME: Gracie Moore DOB: 10/15
AGE: 34 SEX: F
PHYSICIAN: G. Davis

CHEMISTRY

	NORMAL	6 years ago	4 years ago	2 years ago	Present	UNITS
DAY:						
DATE:						
TIME:						
LOCATION:						
Albumin	3.5-5	4.1	4.3	4.0	4.2	g/dL
Total protein	6-8	6.9	6.8	7.1	7.0	g/dL
Prealbumin	16-35	21.5	22.1	21.9	22.1	mg/dL
Transferrin	250-380 (women) 215-365 (men)	205	210	197	235	mg/dL
Sodium	136-145	140	139	137	138	mEq/L
Potassium	3.5-5.5	4.3	4.1	4.2	4.0	mEq/L
Chloride	95-105	103	100	99	103	mEq/L
PO_4	2.3-4.7	2.5	3.1	2.9	3.2	mg/dL
Magnesium	1.8-3	2.1	2.4	2.2	2.3	mg/dL
Osmolality	285-295	290	291	289	292	mmol/kg/H_2O
Total CO_2	23-30	25	27	28	26	mEq/L
Glucose	70-110	90	91	92	93	mg/dL
BUN	8-18	11	14	12	11	mg/dL
Creatinine	0.6-1.2	0.7	0.8	0.9	0.7	mg/dL
Uric acid	2.8-8.8 (women) 4.0-9.0 (men)	6.4	6.5	6.9	6.6	mg/dL
Calcium	9-11	9.7	9.8	9.9	9.6	mg/dL
Bilirubin	≤0.3	0.4 H	0.4 H	0.4 H	0.41 H	mg/dL
Ammonia (NH_3)	9-33	19	20	18	21	µmol/L
ALT	4-36	41 H	43 H	41 H	42 H	U/L
AST	0-35	12	16	18	20	U/L
Alk phos	30-120	118	102	98	80	U/L
CPK	30-135 (women) 55-170 (men)	135	137	134	130	U/L
LDH	208-378	405	410	426	400	U/L
CHOL	120-199	189	187	207 H	197	mg/dL
HDL-C	> 55 (women) > 45 (men)	60	58	52 L	51 L	mg/dL
VLDL	7-32	34	44	42		mg/dL
LDL	< 130	95	85	141 H	132 H	mg/dL
LDL/HDL ratio	< 3.22 (women) < 3.55 (men)	1.6	1.5	2.7	2.5	
TG	35-135 (women) 40-160 (men)	174	224 H	211 H	184	mg/dL
T_4	4-12	11.4	11.2	9.3	10.1	mcg/dL
T_3 Uptake	75-98	24	28	30	32	mcg/dL
HbA_{1C}	3.9-5.2					%
TSH	0.35-5.50	3.50	2.174	2.515	2.68	mcIU/dL
Insulin	< 17		14			mcIU/ml
Testosterone	20-76 (women)	56	75	87 H	25	mg/dL
2-hr GTT						
Glucose administered	75 g					
Fasting glucose	70-115		96			mg/dL
½-hour glucose	< 200		149			mg/dL
1-hour glucose	< 200		134			mg/dL
2-hour glucose	< 200		116			mg/dL

UH UNIVERSITY HOSPITAL

NAME: Gracie Moore
AGE: 34
PHYSICIAN: G. Davis, MD

DOB: 10/15
SEX: F

HEMATOLOGY

DAY:		6 years ago	4 years ago	2 years ago	Present	
DATE:						
TIME:						
LOCATION:						
	NORMAL					UNITS
WBC	4.8-11.8	6.6	6.5	6.7	6.5	$\times 10^3/mm^3$
RBC	4.2-5.4 (women) 4.5-6.2 (men)	4.94	4.89	4.98	4.69	$\times 10^6/mm^3$
HGB	12-15 (women) 14-17 (men)	14.6	14.2	14.3	14.5	g/dL
HCT	37-47 (women) 40-54 (men)	43.4	42.1	43.1	40.6	%
MCV	80-96	88	87	87.5	86.6	μm^3
RETIC	0.8-2.8					%
MCH	26-32	29.6	30.2	29.9	30.9	pg
MCHC	31.5-36	33.7	34.9	33.9	35.7	g/dL
RDW	11.6-16.5	12.4	12.6	12.6	12.8	%
Plt Ct	140-440	262	259	260	248	$\times 10^3/mm^3$
Diff TYPE						
ESR	0-25 (women) 0-15 (men)					mm/hr
% GRANS	34.6-79.2					%
% LYM	19.6-52.7					%
SEGS	50-62					%
BANDS	3-6					%
LYMPHS	24-44	30	35	38.7	40.8	%
MONOS	4-8	6	6.8	6.5	6.8	%
EOS	0.5-4	2	2.2	2.6	2.8	%
Ferritin	20-120 (women) 20-300 (men)					mg/mL
ZPP	30-80					μmol/mol
Vitamin B_{12}	24.4-100					ng/dL
Folate	5-25					μg/dL
Total T cells	812-2,318					mm^3
T-helper cells	589-1,505					mm^3
T-suppressor cells	325-997					mm^3
PT	11-16					sec

Case Questions

I. Understanding the Disease and Pathophysiology

1. Define *PCOS,* and describe the etiology of this disorder.

2. Outline the diagnostic criteria for PCOS.

3. Describe the medical complications associated with PCOS.

4. Using the history and physical for Gracie, identify the signs and symptoms that are consistent with PCOS. Are there any other typical signs and symptoms that Gracie does not have?

5. PCOS is often associated with many of the same signs and symptoms as metabolic syndrome. Define *metabolic syndrome* and outline any differences between this condition and PCOS.

6. What are the long-term complications of PCOS?

II. Understanding the Nutrition Therapy

7. Briefly, what are the primary nutritional treatment goals for PCOS?

III. Nutrition Assessment

A. Evaluation of Weight/Body Composition

8. Gracie's waist measures 36 inches. Calculate her waist-to-hip ratio and explain the implications of this anthropometric measurement in the diagnosis of metabolic syndrome.

9. Calculate and interpret Gracie's BMI.

10. Assess Gracie's weight gain and explain the nutritional implications of the BMI and overall weight gain. Is there a relationship among BMI, PCOS, and metabolic syndrome? If so, explain.

B. Calculation of Nutrient Requirements

11. Calculate total daily energy requirements for Gracie based on a weight within the normal BMI range and on her current weight. Which would you recommend as you plan her nutrition therapy and why?

C. Intake Domain

12. Assess Gracie's 24-hour recall for total kcal, % CHO, % PRO, and % FAT.

13. List four common nutritional recommendations for individuals with PCOS.

14. Compare Gracie's intake to the current nutrition recommendations. List any nutrition problems within the intake domain.

D. Clinical Domain

15. Evaluate Gracie's lab results.

Abnormal Lab	Normal Value	Reason for Abnormality	Nutritional Implication

16. Evaluate each of the medications that Gracie is prescribed. Determine the function of each medication, and identify any nutritional implications.

Medication	Function of Medication	Nutritional Implications

17. Some medications are used "off-label" to treat PCOS. What does "off-label" mean? What are the primary medications used to treat PCOS?

18. List nutrition problems within the clinical domain using the diagnostic term.

E. Behavioral–Environmental Domain

19. From Gracie's history, are there any lifestyle factors that may impact the treatment of her PCOS?

20. While you are interviewing Gracie, she tells you that one of her friends with PCOS takes fenugreek, cinnamon, and ginseng to control her blood glucose levels. Discuss any current research regarading the use of cinnamon, ginseng, or fenugreek to lower blood glucose levels.

21. What information would you recommend to your patient regarding the use of herbal remedies in lowering blood glucose levels?

22. List any nutrition problems within the behavioral–environmental domain using the diagnostic term.

IV. Nutrition Diagnosis

23. Select two high-priority nutrition problems and complete PES statements for each.

V. Nutrition Intervention

24. For each PES statement written, establish an ideal goal (based on signs and symptoms) and an appropriate intervention (based on etiology).

VI. Nutrition Monitoring and Evaluation

25. What nutritional parameters can be used to measure Gracie's response to treatment?

26. When should you schedule your next counseling session with Gracie, and what would you evaluate?

Bibliography

Bruner B, Chad K, Chizen D. Effects of exercise and nutritional counseling in women with polycystic ovary syndrome. *Appl Physiol Nutr Metab.* 2006;31(4):384–391

Farshchi H, Rane A, Love A, Kennedy RL. Diet and nutrition in polycystic ovary syndrome (PCOS): Pointers for nutritional management. *J Obstet Gynaecol.* 2007;27(8):762–773.

Khan MI, Klachko DM. Polycystic ovarian syndrome. *eMedicine.* Available at: http://www.emedicine.com/med/TOPIC2173.htm. Accessed January 2, 2008.

Legro RS. Clinical crossroads. A 27-year-old woman with a diagnosis of polycystic ovary syndrome. *JAMA.* 2007;297:509–519.

Long S. Diseases of the endocrine system. In: Nelms M, Sucher K, Long S. *Nutrition Therapy and Pathophysiology.* Belmont, CA: Thomson/Brooks-Cole; 2007:549–608.

National Women's Health Information Center, Polycystic ovarian syndrome (PCOS) [monograph on the Internet]. Washington (DC): U.S. Department of Health and Human Services Office on Women's Health; 2004. Available at: http://www.womenshealth.gov/faq/pcos.htm. Accessed April 29, 2008.

Norman RJ, Homan G, Moran L, Noakes M. Lifestyle choices, diet, and insulin sensitizers in polycystic ovary syndrome. *Endocrine.* 2006;30(1):35–43.

Pham AQ, Kourlas H, Pham DQ. Cinnamon supplementation in patients with type 2 diabetes mellitus. *Pharmacotherapy.* 2007;27(4):595–599.

Pronsky Z. *Food Medication Interactions,* 14th ed. Pottstown, PA: Food Medication Interactions, 2008.

Solomon TP, Blannin AK. Effects of short-term cinnamon ingestion on in vivo glucose tolerance. *Diabetes Obes Metab.* 2007;9(6):895–901.

Thatcher SS. *PCOS. The Hidden Epidemic.* Indianapolis, IN: Perspectives Press, 2000.

Internet Resources

American College of Obstetricians and Gynecologists. http://www.acog.org/publications/patient_education/bp121.cfm

Familydoctor.org: Polycystic Ovary Syndrome. http://familydoctor.org/online/famdocen/home/women/reproductive/gynecologic/620.html

The Hormone Foundation. http://www.hormone.org

International Council of Infertility Information Dissemination, Inc.: Polycystic Ovary Syndrome (PCOS) FAQ. http://www.inciid.org/faq.php?cat=infertility101&id=2

MayoClinic.com: Polycystic Ovary Syndrome. http://www.mayoclinic.com/health/polycystic-ovary-syndrome/DS00423/DSECTION=7

MedlinePlus: Polycystic Ovary Disease. http://www.nlm.nih.gov/medlineplus/ency/article/000369.htm

The University of Chicago Medical Center: Center for Polycystic Ovary Syndrome. http://www.uchospitals.edu/specialties/pcos/

Unit Eight

NUTRITION THERAPY FOR RENAL DISORDERS

There has been a noticeable growth in the field of medical nutrition therapy for patients with chronic kidney disease (CKD). The importance of nutrition in the care of patients with CKD is illustrated by the fact that indicators of nutritional status effectively predict morbidity and mortality in these patients. In 2005, over 15 million Americans had clinical evidence of kidney disease. Of this number, more than 85,000 Americans die each year because of kidney disease, and more than 485,000 Americans suffer from advanced CKD and need renal replacement therapy to stay alive. Kidney disease is one of the costliest illnesses. In 2005, more than $30 billion was spent on renal replacement therapy (available at: http://kidney.niddk.nih.gov/kudiseases/pubs/kustats/index.htm#kp; accessed April 28, 2008).

The primary cause of CKD is diabetes mellitus, which accounts for about 43 percent of all new cases each year. Uncontrolled hypertension is the second leading cause of CKD in the United States, accounting for approximately 26 percent of U.S. cases.

The cases presented in this section can stand alone or be used in tandem to illustrate progression of renal disease from impaired renal function to CKD. Integrated into each case are aspects that predispose an individual to CKD, such as diabetes mellitus and ethnicity. Researchers from the National Institutes of Health have established that CKD caused by diabetes mellitus is anywhere from 10 to 75 times more prevalent in Native Americans than in whites, and the prevalence differs among tribes; 50 percent of Pima Indians age 35 years and over have type 2 diabetes mellitus—the highest rate in the world. Fundamental principles, such as modification of nutrient composition in impaired renal function, CKD, and renal replacement therapy, are included. Type 2 DM in adolescents is touched on in these cases, as well as results from the Modification of Diet in Renal Disease Study, the largest randomized, multicenter clinical trial designed to evaluate effects of dietary protein and phosphorus, and of blood pressure control on the progression rate of chronic renal insufficiency.

Case 26

Chronic Kidney Disease (CKD) Treated with Dialysis

Objectives

After completing this case, the student will be able to:

1. Describe the pathophysiology of chronic kidney disease (CKD).
2. Describe the stages of CKD.
3. Differentiate the physiology of peritoneal dialysis and hemodialysis.
4. Identify and explain common nutritional problems associated with CKD.
5. Interpret laboratory parameters for nutritional implications and significance.
6. Analyze nutrition assessment data to evaluate nutritional status and identify specific nutrition problems.
7. Determine nutrition diagnoses and write appropriate PES statements.
8. Develop a nutrition care plan with appropriate measurable goals, interventions, and strategies for monitoring and evaluation that addresses the nutrition diagnoses of this case.
9. Integrate sociocultural and ethnic food consumption issues within a nutrition care plan.
10. Make appropriate documentation in the medical record.

Enez Joaquin is a 24-year-old Pima Indian who has had type 2 diabetes mellitus since age 13. Mrs. Joaquin has experienced a declining glomerular filtration rate for the past 2 years. She is being admitted in preparation for kidney replacement therapy.

Name: Enez Joaquin
DOB: 4/13 (age 24)
Physician: L. Nila, MD

ADMISSION DATABASE

BED #	DATE:	TIME:	TRIAGE STATUS (ER ONLY):
2	3/5	1830	☐ Red ☐ Yellow ☐ Green ☐ White

Initial Vital Signs

TEMP:	RESP:	SAO$_2$:	
98.6	25		
HT: 5'0"	WT (lb): 170	B/P: 220/80	PULSE: 84
LAST TETANUS: 4 years ago		LAST ATE: 2 days ago	LAST DRANK: 4 hours ago–water

PRIMARY PERSON TO CONTACT:
Name: Eddie Joaquin (husband)
Home #: 555-3947
Work #: 554-2100

ORIENTATION TO UNIT: ☒ Call light ☒ Television/telephone ☒ Bathroom ☒ Visiting ☒ Smoking ☒ Meals ☒ Patient rights/responsibilities

CHIEF COMPLAINT/HX OF PRESENT ILLNESS

N/A

PERSONAL ARTICLES: (Check if retained/describe)
☐ Contacts ☐ R ☐ L ☐ Dentures ☐ Upper ☐ Lower
☐ Jewelry:
☐ Other:

ALLERGIES: Meds, Food, IVP Dye, Seafood: Type of Reaction

none

VALUABLES ENVELOPE:
☐ Valuables instructions

PREVIOUS HOSPITALIZATIONS/SURGERIES

childbirth 7 years ago

INFORMATION OBTAINED FROM:
☒ Patient ☐ Previous record
☐ Family ☐ Responsible party

Signature *Enez Joaquin*

Home Medications (including OTC) **Codes: A=Sent home** **B=Sent to pharmacy** **C=Not brought in**

Medication	Dose	Frequency	Time of Last Dose	Code	Patient Understanding of Drug
Glucophage	850 mg	twice daily	?	C	no
Vasotec	20 mg	three X daily	?	C	no

Do you take all medications as prescribed? ☐ Yes ☒ No If no, why?

PATIENT/FAMILY HISTORY

☐ Cold in past two weeks
☐ Hay fever
☐ Emphysema/lung problems
☐ TB disease/positive TB skin test
☐ Cancer
☐ Stroke/past paralysis
☐ Heart attack
☐ Angina/chest pain
☐ Heart problems
☒ High blood pressure Patient
☐ Arthritis
☐ Claustrophobia
☐ Circulation problems
☐ Easy bleeding/bruising/anemia
☐ Sickle cell disease
☐ Liver disease/jaundice
☐ Thyroid disease
☒ Diabetes Patient
☒ Kidney/urinary problems Patient
☒ Gastric/abdominal pain/heartburn Patient
☐ Hearing problems
☐ Glaucoma/eye problems
☐ Back pain
☐ Seizures
☐ Other

RISK SCREENING

Have you had a blood transfusion? ☐ Yes ☒ No
Do you smoke? ☐ Yes ☒ No
If yes, how many pack(s)?
Does anyone in your household smoke? ☒ Yes ☐ No
Do you drink alcohol? ☒ Yes ☐ No
If yes, how often? daily How much? 12 oz beer
When was your last drink? 3/4
Do you take any recreational drugs? ☐ Yes ☒ No
If yes, type:______ Route:
Frequency:______ Date last used:______/______/______

FOR WOMEN Ages 12–52

Is there any chance you could be pregnant? ☐ Yes ☒ No
If yes, expected date (EDC):
Gravida/Para: 1/1

ALL WOMEN

Date of last Pap smear: 1/25
Do you perform regular breast self-exams? ☐ Yes ☒ No

ALL MEN

Do you perform regular testicular exams? ☐ Yes ☐ No

Additional comments:

x *Liz Romero, RN*
Signature/Title

Client name: Enez Joaquin
DOB: 4/13
Age: 24
Sex: Female
Education: High school
Occupation: Secretary
Hours of work: 9 AM–5 PM
Household members: Husband age 26, type 2 diabetes under control; daughter age 7, in good health
Ethnic background: Pima Indian
Religious affiliation: Catholic
Referring physician: Lourdes Nila, MD (nephrology)

Chief complaint:
Patient complains of anorexia; N/V; 4 kg weight gain in the past 2 weeks, edema in extremities, face, and eyes; malaise; progressive SOB with 3-pillow orthopnea; pruritus; muscle cramps; and inability to urinate.

Patient history:
Mrs. Joaquin is a 24-year-old Native American woman who was diagnosed with type 2 DM when she was 13 years old and has been poorly compliant with prescribed treatment. She is from the Pima Indian tribe of southern Arizona. She lives with her husband and 7-year-old daughter. Her husband also has type 2 DM. He was diagnosed at the age of 18. Her renal function has been monitored for the past 7 years. Progressive decompensation of kidney function has been documented by declining GFR, increasing creatinine and urea concentrations, elevated serum phosphate, and normochromic, normocytic anemia. She is being admitted for preparation for kidney replacement therapy.
Onset of disease: Diagnosed with Stage 3 chronic kidney disease 2 years ago. Her acute symptoms have developed over the last 2 weeks.
Type of Tx: Control BP; prepare for kidney replacement therapy; nutrition consult.
PMH: Gravida 1/para 1. Infant weighed 10 lbs at birth 7 years ago. Patient admits she recently stopped taking a prescribed hypoglycemic agent, and she has never filled her prescription for antihypertensive medication.
Meds: Glucophage (metformin) 850 mg bid
Smoker: No
Family Hx: Both mother and father diagnosed with DM

Physical exam:
General appearance: Overweight Native American female who appears her age. Lethargic, complaining of N/V.
Vitals: Temp 98.6°F, BP 220/80 mm Hg, HR 86 bpm, RR 25 bpm
Heart: S4, S1, and S2, regular rate and rhythm. I/VI systolic ejection murmur, upper left sternal border.
HEENT: Normocephalic, equal carotid pulses, neck supple, no bruits
  Eyes: PERRLA
  Ears: Noncontributory

Nose: Noncontributory
Throat: Noncontributory
Genitalia: Normal female
Neurologic: Oriented to person, place, and time; intact, mild asterixis
Extremities: Muscle weakness; 3+ pitting edema to the knees, no cyanosis
Skin: Dry and yellowish brown
Chest/lungs: Generalized rhonchi with rales that are mild at the bases (patient breathes with poor effort)
Peripheral vascular: Normal pulse (3+) bilaterally
Abdomen: Bowel sounds positive, soft; generalized mild tenderness; no rebound

Nutrition Hx:
General: Intake has been poor due to anorexia, N & V. Patient states that she tried to follow the diet that she was taught 2 years ago. "It went pretty well for awhile, but it was hard to keep up with."

Usual dietary intake:
Breakfast: Cold cereal, bread or fried potatoes, fried egg (occasionally)
Lunch: Bologna sandwich, potato chips, Coke
Dinner: Chopped meat, fried potatoes
Snacks: Crackers and peanut butter

Food allergies/intolerances/aversions: None
Previous nutrition therapy? Yes *If yes, when:* 2 years ago when patient dx with Stage 3 chronic kidney disease *Where?* Reservation Health Service
Food purchase/preparation: Self
Vit/min intake: None
Current diet order: 30 kcal/kg, 0.8 g protein/kg, 8–12 mg phosphorus/kg, 2–3 g Na
24-hour recall: N/A

Dx:
Chronic kidney disease; type 2 DM

Tx plan:
Evaluate for kidney replacement therapy
Capoten/captopril
Erythropoietin (r-HuEPO) 30 units/kg
Vitamin/mineral supplement
Hectorol 2.5 μg four times daily 3 × week
35 kcal/kg, 1.2 g protein/kg, 2 g K, 1 g phosphorus, 2 g Na, 1,000 mL fluid + urine output per day
Glucophage (metformin) 850 mg twice daily
CBC, chemistry
Phos Lo
Stool softener
Sodium bicarbonate, 2 g every day
Occult fecal blood

UH UNIVERSITY HOSPITAL

NAME: Enez Joaquin DOB: 4/13
AGE: 24 SEX: F
PHYSICIAN: L. Nila, MD

CHEMISTRY

DAY:		Admit	d/c	
DATE:				
TIME:				
LOCATION:				
	NORMAL			UNITS
Albumin	3.5–5	3.7	3.4 L	g/dL
Total protein	6–8	6.2	6.0	g/dL
Prealbumin	16–35			mg/dL
Transferrin	250–380 (women) 215–365 (men)			mg/dL
Sodium	136–145	130 L	134 L	mEq/L
Potassium	3.5–5.5	5.8 H	5.6 H	mEq/L
Chloride	95–105	91 L	100	mEq/L
PO_4	2.3–4.7	9.5 H	7.3 H	mg/dL
Magnesium	1.8–3	2.9	2.7	mg/dL
Osmolality	285–295			mmol/kg/H_2O
Total CO_2	23–30	20 L	23	mEq/L
Glucose	70–110	282 H	200 H	mg/dL
BUN	8–18	69 H	55 H	mg/dL
Creatinine	0.6–1.2	12.0 H	8.5 H	mg/dL
Uric acid	2.8–8.8 (women) 4.0–9.0 (men)			mg/dL
Calcium	9–11	8.2 L	8.6 L	mg/dL
Bilirubin	≤0.3			mg/dL
Ammonia (NH_3)	9–33			μmol/L
ALT	4–36	26		U/L
AST	0–35	28		U/L
Alk phos	30–120	131		U/L
CPK	30–135 (women) 55–170 (men)			U/L
LDH	208–378	315		U/L
CHOL	120–199	220 H		mg/dL
HDL-C	>55 (women) >45 (men)			mg/dL
VLDL	7–32			mg/dL
LDL	<130			mg/dL
LDL/HDL ratio	<3.22 (women) <3.55 (men)			
Apo A	101–199 (women) 94–178 (men)			mg/dL
Apo B	60–126 (women) 63–133 (men)			mg/dL
TG	35–135 (women) 40–160 (men)	200 H		mg/dL
T_4	4–12			mcg/dL
T_3	75–98			mcg/dL
HbA_{1C}	3.9–5.2	8.9 H		%

UH UNIVERSITY HOSPITAL

NAME: Enez Joaquin DOB: 4/13
AGE: 24 SEX: F
PHYSICIAN: L. Nila, MD

URINALYSIS

DAY:		Admit	Postop	d/c	
DATE:					
TIME:					
LOCATION:					
	NORMAL				UNITS
Coll meth		Random specimen	First morning	First morning	
Color		Straw	Straw	Pale yellow	
Appear		Hazy	Slightly hazy	Slightly hazy	
Sp grv	1.003-1.030	1.010			
pH	5-7	7.9			
Prot	NEG	2+			mg/dL
Glu	NEG				mg/dL
Ket	NEG				
Occ bld	NEG				
Ubil	NEG				
Nit	NEG				
Urobil	<1.1				EU/dL
Leu bst	NEG				
Prot chk	NEG				
WBCs	0-5	20			/HPF
RBCs	0-5				/HPF
EPIs	0				/LPF
Bact	0				
Mucus	0				
Crys	0				
Casts	0				/LPF
Yeast	0				

Name: Enez Joaquin
Physician: L. Nila, MD

PATIENT CARE SUMMARY SHEET

Date: 3/5 Room: 324 Wt Yesterday: Today: 170

Temp °F	NIGHTS								DAYS								EVENINGS							
	00	01	02	03	04	05	06	07	08	09	10	11	12	13	14	15	16	17	18	19	20	21	22	23
105																								
104																								
103																								
102																								
101																								
100																								
99																								
98																								
97																								
96																								
Pulse																			84					82
Respiration																			25					24
BP																			220/80					210/78
Blood Glucose																			210					
Appetite/Assist																			0					
INTAKE																								
Oral																			0	50				
IV																			0					
TF Formula/Flush																			0					
Shift Total																								
OUTPUT																								
Void																			N/A					100
Cath.																								
Emesis																					50			
BM																								
Drains																								
Shift Total																	50 cc							
Gain																	150 cc							
Loss																								
Signatures																	Sandy Dunn, RN							

Name: Enez Joaquin
Physician: L. Nila, MD

PATIENT CARE SUMMARY SHEET

Date: 3/6 Room: 324 Wt Yesterday: 170 Today: 165 Postdialysis: 165

Temp °F	NIGHTS								DAYS								EVENINGS							
	00	01	02	03	04	05	06	07	08	09	10	11	12	13	14	15	16	17	18	19	20	21	22	23
105																								
104																								
103																								
102																								
101																								
100																								
99								99																
98																								
97																								
96																								
Pulse								80			84													
Respiration								23			25													
BP								200/75			220/80													
Blood Glucose								170			200													
Appetite/Assist								NPO			NPO													
INTAKE																								
Oral								0			0													
IV																								
TF Formula/Flush																								
Shift Total																								
OUTPUT																					300			
Void								200																
Cath.																								
Emesis				100												50								
BM								×1																
Drains																								
Shift Total																								
Gain	NPO																							
Loss	300 cc																							
Signatures	Bill Larga, RN								Sandy Dunn, RN								Michele Barker, RN							

Case Questions

I. Understanding the Disease and Pathophysiology

1. Describe the physiological functions of the kidneys.

2. What diseases/conditions can lead to chronic kidney disease (CKD)?

3. Explain how type 2 diabetes mellitus can lead to CKD.

4. Outline the stages of CKD, including the distinguishing signs and symptoms.

5. From your reading of Mrs. Joaquin's history and physical, what signs and symptoms did she have?

6. What are the treatment options for Stage 5 CKD?

7. Describe the differences between hemodialysis and peritoneal dialysis.

II. Understanding the Nutrition Therapy

8. Explain the reasons for the following components of Mrs. Joaquin's medical nutrition therapy:

Nutrition Therapy	Rationale
35 kcal/kg	
1.2 g protein/kg	
2 g K	
1 g phosphorus	
2 g Na	
1,000 mL fluid + urine output	

III. Nutrition Assessment

A. Evaluation of Weight/Body Composition

9. Calculate and interpret Mrs. Joaquin's BMI. How does edema affect your interpretation?

10. What is edema-free weight? The following equation can be used to calculate the edema-free adjusted body weight (aBW_{ef}):

$$aBW_{ef} = BW_{ef} + [(SBW - BW_{ef}) \times 0.25]$$

where BW_{ef} is the actual edema-free body weight and SBW is the standard body weight as determined from the NHANES II data.

Calculate Mrs. Joaquin's edema-free weight. Is this the same as dry weight?

B. Calculation of Nutrient Requirements

11. What are the energy requirements for CKD?

12. Calculate what Mrs. Joaquin's energy needs will be once she begins hemodialysis.

13. What are Mrs. Joaquin's protein requirements when she begins hemodialysis?

14. What is the rationale? How would these change if she were on peritoneal dialysis?

C. Intake Domain

15. Are there any potential benefits of using different types of protein, such as plant protein rather than animal protein, in the diet for a patient with CKD? Explain.

16. Mrs. Joaquin has a PO_4 restriction. Why?

17. What foods have the highest levels of phosphorus?

18. Mrs. Joaquin tells you that one of her friends can drink only certain amounts of liquids and wants to know if that is the case for her. What foods are considered to be fluids? What recommendations can you make for Mrs. Joaquin?

19. If a patient must follow a fluid restriction, what can be done to help reduce his or her thirst?

20. Identify nutrition problems within the intake domain using the appropriate diagnostic term.

D. Clinical Domain

21. Several biochemical indices are used to diagnose chronic kidney disease. One is glomerular filtration rate (GFR). What does GFR measure?

22. What test is usually done to estimate glomerular filtration rate?

23. Mrs. Joaquin's GFR is 28 mL/min. What does this tell you about her kidney function?

24. Evaluate Mrs. Joaquin's chemistry report. What labs support the diagnosis of Stage 4 CKD?

25. Examine the patient care summary sheet for hospital day 2. What was Mrs. Joaquin's weight postdialysis? Why did it change?

26. Which of Mrs. Joaquin's other symptoms would you expect to begin to improve?

27. Explain why the following medications were prescribed by completing the table.

Medication	Indications/Mechanism	Nutritional Concerns
Vasotec		
Erythropoietin		
Vitamin/mineral supplement		
Calcitriol		
Glucophage		
Sodium bicarbonate		
Phos Lo		

28. Identify nutrition problems within the clinical domain using the appropriate diagnostic term.

E. Behavioral–Environmental Domain

29. What health problems have been identified in the Pima Indians through epidemiological data?

30. Explain what is meant by the "thrifty gene" theory.

31. How does nephropathy affect Pima Indians?

IV. Nutrition Diagnosis

32. Choose two high-priority nutrition problems and complete a PES statement for each.

V. Nutrition Intervention

33. For each PES statement, establish an ideal goal (based on the signs and symptoms) and appropriate intervention (based on the etiology).

34. When Mrs. Joaquin begins dialysis, energy and protein recommendations will increase. Explain why.

35. Why is it recommended for patients to have at least 50% of their protein from sources that have high biological value?

36. The MD ordered daily use of a multivitamin/mineral supplement containing B-complex, but not fat-soluble vitamins. Why are these restrictions specified?

37. What resources would you use to teach Mrs. Joaquin about her diet?

38. Using Mrs. Joaquin's typical intake and the prescribed diet, write a sample menu. Make sure you can justify your changes and that it is consistent with her nutrition prescription.

Diet PTA		**Sample Menu**
Breakfast:	Cold cereal (¾ c unsweetened)	
	Bread (2 slices) or fried potatoes (1 med potato)	
	1 fried egg (occasionally)	
Lunch:	Bologna sandwich (2 slices white bread, 2 slices bologna, mustard)	
	Potato chips (1 oz)	
	1 can Coke	
Dinner:	Chopped meat (3 oz beef)	
	Fried potatoes (1½ medium)	
HS Snack:	Crackers (6 saltines) and peanut butter (2 tbsp)	

39. Using the renal exchange list, plan a 1-day diet that complies with your diet order. Provide a nutrient analysis to assure consistency with all components of the prescription.

40. Write an initial medical record note for your consultation with Mrs. Joaquin.

Bibliography

American Dietetic Association. *Nutrition Diagnosis and Intervention: Standardized Language for the Nutrition Care Process.* Chicago, IL: American Dietetic Association; 2007.

Brown TL. Ethnic populations. In Ross TA, Boucher JL, O'Connell BS. *American Dietetic Association Guide to Diabetes Medical Nutrition Therapy and Education.* Chicago, IL: American Dietetic Association; 2005:227–238.

Byham-Gray L, Wiesen K. *A Clinical Guide to Nutrition Care in Kidney Disease.* Chicago, IL: American Dietetic Association; 2004.

Escott-Stump S. Renal disorders. In: Escott-Stump S. *Nutrition and Diagnosis-Related Care,* 6th ed. Baltimore, MD: Lippincott Williams & Wilkins; 2008:785–819.

Freedman BI, DuBose TD. Chronic kidney disease: Cause and consequence of cardiovascular disease. *Arch Int Med.* 2007;167:1113–1115.

Hill L, Goeddeke-Merickel CM. Chronic kidney disease—Nondialysis. In Ross TA, Boucher JL, O'Connell BS. *American Dietetic Association Guide to Diabetes Medical Nutrition Therapy and Education.* Chicago, IL: American Dietetic Association; 2005:264–275.

Karalis M, Pavlinac JM, Goldstein-Fuchs J. Diseases of the renal system. In: Nelms M, Sucher K, Long S. *Nutrition Therapy and Pathophysiology.* Belmont, CA: Thomson/Brooks-Cole; 2007:609–650.

Kittler PG, Sucher KP. *Food and Culture.* 4th ed. Belmont, CA: Wadsworth Thompson Learning; 2004.

Lacey K. The nutrition care process. In: Nelms M, Sucher K, Long S. *Nutrition Therapy and Pathophysiology.* Belmont, CA: Thomson/Brooks-Cole; 2007:39–64.

National Institute of Diabetes and Digestive and Kidney Diseases (NIDDK). American Indians, Alaska Natives, and Diabetes. Available at: http://diabetes.niddk.nih.gov/dm/pubs/americanindian/. Accessed July 24, 2007.

National Institute of Diabetes and Digestive and Kidney Diseases (NIDDK). *The Pima Indians: Pathfinders for Health.* Available at: http://diabetes.niddk.nih.gov/dm/pubs/pima/. Accessed July 24, 2007.

Nelms MN. Assessment of nutrition status and risk. In: Nelms M, Sucher K, Long S. *Nutrition Therapy and Pathophysiology.* Belmont, CA: Thomson/Brooks-Cole; 2007:101–135.

Nelson RG, Bennett PH, Beck GJ, Tan M, Knowler WC, Mitch WE, Hirschman GH, Myers BD. Development and progression of renal disease in Pima Indians with non-insulin-dependent diabetes mellitus. Diabetic Renal Disease Study Group. *N Engl J Med.* 1996;335(22):1636–1642.

O'Connell BS. Early renal disease in diabetes: A brief review. *Diabetes Care Educ.* 2001;22(1):7–11.

Pronsky ZM. *Powers and Moore's Food and Medication Interaction,* 14th ed. Birchrunville, PA: Food–Medication Interactions; 2006.

Internet Resources

American Association of Kidney Patients. http://www.aakp.org/

Atlas of Diseases of the Kidney. http://www.kidneyatlas.org/

Cook's Thesaurus. http://www.foodsubs.com/

Culinary Kidney Cooks. http://www.culinarykidneycooks.com/

eMedicineHealth: Chronic Kidney Disease. http://www.emedicinehealth.com/chronic_kidney_disease/article_em.htm

Kidney School. http://www.kidneyschool.org/

National Institute of Diabetes, Digestive and Kidney Diseases (NIDDK). http://www2.niddk.nih.gov/

National Kidney Foundation. http://www.kidney.org/atoz/atozTopic.cfm?topic=4

National Kidney Foundation K/DOQI Guidelines: Evaluation, Classification, and Stratification. http://www.kidney.org/professionals/kdoqi/guidelines_ckd/toc.htm

The Nephron Information Center. http://www.nephron.com/

The Nephron Information Center: Food Values. http://foodvalues.us/

Renal Web. http://www.renalweb.com/

San Jose State University: Renal Dialysis—A Team Effort. http://www.nufs.sjsu.edu/renaldial/index.html

United States Renal Data System (USRDS). http://www.usrds.org/

Case 27

Renal Transplant

Objectives

After completing this case, the student will be able to:

1. Describe the physiology of organ transplantation.
2. Explain how transplant recipients are matched to donors.
3. Compare and contrast nutrition therapy for post-transplant patients during the acute and chronic phases.
4. Identify nutritional implications associated with the pharmacotherapy of renal transplantation.
5. Interpret laboratory parameters for nutritional implications and significance.
6. Determine nutrition diagnoses and write appropriate PES statements.
7. Develop a nutrition care plan with appropriate measurable goals, interventions, and strategies for monitoring and evaluation that addresses the nutrition diagnoses of this case.
8. Integrate sociocultural and ethnic food consumption issues within a nutrition care plan.

Enez Joaquin is a 26-year-old Pima Indian who has had type 2 diabetes mellitus since age 13. Mrs. Joaquin was placed on a transplant list 2 years ago when she began hemodialysis to treat her CKD. She has been matched to a kidney donor and enters the hospital for a kidney transplant.

Name: Enez Joaquin
DOB: 4/13 (age 26)
Physician: L. Nila, MD

ADMISSION DATABASE

BED #	DATE:	TIME:	TRIAGE STATUS (ER ONLY):
2	9/5	830	☐ Red ☐ Yellow ☐ Green ☐ White

Initial Vital Signs

TEMP:	RESP:	SAO$_2$:	
98.6	25		
HT:	WT (lb):	B/P:	PULSE:
5′0″	165	130/85	87
LAST TETANUS		LAST ATE	LAST DRANK
6 years ago		this AM	1 hour ago-water

CHIEF COMPLAINT/HX OF PRESENT ILLNESS

N/A

ALLERGIES: Meds, Food, IVP Dye, Seafood: Type of Reaction

None

PREVIOUS HOSPITALIZATIONS/SURGERIES

childbirth 9 years ago

AVF for hemodialysis 2 years ago

PRIMARY PERSON TO CONTACT:
Name: Eddie Joaquin (husband)
Home #: 555-3947
Work #: 554-2100

ORIENTATION TO UNIT: ☒ Call light ☒ Television/telephone ☒ Bathroom ☒ Visiting ☒ Smoking ☒ Meals ☒ Patient rights/responsibilities

PERSONAL ARTICLES: (Check if retained/describe)
☐ Contacts ☐ R ☐ L ☐ Dentures ☐ Upper ☐ Lower
☐ Jewelry:
☐ Other:

VALUABLES ENVELOPE:
☐ Valuables instructions

INFORMATION OBTAINED FROM:
☒ Patient ☐ Previous record
☐ Family ☐ Responsible party

Signature *Enez Joaquin*

Home Medications (including OTC) **Codes: A=Sent home** **B=Sent to pharmacy** **C=Not brought in**

Medication	Dose	Frequency	Time of Last Dose	Code	Patient Understanding of Drug
Glucophage	850 mg	twice daily	0700	C	yes
Vasotec	20 mg	three X daily	0700	C	yes
Phos Lo	3 tablets	daily w/meals	0700	C	yes
erythropoietin	30 units/kg		yesterday	C	yes
vitamin/mineral supplement	N/A	daily	0700	C	yes
calcitriol	0.25 mcg	daily	0700	C	yes
sodium bicarbonate	2g	daily	0700	C	yes

Do you take all medications as prescribed? ☒ Yes ☐ No If no, why?

PATIENT/FAMILY HISTORY

☐ Cold in past two weeks
☐ Hay fever
☐ Emphysema/lung problems
☐ TB disease/positive TB skin test
☐ Cancer
☐ Stroke/past paralysis
☐ Heart attack
☐ Angina/chest pain
☐ Heart problems
☒ High blood pressure Patient
☐ Arthritis
☐ Claustrophobia
☐ Circulation problems
☐ Easy bleeding/bruising/anemia
☐ Sickle cell disease
☐ Liver disease/jaundice
☐ Thyroid disease
☒ Diabetes Patient
☒ Kidney/urinary problems Patient
☒ Gastric/abdominal pain/heartburn Patient
☐ Hearing problems
☐ Glaucoma/eye problems
☐ Back pain
☐ Seizures
☐ Other

RISK SCREENING

Have you had a blood transfusion? ☐ Yes ☒ No
Do you smoke? ☐ Yes ☒ No
If yes, how many pack(s)?
Does anyone in your household smoke? ☒ Yes ☐ No
Do you drink alcohol? ☒ Yes ☐ No
If yes, how often? 1 x/week How much? 12 oz beer
When was your last drink? 9/4
Do you take any recreational drugs? ☐ Yes ☒ No
If yes, type:______ Route:
Frequency:______ Date last used:______/______/______

FOR WOMEN Ages 12–52

Is there any chance you could be pregnant? ☐ Yes ☒ No
If yes, expected date (EDC):
Gravida/Para: 1/1

ALL WOMEN

Date of last Pap smear: 1/25
Do you perform regular breast self-exams? ☐ Yes ☒ No

ALL MEN

Do you perform regular testicular exams? ☐ Yes ☐ No

Additional comments:

x *Liz Romero, RN*
Signature/Title

Client name: Enez Joaquin
DOB: 4/13
Age: 26
Sex: Female
Education: High school
Occupation: Secretary
Hours of work: 9 AM–5 PM
Household members: Husband age 28, type 2 diabetes under control; daughter age 9, in good health
Ethnic background: Pima Indian
Religious affiliation: Catholic
Referring physician: Lourdes Nila, MD (nephrology)

Chief complaint:
Patient admitted for deceased donor kidney transplant.

Patient history:
Mrs. Joaquin is a 26-year-old Native American woman who was diagnosed with type 2 DM when she was 13 years old. She is from the Pima Indian tribe of southern Arizona, and lives with her husband and 9-year-old daughter. Her husband also has type 2 DM, and was diagnosed at the age of 18. Her renal function progressively decompensated over 7 years, and she was placed on hemodialysis for kidney replacement. A transplant evaluation was done, and she was placed on a transplant list 2 years ago. A kidney from a deceased donor has become available, so she is being admitted for preparation for kidney transplantation.
Onset of disease: Diagnosed with Stage 5 chronic kidney disease 2 years ago when she was placed on hemodialysis
Type of Tx: Control BP; prepare for kidney transplantation; nutrition consult.
PMH: Gravida 1/para 1. Infant weighed 10 lbs at birth 9 years ago. Patient has been compliant with medication and kidney replacement regimens since onset of hemodialysis.
Meds: Glucophage (metformin) 850 mg bid, Vasotec, erythropoietin, calcitriol, sodium bicarbonate, Phos Lo, vitamin/mineral supplement
Smoker: No
Family Hx: Both mother and father diagnosed with DM

Physical exam:
General appearance: Overweight Native American female who appears her age
Vitals: Temp 98.6°F, BP 130/85 mm Hg, HR 87 bpm, RR 25 bpm
Heart: S4, S1, and S2, regular rate and rhythm; I/VI systolic ejection murmur, upper left sternal border
HEENT: Normocephalic, equal carotid pulses, neck supple, no bruits
Eyes: PERRLA
Ears: Noncontributory
Nose: Noncontributory
Throat: Noncontributory
Genitalia: Normal female
Neurologic: Oriented to person, place, and time

Extremities: No cyanosis
Skin: Dry. AVF in right forearm.
Chest/lungs: Generalized rhonchi with rales that are mild at the bases (patient breathes with poor effort)
Peripheral vascular: Normal pulse (3+) bilaterally
Abdomen: Bowel sounds positive, soft; generalized mild tenderness; no rebound

Nutrition Hx:
General: States appetite is good and she has been following the diet prescribed when she began hemodialysis. She follows up with the RD at the dialysis center at least every few months.

Usual dietary intake:
Breakfast: 1 soft-cooked egg, 2 slices wheat toast with 1 tsp low-fat margarine, 1 c artificially sweetened cranberry juice
Lunch: 2 beef tamales with ¼ c chili con carne, 1 can Diet Coke
Dinner: 2 soft-shell tacos made with ½ c black beans, 2 flour tortillas, ½ c shredded lettuce, ¼ c chopped tomatoes, ¼ c chopped onions; 1 can Diet Coke
Snacks: 6 vanilla wafers

Food allergies/intolerances/aversions: None
Previous nutrition therapy? Yes *If yes, when:* 2 years ago when placed on hemodialysis *Where?* Reservation Health Service
Food purchase/preparation: Self
Vit/min intake: None
Current diet order: 35 kcal/kg, 1.2 g protein/kg, 2 g K, 1 g phosphorous, 2 g Na, 1,000 mL fluid + urine output
24-hour recall: N/A

Dx:
Chronic kidney disease

Tx plan:
Prep for surgery. Medications after surgery: Neoral 450 mg q 12 hrs, Imuran 150 mg q d, prednisone 90 mg q d, magnesium oxide 400 mg three times daily, Bactrim, Neutra-phos, Persantine, omeprazole, Glucophage

UH UNIVERSITY HOSPITAL

NAME: Enez Joaquin
AGE: 26
PHYSICIAN: L. Nila, MD

DOB: 4/13
SEX: F

CHEMISTRY

DAY:		Admit	d/c	
DATE:				
TIME:				
LOCATION:				
	NORMAL			UNITS
Albumin	3.5–5	3.8	3.9	g/dL
Total protein	6–8			g/dL
Prealbumin	16–35			mg/dL
Transferrin	250–380 (women) 215–365 (men)			mg/dL
Sodium	136–145	136	138	mEq/L
Potassium	3.5–5.5	5.5	5.4	mEq/L
Chloride	95–105	95	100	mEq/L
PO_4	2.3–4.7	6.2 H	4.5	mg/dL
Magnesium	1.8–3	2.9	2.7	mg/dL
Osmolality	285–295			mmol/kg/H_2O
Total CO_2	23–30	25	26	mEq/L
Glucose	70–110	282 H	200 H	mg/dL
BUN	8–18	69 H	55 H	mg/dL
Creatinine	0.6–1.2	12.0 H	8.5 H	mg/dL
Uric acid	2.8–8.8 (women) 4.0–9.0 (men)			mg/dL
Calcium	9–11	8.9 L	9.1	mg/dL
Bilirubin	≤0.3			mg/dL
Ammonia (NH_3)	9–33			μmol/L
ALT	4–36	26		U/L
AST	0–35	28		U/L
Alk phos	30–120	131 H		U/L
CPK	30–135 (women) 55–170 (men)			U/L
LDH	208–378	315		U/L
CHOL	120–199	200 H		mg/dL
HDL-C	>55 (women) >45 (men)			mg/dL
VLDL	7–32			mg/dL
LDL	<130			mg/dL
LDL/HDL ratio	<3.22 (women) <3.55 (men)			
Apo A	101–199 (women) 94–178 (men)			mg/dL
Apo B	60–126 (women) 63–133 (men)			mg/dL
TG	35–135 (women) 40–160 (men)	195 H		mg/dL
T_4	4–12			mcg/dL
T_3	75–98			mcg/dL
HbA_{1C}	3.9–5.2	7.1 H		%

UH UNIVERSITY HOSPITAL

NAME: Enez Joaquin DOB: 4/13
AGE: 26 SEX: F
PHYSICIAN: L. Nila, MD

HEMATOLOGY

DAY: Admit
DATE:
TIME:
LOCATION:

	NORMAL	Admit	UNITS
WBC	4.8-11.8	11.1	$\times 10^3/mm^3$
RBC	4.2-5.4 (women) 4.5-6.2 (men)	4.0 L	$\times 10^6/mm^3$
HGB	12-15 (women) 14-17 (men)	10.9 L	g/dL
HCT	37-47 (women) 40-54 (men)	35 L	%
MCV	80-96	81	μm^3
RETIC	0.8-2.8	0.9	%
MCH	26-32	28	pg
MCHC	31.5-36	33.1	g/dL
RDW	11.6-16.5	12.1	%
Plt Ct	140-440	143	$\times 10^3/mm^3$
Diff TYPE			
ESR	0-25 (women) 0-15 (men)	20	mm/hr
% GRANS	34.6-79.2	45.6	%
% LYM	19.6-52.7	26.5	%
SEGS	50-62	53	%
BANDS	3-6	4.1	%
LYMPHS	24-44	26	%
MONOS	4-8	6	%
EOS	0.5-4	2.6	%
Ferritin	20-120 (women) 20-300 (men)	20	mg/mL
ZPP	30-80	55	µmol/mol
Vitamin B_{12}	24.4-100	56	ng/dL
Folate	5-25	7.4	µg/dL
Total T cells	812-2,318	1,957	mm^3
T-helper cells	589-1,505	1,436	mm^3
T-suppressor cells	325-997	550	mm^3
PT	11-16	15.5	sec

Case Questions

I. Understanding the Disease and Pathophysiology

1. Describe the physiological functions of the kidneys.

2. What diseases/conditions can lead to chronic kidney disease (CKD)?

3. What was the likely cause of Mrs. Joaquin's CKD?

4. Mrs. Joaquin's transplant evaluation took place 2 years ago and included each of the following. What were each of these procedures used to evaluate?

Procedure	Used to Evaluate
Abdominal and renal ultrasound	
EKG and echocardiogram	
Chest X-ray	
Meeting with transplant nurse, social worker, surgeon, and financial counselor	
Blood typing and tissue typing	
Dental exam	
Viral testing on blood	
Mammogram and PAP test	

5. Describe why the immunological characteristics of the donated organ must match with the recipient's medical and immunological characteristics.

6. Explain the role of the major histocompatibility complex (MHC).

II. Understanding the Nutrition Therapy

7. What are the differences between nutrition therapy during the acute phase (up to 8 weeks following transplant) and during the chronic phase (starting ninth week following transplantation) post-transplantation? Explain the rationale for each.

Nutrient	Acute Phase	Chronic Phase	Rationale
Protein			
Energy			
Carbohydrates			
Fats			
Cholesterol			
Potassium			
Sodium			
Calcium			
Phosphorus			
Vitamins/minerals			
Fluids			

III. Nutrition Assessment

A. Evaluation of Weight/Body Composition

8. Calculate Mrs. Joaquin's BMI.

9. How would you interpret Mrs. Joaquin's BMI? Explain your rationale.

B. Calculation of Nutrient Requirements

10. What are the recommendations for estimating energy requirements for (post) renal transplantation? Calculate Mrs. Joaquin's energy needs accordingly.

11. What will Mrs. Joaquin's protein requirements be after the transplant?

12. Compare her energy and protein needs prior to and post-transplant. Explain how and why they are different.

C. Intake Domain

13. Explain the importance of food safety education for transplant patients.

D. Clinical Domain

14. On POD #2, Mrs. Joaquin was doing well and transferred to the medical floor. Her recovery and course were uneventful. She was taken to Nuclear Medicine for a scan. Results showed good perfusion and function of the kidney. Her intake and output were good. During the remainder of her hospitalization, Mrs. Joaquin received detailed instructions about postoperative care and medications. The instructions were:

- Keep incision clean and dry
- Staples will be removed in 3 weeks
- Avoid lifting over 5 pounds
- Can resume driving and sexual activity in 2–4 weeks or when pain free
- Follow prescribed diet

Explain why the following medications were prescribed, and indicate any nutrition implications.

Medication	Indications/Mechanism	Nutritional Implications
Neoral		
Imuran		
Prednisone		
Magnesium oxide		
Bactrim		
Neutral-phos		
Persantine		
Omeprazole		
Glucophage		

15. Explain the role of immunosuppression in organ transplantation.

16. How long will Mrs. Joaquin require immunosuppression?

17. How will taking prednisone for her transplant affect her glycemic control?

18. Mrs. Joaquin is also instructed to watch for signs of rejection. Explain what is meant by rejection and list at least three signs of transplant rejection.

19. What will happen if Mrs. Joaquin does reject her transplanted kidney?

E. Behavioral–Environmental Domain

20. Mrs. Joaquin tells you that she's heard transplant patients gain weight after surgery, and she wants to know if this will happen to her. How do you answer her question?

IV. Nutrition Diagnosis

21. Prioritize the nutrition diagnoses by listing them in the order in which you would expect interventions to be developed.

22. Select two high-priority nutrition problems and complete the PES statement for each.

V. Nutrition Intervention

23. Using your PES statement, establish an ideal goal (based on the signs and symptoms) and appropriate intervention (based on the etiology).

24. Using Mrs. Joaquin's typical intake and the prescribed diet, write a sample menu for her post-transplant nutritional needs.

Diet PTA	Sample Menu
1 soft-cooked egg	
2 slices wheat toast with 1 tsp low-fat margarine	
1 c artificially sweetened cranberry juice	
2 beef tamales with ¼ c chili con carne	
1 can Diet Coke	
2 soft-shell tacos made with ½ c black beans	
2 flour tortillas	
½ c shredded lettuce	
¼ c chopped tomatoes	
¼ c chopped onions	
1 can Diet Coke	
6 vanilla wafers	

25. Write an initial medical record note for your consultation with Mrs. Joaquin.

Bibliography

American Dietetic Association. *Nutrition Diagnosis and Intervention: Standardized Language for the Nutrition Care Process.* Chicago, IL: American Dietetic Association; 2007.

Beisswenger PJ, Drummond KS, Nelson RG, Howell SK, Szwergold BS, Mauer M. Susceptibility to diabetic nephropathy is related to dicarbonyl and oxidative stress. *Diabetes.* 2005;54(11):3274–3278.

Bernardi A, Biasia F, Pati T, Piva M, Scaramuzzo P, Stoppa F, Bucciante G. Factors affecting nutritional status, response to exercise, and progression of chronic rejection in kidney transplant recipients. *J Ren Nutr.* 2005;15(1):54–57.

Brown TL. Ethnic populations. In Ross TA, Boucher JL, O'Connell BS. *American Dietetic Association Guide to Diabetes Medical Nutrition Therapy and Education.* Chicago, IL: American Dietetic Association; 2005:227–238.

Escott-Stump S. *Nutrition and Diagnosis-Related Care,* 6th ed. Baltimore, MD: Lippincott Williams & Wilkins; 2008.

Karalis M, Pavlinac JM, Goldstein-Fuchs J. Diseases of the renal system. In: Nelms M, Sucher K, Long S. *Nutrition Therapy and Pathophysiology.* Belmont, CA: Thomson/Brooks-Cole; 2007:609–649.

Kasiske BL, Vazquez MA, Harmon WE, Brown RS, Danovitch GM, Gaston RS, Roth D, Scandling JD, Singer GG. Recommendations for the outpatient surveillance of renal transplant recipients. *J Am Soc Nephrol.* 2000;11(Suppl 15):S1–S86.

Kasiske BL, Magdalena AA. Nutritional management of renal transplantation. In Kopple J, Massry S, eds. *Nutritional Management of Renal Disease,* 2nd ed. Philadelphia, PA: Lippincott Williams & Wilkins; 2004:513–525.

Kidney Disease Outcomes Quality Initiative (K/DOQI) Group. K/DOQI clinical practice guidelines for managing dyslipidemias in chronic kidney disease. *Am J Kidney Dis.* 2003;41(4 Suppl 3):I–IV, S1–S91.

Kittler PG, Sucher KP. *Food and Culture.* 4th ed. Belmont, CA: Wadsworth Thomson Learning; 2004.

Lacey K. The nutrition care process. In: Nelms M, Sucher K, Long S. *Nutrition Therapy and Pathophysiology.* Belmont, CA: Thomson/Brooks-Cole; 2007:39–64.

Mantoo S, Abraham G, Pratap GB, Jayanthi V, Obulakshmi S, Bhaskar SS, Lesley N. Nutritional status in renal transplant recipients. *Saudi J Kidney Dis Transpl.* 2007;18(3):382–386.

McCarthy MR. Nutrition and transplant—How to help patients on dialysis prepare. *Nephrol Nurs J.* 2006;33(5):570–572.

Nakai I, Omori Y, Aikawa I, Yasumura T, Suzuki S, Yoshimura N, Arakawa K, Matsui S, Oka T. Effect of cyclosporine on glucose metabolism in kidney transplant recipients. *Transplant Proc.* 1988;20(3 Suppl 3):969–978.

National Kidney Foundation. K/DOQI clinical practice guidelines for chronic kidney disease: Evaluation, classification, and stratification. *Am J Kidney Dis.* 2002;39(2 Suppl 1):S1–S266.

Nelms MN. Assessment of nutrition status and risk. In: Nelms M, Sucher K, Long S. *Nutrition Therapy and Pathophysiology.* Belmont, CA: Thomson/Brooks-Cole; 2007:101–135.

Nelson RG, Bennett PH, Beck GJ, Tan M, Knowler WC, Mitch WE, Hirschman GH, Myers BD. Development and progression of renal disease in Pima Indians with non-insulin-dependent diabetes mellitus. Diabetic Renal Disease Study Group. *N Engl J Med.* 1996;335(22):1636–1642.

Nelson RG, Meyer TW, Myers BD, Bennett PH. Course of renal disease in Pima Indians with non-insulin-dependent diabetes mellitus. *Kidney Int Suppl.* 1997;63:S45–S48.

Patel MG. The effect of dietary intervention on weight gain after renal transplantation. *J Renal Nutr.* 1998;8(3):137–141.

Pronsky ZM. *Powers and Moore's Food and Medication Interaction,* 15th ed. Birchrunville, PA: Food–Medication Interactions; 2008.

Pupim LB, Cuppari L, Ikizler TA. Nutrition and metabolism in kidney disease. *Semin Nephrol.* 2006;26(2):134–157.

Sancho A, Gavela E, Avila A, Morales A, Fernández-Nájera JE, Crespo JF, Pallardo LM. Risk factors and prognosis for proteinuria in renal transplant recipients. *Transplant Proc.* 2007;39(7):2145–2147.

Sezer S, Ozdemir FN, Afsar B, Colak T, Kizay U, Haberal M. Subjective global assessment is a useful method to detect malnutrition in renal transplant patients. *Transplant Proc.* 2006;38(2):517–512.

Sullivan SS, Anderson EJ, Best S, Sonnenberg LM, Williams WW. The effect of diet on hypercholesterolemia in renal transplant recipients. *J Renal Nutr.* 1996;6(3):141–151.

United Network for Organ Sharing. U.S. Transplantation Data. Organ Procurement and Transplantation Network, 2006. Available at: http://www.unos.org/data/default.asp?displayType=usData. Accessed February 21, 2008.

Wiggins KL. *Guidelines for Nutrition Care of Renal Patients,* 3rd ed. Chicago, IL: American Dietetic Association; 2002.

Internet Resources

Donate Life America. http://donatelife.net/

The National Institute of Diabetes and Digestive and Kidney Diseases (NIDDK). http://www2.niddk.nih.gov/

National Kidney Disease Education Program. http://www.nkdep.nih.gov/

Transplant Living. http://www.transplantliving.org/

The University of Texas Medical Branch: Glossary of transplant terms. http://www.utmb.edu/renaltx/gloss.htm

U.S. Renal Data System. http://www.usrds.org/

Unit Nine

NUTRITION THERAPY FOR HYPERMETABOLISM, INFECTION, AND TRAUMA

The physiological response to stress, trauma, and infection has been an important area of nutrition research for the past several decades. This metabolic response is characterized by catabolism of stored nutrients to meet the increased energy requirements.

Unlike other situations in which the body faces increased energy requirements, the stress response demands a preferential use of glucose for fuel. Because glycogen stores are quickly depleted, the body turns to lean body mass for glucose produced via gluconeogenesis.

Under the influence of counterregulatory hormones such as glucagon, epinephrine, norepinephrine, and cortisol, as well as cytokines such as interleukin and tumor necrosis factor, the body shifts from its normal state of anabolism to catabolism. All sources of fuel metabolism are affected by the stress response and the subsequent control of counterregulatory hormones. Despite increased lipolysis, there is not an effective use of fatty acids and glycerol as sources of fuel.

The body's inability to keep up with the rate of protein catabolism results in significant loss of skeletal muscle and high urinary losses of nitrogen. The liver's rate of gluconeogenesis is increased, and hyperglycemia is common. In addition, many tissues—especially skeletal tissue—develop insulin resistance, which contributes to the hyperglycemic state.

Nutrition support during these conditions is challenging, to say the least. Research indicates that both overfeeding and underfeeding can harm the patient. Advances in enteral and parenteral feeding have allowed refinement of this nutrition support practice, and today, medical nutrition therapy can certainly support the trauma patient appropriately and adequately.

Case 28 allows you to assess a patient with a closed head injury from a motor vehicle accident. Closed head injuries are an excellent example of the post-traumatic, hypermetabolic state. Determining nutritional needs, prescribing appropriate nutrition support, and monitoring daily progress are all addressed in this case. Other conditions resulting in metabolic stress include trauma, open wounds, and sepsis. These situations also demand close attention to nutrition support to minimize complications of protein-calorie malnutrition and to optimize recovery through medical nutrition therapy. You can easily transfer the same concepts for nutrition assessment and support to other individual cases you may encounter.

The second case in this section involves acquired immunodeficiency syndrome (AIDS). This condition may seem very different from closed head injury or trauma, but in many ways, the metabolic response is similar. Viral load, opportunistic infections, and the presence of wasting syndrome all can increase energy expenditure and shift substrate metabolism. Other issues for HIV and AIDS are included in this case. Drug–nutrient interactions, biochemical indices of viral load, and appropriate nutrition education are all crucial aspects of medical nutrition therapy for the patient with HIV and AIDS.

The final case in this section provides the opportunity to understand metabolic stress in the context of an acute trauma and the subsequent complications of an open abdominal wound. The complexities of hypermetabolism, nutrition support, and wound healing are essential components of this case.

Case 28

Traumatic Brain Injury: Metabolic Stress with Nutrition Support

Objectives

After completing this case, the student will be able to:

1. Discuss the pathophysiology of a closed head injury.
2. Describe the metabolic response to stress and trauma.
3. Determine nutrient, fluid, and electrolyte requirements for children.
4. Analyze nutrition assessment data to evaluate nutritional status and identify specific nutrition problems.
5. Determine nutrition diagnoses and write appropriate PES statements.
6. Calculate enteral nutrition formulations.
7. Evaluate a standard enteral nutrition regimen.

Chelsea Montgomery is a 9-year-old girl admitted through the emergency room after being injured as a restrained front-seat passenger in a motor vehicle accident. She is transferred to the neurointensive care unit with a traumatic brain injury.

Name: Chelsea Montgomery
DOB: 1/12 (age 9)
Physician: E. Mantio, MD

ADMISSION DATABASE

BED #	DATE:	TIME:	TRIAGE STATUS (ER ONLY):
2	5/24	1400	☒ Red ☐ Yellow ☐ Green ☐ White

Initial Vital Signs

TEMP:	RESP:	SAO$_2$:	
97	27		
HT:	**WT (lb):**	**B/P:**	**PULSE:**
4′4″	61	138/90	100
LAST TETANUS		**LAST ATE**	**LAST DRANK**
6 months ago		lunch today	?

PRIMARY PERSON TO CONTACT:
Name: Jacob and Melanie Montgomery
Home #: 334-555-5689
Work #: 334-555-3200

ORIENTATION TO UNIT: ☐ Call light ☐ Television/telephone ☐ Bathroom ☒ Visiting ☐ Smoking ☐ Meals ☐ Patient rights/responsibilities

CHIEF COMPLAINT/HX OF PRESENT ILLNESS

Admitted through ER–victim of high-speed MVA with head-on collision with truck. Victim was a restrained front seat passenger.

ALLERGIES: Meds, Food, IVP Dye, Seafood: Type of Reaction

NKA

PREVIOUS HOSPITALIZATIONS/SURGERIES

PERSONAL ARTICLES: (Check if retained/describe)
☐ Contacts ☐ R ☐ L ☐ Dentures ☐ Upper ☐ Lower
☐ Jewelry:
☐ Other:

VALUABLES ENVELOPE:
☐ Valuables instructions

INFORMATION OBTAINED FROM:
☐ Patient ☐ Previous record
☒ Family ☒ Responsible party

Signature: Melanie Montgomery

Home Medications (including OTC) **Codes: A=Sent home** **B=Sent to pharmacy** **C=Not brought in**

Medication	Dose	Frequency	Time of Last Dose	Code	Patient Understanding of Drug
multivitamin	1	daily			

Do you take all medications as prescribed? ☐ Yes ☒ No If no, why? "Sometimes I can't afford to buy them."

PATIENT/FAMILY HISTORY

☐ Cold in past two weeks
☐ Hay fever
☐ Emphysema/lung problems
☐ TB disease/positive TB skin test
☐ Cancer
☐ Stroke/past paralysis
☒ Heart attack Paternal grandfather
☐ Angina/chest pain
☒ Heart problems Paternal grandfather
☒ High blood pressure Father
☐ Arthritis
☐ Claustrophobia
☐ Circulation problems
☐ Easy bleeding/bruising/anemia
☐ Sickle cell disease
☐ Liver disease/jaundice
☐ Thyroid disease
☒ Diabetes Brother
☐ Kidney/urinary problems
☐ Gastric/abdominal pain/heartburn
☐ Hearing problems
☐ Glaucoma/eye problems
☐ Back pain
☐ Seizures
☐ Other

RISK SCREENING

Have you had a blood transfusion? ☒ Yes ☐ No
Do you smoke? ☐ Yes ☒ No
If yes, how many pack(s)?
Does anyone in your household smoke? ☐ Yes ☒ No
Do you drink alcohol? ☐ Yes ☒ No
If yes, how often?_______ How much?
When was your last drink? _______/_______/_______
Do you take any recreational drugs? ☐ Yes ☒ No
If yes, type:_______ Route:
Frequency:_______ Date last used:_______/_______/_______

FOR WOMEN Ages 12–52

Is there any chance you could be pregnant? ☐ Yes ☐ No
If yes, expected date (EDC):
Gravida/Para:

ALL WOMEN

Date of last Pap smear:
Do you perform regular breast self-exams? ☐ Yes ☐ No

ALL MEN

Do you perform regular testicular exams? ☐ Yes ☐ No

Additional comments:

✘ Ginger Syler, RN
Signature/Title

Client name: Chelsea Montgomery
DOB: 1/12
Age: 9
Sex: Female
Education: Less than high school *What grade/level?* 3rd grade
Occupation: Student
Hours of work: N/A
Household members: Mother age 36; father age 37; brother age 11 with type 1 DM
Ethnic background: Caucasian
Religious affiliation: Catholic
Referring physician: Elizabeth Mantio, MD (intensive care)

Chief complaint:
Admitted through ER after high-speed MVA—head-on collision with truck. Chelsea was a restrained front seat passenger.

Patient history:
Onset of disease: N/A
Type of Tx: N/A
PMH: Full-term infant weighing 9 lbs 1 oz, delivered via cesarean. Healthy except for severe myopia. Good student; competitive gymnast, softball player, and participant in Girl Scouts and after-school program.
Meds: None
Smoker: No
Family Hx: What? CAD *Who?* Paternal grandfather; Diabetes—older brother

Physical exam:
General appearance: 9-year-old female child alternating between crying and unconsciousness
Vitals: Temp: 97°F, BP 138/90 mm Hg, HR 100 bpm, RR 27 bpm
Heart: RRR, nl S1–S2, tachycardia, no murmur
HEENT:
Eyes: Pupils 4 mm reactive; no battle/raccoon signs
Ears: WNL
Nose: WNL
Throat: WNL
Genitalia: +Rectal tone, heme negative
Neurologic: GCS = 10 E4 V2 M4. Obtundation and L-sided hemiparesis. No verbal responses. Withdrawal and moaning when touched.
Extremities: DTR symmetric, WNL. 3+ lower extremities; 2+ R biceps; 1+ biceps. 2-cm laceration on R knee.
Skin: WNL
Chest/lungs: Breath sounds bilaterally
Peripheral vascular: No ankle edema
Abdomen: Soft; bowel sounds diminished, linear mark in LUQ, + guarding throughout

Nutrition Hx:
General: Parents indicate that patient had normal growth and appetite PTA.

Usual dietary intake:
Breakfast: Cereal, juice, milk, toast
Lunch: At school cafeteria
Snacks: Prior to gymnastics or softball practice: cookies, fruit, juice, or milk
Dinner: Meat, pasta or potatoes, rolls or bread. Likes only green beans, corn, and salad as vegetables. Will eat any fruit.

24-hour recall: NPO
Food allergies/intolerances/aversions: NKA
Previous nutrition therapy? No
Food purchase/preparation: Parents
Vit/min intake: General multivitamin with iron

Nutrition consult (excerpt from nutrition assessment note):
Recommendations for enteral feeding: Nutren Jr. with fiber @ 25 cc/hr. ↑ 10 cc every 4–6 hrs to goal rate of 85 cc/hr via continuous drip × 16 hrs then gradually switch to bolus as patient tolerates. Start bolus q 4 hrs @ 60 cc; then ↑ 120 cc; then ↑ 340 cc. Suggest to ↓ IVF as TF ↑.
(Signed) P. Marietta, MS, RD

Dx:
Closed head injury secondary to MVA

Tx plan:
Admit to Neurointensive Care Unit
D_5 0.9 NS with 10 mEq KCl
Zantac 25 mg every 6 hrs; Tylenol 450 mg every 6 hrs; ibuprofen 200 mg every 6 hrs; Zofran 2 mg IV every 6 hrs.
NPO
Nasogastric tube to low wall suction
O_2 to keep sat > 95%
I/O
Foley to gravity

Hospital course:
By day 4, aroused easily—automatic speech of "No-no-no." One-level commands followed. Oriented to parents, but not place or time. CT and MRI completed. Rehabilitation consult. Nutrition consult on day 3 for nutrition support recommendations. Patient began PO on hospital day 14. Weaned from enteral feeding completely on hospital day 17. During hospitalization, patient had extensive physical, speech, and occupational therapy. Patient discharged on hospital day 21 with orders for patient direct supervision 24 hrs/day, 7 days a week, with gradual removal of restrictions as clinically indicated. Patient to receive PT weekly; OT 3–5 × week and speech therapy 3–5 × week.

DEPARTMENT OF RADIOLOGY

CT Report

Date: 5/24

Patient: Chelsea Montgomery

DOB: 1/12 (age 9)

Physician: Elizabeth Mantio, MD

Two areas of increased density in L frontal lobe near vertex and possibly left central modality.

Victoria Roundtree, MD

Department of Radiology

DEPARTMENT OF RADIOLOGY

MRI Report

Date: 5/29

Patient: Chelsea Montgomery

DOB: 1/12 (age 9)

Physician: Elizabeth Mantio, MD

MRI showed areas of hemorrhagic edema in deep white matter of L frontal lobe anteriorly. Additionally, heme and edema found in the splenium of corpus callosum. 3.4 cm × 4.2 cm × 1.0 cm representing areas of shearing injury.

James Morgan, MD

Department of Radiology

5/29

DEPARTMENT OF SPEECH PATHOLOGY

RE: Interpretation of video fluoroscopy and speech/swallowing evaluation

Date: 6/3 Hospital day 10

Patient: Chelsea Montgomery

DOB: 1/12 (age 9)

Physician: Elizabeth Mantio, MD

Patient accepted macaroni and cheese with appropriate tongue lateralization and chewing skills but choked after 5–7 ice chips. Oral skills appropriate. Showed significant signs of fatigue and decreased cooperation after a few swallows, which therefore inhibited PO feeding. Video swallow studies showed no evidence of penetration or aspiration.

Carol Davie, MS, SLP

UH UNIVERSITY HOSPITAL

NAME: Chelsea Montgomery　　DOB: 1/12
AGE: 9　　SEX: F
PHYSICIAN: E. Mantio, MD

CHEMISTRY

	NORMAL			UNITS
DAY:		1	10	
DATE:		5/24	6/3	
TIME:				
LOCATION:				
Albumin	3.5-5	3.7	3.3 L	g/dL
Total protein	6-8	6.4		g/dL
Prealbumin	16-35		15 L	mg/dL
Transferrin	250-380 (women) 215-365 (men)			mg/dL
Sodium	136-145	142	139	mEq/L
Potassium	3.5-5.5	3.9	3.6	mEq/L
Chloride	95-105	101	105	mEq/L
PO_4	2.3-4.7	4.2	3.8	mg/dL
Magnesium	1.8-3	2.1	2.2	mg/dL
Osmolality	285-295	291	289	mmol/kg/H_2O
Total CO_2	23-30	28	29	mEq/L
Glucose	70-110	145 H	109	mg/dL
BUN	8-18	8	10	mg/dL
Creatinine	0.6-1.2	0.6	0.7	mg/dL
Uric acid	2.8-8.8 (women) 4.0-9.0 (men)			mg/dL
Calcium	9-11	9.1	9.9	mg/dL
Bilirubin	≤0.3	0.29	0.3	mg/dL
Ammonia (NH_3)	9-33			μmol/L
ALT	4-36	105 H	34	U/L
AST	0-35	111 H	36 H	U/L
Alk phos	30-120	261 H	119	U/L
CPK	30-135 (women) 55-170 (men)			U/L
LDH	208-378			U/L
CHOL	120-199			mg/dL
HDL-C	>55 (women) >45 (men)			mg/dL
VLDL	7-32			mg/dL
LDL	<130			mg/dL
LDL/HDL ratio	<3.22 (women) <3.55 (men)			
Apo A	101-199 (women) 94-178 (men)			mg/dL
Apo B	60-126 (women) 63-133 (men)			mg/dL
TG	35-135 (women) 40-160 (men)			mg/dL
T_4	4-12			mcg/dL
T_3	75-98			mcg/dL
HbA_{1C}	3.9-5.2			%

Name: Chelsea Montgomery
Physician: E. Mantio, MD

PATIENT CARE SUMMARY SHEET

Date: 6/5 Room: NICU Bed 3 Wt Yesterday: 25.5 kg Today: 25.2 kg

Temp °F	NIGHTS								DAYS								EVENINGS							
	00	01	02	03	04	05	06	07	08	09	10	11	12	13	14	15	16	17	18	19	20	21	22	23
105																								
104																								
103																								
102																								
101																								
100																								
99																								
98																								
97																								
96																								
Pulse	108								94								100							
Respiration	20								20								24							
BP	100/51								121/62								124/72							
Blood Glucose	NG								NG								NG							
Appetite/Assist																								
INTAKE																								
Oral																								
IV																								
TF Formula/Flush	85	85	85	85	85	85	85	85/30	85	85	85	85	85	85	85	85	85	85	*	*	*	*	*	50
Shift Total	680 TF + 30 flush								220															
OUTPUT								INC																
Void		50						50							100		70		100					
Cath.																								
Emesis																								
BM								1—soft																
Drains																								
Shift Total	incontinent								incontinent								incontinent							
Gain																								
Loss																								
Signatures	J. Welter, RN								A. Koch, RN								K. Vangilder, RN							

Case Questions

I. Understanding the Disease and Pathophysiology

1. What is the Glasgow Coma Scale (GCS)?

2. What was Chelsea's initial GCS score? Is anything in the initial physical assessment consistent with this score? Explain.

3. Define the following terms found in the admitting history and physical.

   a. *Intensivist:*

   b. *L-sided hemiparesis:*

4. Read the CT scan and MRI report. The CT scan report was very general, noting density in the frontal lobe. The MRI indicated more-localized areas of edema and blood in the frontal lobe. It also discusses a shearing injury.

   a. What causes edema and bleeding in a traumatic brain injury?

   b. What general functions occur in the frontal lobe? How might Chelsea's injury affect her in the long term?

5. Define *metabolic stress.*

6. The stress response has been described as a progression through three phases. Describe each.

II. Understanding the Nutrition Therapy

7. Explain the metabolic changes in nutrient metabolism (energy requirements, carbohydrate, protein, and lipid) that occur with a traumatic brain injury as a result of the stress response.

8. What additional factors may place the patient with traumatic brain injury at nutritional risk?

III. Nutrition Assessment

A. Evaluation of Weight/Body Composition

9. Chelsea's height is 132 cm, and her weight on admission is 27.7 kg. At 9 years of age, what is the most appropriate method to evaluate her height and weight? Assess her height and weight.

B. Calculation of Nutrient Requirements

10. What method should you use to determine Chelsea's energy and protein requirements? Be sure to consider her age and injury. After specifying your method, determine her energy and protein needs.

C. Intake Domain

11. Chelsea was to receive a goal rate of Nutren Jr. with fiber @ 85 cc/hr. How much energy and protein would this provide? Show your calculations. Does it meet her needs?

12. What is the rationale for using enteral nutrition rather than parenteral nutrition support for Chelsea?

13. Using the patient care summary sheet, answer the following:

    a. What was the total volume of feeding she received on June 5?

    b. What was the nutritional value of her feeding for that day? Calculate the total energy and protein.

    c. What percentage of her needs was met?

    d. There is a note on the evening shift that the feeding was held for high residual. What does that mean?

e. What is aspiration? What are the potential consequences?

f. What is the usual procedure for handling a high gastric residual? How do you think Chelsea's situation was handled?

g. What other information on the patient care summary sheet would you assess to determine her tolerance to the enteral feeding?

h. Look at the additional information on the patient care summary sheet. Are there any factors of concern? Explain.

14. From the information gathered within the intake domain, list possible nutrition problems using the diagnostic term.

D. Clinical Domain

15. Evaluate Chelsea's laboratory data. Note any changes from admission day labs to June 3. Are any changes of nutritional concern?

16. On June 6, a 24-hour urine sample was collected for nitrogen balance. On this day, she received 1,650 cc of Nutren Jr. Her total nitrogen output was 14 g.

a. Calculate her nitrogen balance from this information. Show all your calculations.

b. How would you assess this information? Explain your response in the context of her hypermetabolism.

c. Are there any factors that may affect the accuracy of this test?

d. The intern taking care of Chelsea pages you when he reads your note regarding her negative nitrogen balance. He asks whether he should change the enteral formula to one higher in nitrogen. Explain the results in the context of the metabolic stress response.

17. From the information gathered within the clinical domain, list possible nutrition problems using the diagnostic term.

IV. Nutrition Diagnosis

18. Select two nutrition problems and complete the PES statement for each.

V. Nutrition Intervention

19. For each of the PES statements that you have written, establish an ideal goal (based on the signs and symptoms) and an appropriate intervention (based on the etiology).

VI. Nutrition Monitoring and Evaluation

20. Chelsea has worked with an occupational therapist, speech therapist, and physical therapist. Summarize the training that each of these professionals receives and what their role might be for Chelsea's rehabilitation.

21. The speech pathologist saw Chelsea for a swallowing evaluation on hospital day 10.

a. What is a video fluoroscopy?

b. What factors were noted that support the need for enteral feeding at this time?

22. Are there additional nutrients that may assist Chelsea in her recovery? Should any specialized enteral products be used?

23. As Chelsea's recovery proceeds, she begins a PO mechanical soft diet. Her calorie counts are as follows: (10/14) oatmeal ¼ c; brown sugar 2 tbsp; whole milk 1 c; 240 cc Carnation Instant Breakfast (CIB) prepared with 2% milk; mashed potatoes 1 c; gravy 2 tbsp (10/15) Cheerios 1 c; whole milk 1 c; 240 cc CIB prepared with 2% milk; grilled cheese sandwich (2 slices bread, 1 oz American cheese, 1 tsp margarine); Jell-O 1 c; 240 cc CIB prepared with 2% milk

a. Calculate her intake and average for these 2 days of calorie counts.

b. What recommendations would you make regarding her enteral feeding?

Bibliography

ASPEN Board of Directors and the Clinical Guidelines Task Force. Guidelines for the use of parenteral and enteral nutrition in adult and pediatric patients. *J Parenter Enteral Nutr.* 2002;26(1 Suppl):1SA–138SA.

American Dietetic Association. Critical Illness Evidence-Based Nutrition Guideline. Available at: http://www.adaevidencelibrary.com/topic.cfm?cat=3016. Accessed November 9, 2007.

Brain Trauma Task Force. Management and prognosis of severe traumatic brain injury. Available at: http://www.braintrauma.org/site/PageServer?pagename=Guidelines&JServSessionIdr007=2nsf2flu95.app6b. Accessed April 30, 2008.

Ghajar J. Traumatic brain injury. *Lancet.* 2000;356(9233):923–929.

Jumisko E, Lexell J, Soderberg S. The meaning of living with traumatic brain injury in people with moderate or severe traumatic brain injury. *J Neurosci Nurs.* 2005;37(1):42–50.

Marion D. Evidenced-based guidelines for traumatic brain injuries. In: Pollock BE, ed. *Guiding Neurosurgery by Evidence. Progress in Neurological Surgery.* Basel, Switzerland: Karger; 2006:171–196.

Marshall LF, Gautille T, Klauber MR, Eisenberg HM, Jane JA, Luerssen TG, Marmarou A, Foulkes MA. The outcome of severe closed head injury. *Neurosurgery.* 1991;75:S28–S36.

Nelms MN. Assessment of nutrition status and risk. In: Nelms M, Sucher K, Long S. *Nutrition Therapy and Pathophysiology.* Belmont, CA: Thomson/Brooks-Cole; 2007:101–135.

Nelms MN. Diseases of the neurological system. In: Nelms M, Sucher K, Long S. *Nutrition Therapy and Pathophysiology.* Belmont, CA: Thomson/Brooks-Cole; 2007:687–714.

Nelms MN. Metabolic stress. In: Nelms M, Sucher K, Long S. *Nutrition Therapy and Pathophysiology.* Belmont, CA: Thomson/Brooks-Cole; 2007:785–804.

Skipper A, Nelms MN. Methods of nutrition support. In: Nelms M, Sucher K, Long S. *Nutrition Therapy and Pathophysiology.* Belmont, CA: Thomson/Brooks-Cole; 2007:149–179.

Teasdale G, Jennett B. Assessment of coma and impaired consciousness: A practical scale. *Lancet.* 1974;2(7872):81–84.

Vincent JL, Berre J. Primer on management of severe brain injury. *Critical Care Med.* 2005;33:1392–1399.

Internet Resources

Brain Trauma Foundation. http://www.braintrauma.org

National Center for Injury Prevention and Control: What Is Traumatic Brain Injury? http://www.cdc.gov/ncipc/tbi/TBI.htm

National Institute of Neurological Disorders and Stroke: NINDS Traumatic Brain Injury Information Page. http://www.ninds.nih.gov/disorders/tbi/tbi.htm

Traumatic Brain Injury: What Is Traumatic Brain Injury (TBI)? http://www.traumaticbraininjury.com

Case 29

AIDS

Objectives

After completing this case, the student will be able to:

1. Apply knowledge of the pathophysiology of HIV infection to identify and explain common nutritional problems associated with HIV and acquired immunodeficiency syndrome (AIDS).
2. Identify nutritional risk factors for the patient with AIDS.
3. Analyze nutrition assessment data to evaluate nutritional status and identify specific nutrition problems.
4. Determine nutrition diagnoses and write appropriate PES statements
5. Develop a nutrition care plan with appropriate measurable goals, interventions, and strategies for monitoring and evaluation that addresses the nutrition diagnoses of this case.
6. Determine potential drug–nutrient interactions and appropriate interventions.
7. Identify key components of nutrition education for the patient with AIDS.
8. Evaluate risks and current recommendations for nutritional supplementation.

Mr. Terry Long, a 32-year-old African American male, is admitted with probable pneumonia and progression to AIDS. Mr. Long was diagnosed HIV positive 4 years ago and has not received any treatment previously.

Name: Terry Long
DOB: 5/12 (age 32)
Physician: A. Fremont, MD

ADMISSION DATABASE

BED #	DATE:	TIME:	TRIAGE STATUS (ER ONLY):
2	10/17	1400	☐ Red ☐ Yellow ☐ Green ☐ White

Initial Vital Signs

TEMP:	RESP:	SAO$_2$:	
98.6	18		
HT: 6′1″	WT (lb): 151	B/P: 120/84	PULSE: 92
LAST TETANUS: > 10 years ago		LAST ATE: 12:00 today	LAST DRANK: 12:00 today

PRIMARY PERSON TO CONTACT:
Name: Fred and Marie Long
Home #: 312-555-4456
Work #: N/A

ORIENTATION TO UNIT: ☒ Call light ☒ Television/telephone ☒ Bathroom ☒ Visiting ☒ Smoking ☒ Meals ☒ Patient rights/responsibilities

CHIEF COMPLAINT/HX OF PRESENT ILLNESS

"I was diagnosed with HIV 4 years ago when I was living in St. Louis. I just recently moved back home because I am not able to work right now."

PERSONAL ARTICLES: (Check if retained/describe)
☐ Contacts ☐ R ☐ L ☐ Dentures ☐ Upper ☐ Lower
☐ Jewelry:
☐ Other:

ALLERGIES: Meds, Food, IVP Dye, Seafood: Type of Reaction

NKA

VALUABLES ENVELOPE:
☐ Valuables instructions

PREVIOUS HOSPITALIZATIONS/SURGERIES

tonsillectomy-age 6
appendectomy-age 18
hernia repair-age 26

INFORMATION OBTAINED FROM:
☒ Patient ☐ Previous record
☐ Family ☐ Responsible party

Signature: Terry Long

Home Medications (including OTC) **Codes: A=Sent home** **B=Sent to pharmacy** **C=Not brought in**

Medication	Dose	Frequency	Time of Last Dose	Code	Patient Understanding of Drug
multivitamin	1	daily	1 week ago	C	yes
ginseng	500 mg	daily	1 week ago	C	no
milk thistle	200 mg	twice daily	1 week ago	C	no
echinacea	88.5 mg capsule	three X daily	1 week ago	C	no
vitamin E	1500 IU	daily	1 week ago	C	no
vitamin C	500 mg	four X daily	1 week ago	C	no
St. John's wort	300 mg	daily	1 week ago	C	no

Do you take all medications as prescribed? ☒ Yes ☒ No If no, why? "Sometimes I can't afford to buy them."

PATIENT/FAMILY HISTORY

☐ Cold in past two weeks
☐ Hay fever
☐ Emphysema/lung problems
☐ TB disease/positive TB skin test
☐ Cancer
☐ Stroke/past paralysis
☐ Heart attack
☒ Angina/chest pain Father
☒ Heart problems Father
☒ High blood pressure Father
☐ Arthritis
☐ Claustrophobia
☐ Circulation problems
☐ Easy bleeding/bruising/anemia
☐ Sickle cell disease
☐ Liver disease/jaundice
☐ Thyroid disease
☐ Diabetes
☐ Kidney/urinary problems
☐ Gastric/abdominal pain/heartburn
☐ Hearing problems
☐ Glaucoma/eye problems
☐ Back pain
☐ Seizures
☒ Other HIV positive-Patient

RISK SCREENING

Have you had a blood transfusion? ☒ Yes ☐ No
Do you smoke? ☐ Yes ☒ No
If yes, how many pack(s)?
Does anyone in your household smoke? ☐ Yes ☒ No
Do you drink alcohol? ☒ Yes ☐ No
If yes, how often? 3-4 x/week How much? 2-3 drinks
When was your last drink? 10/12
Do you take any recreational drugs? ☒ Yes ☐ No
If yes, type: marijuana Route:
Frequency: occasionally Date last used: last summer

FOR WOMEN Ages 12–52

Is there any chance you could be pregnant? ☐ Yes ☐ No
If yes, expected date (EDC):
Gravida/Para:

ALL WOMEN

Date of last Pap smear:
Do you perform regular breast self-exams? ☐ Yes ☐ No

ALL MEN

Do you perform regular testicular exams? ☐ Yes ☐ No

Additional comments:

× Jessie Farmer, RN
Signature/Title

Client name: Terry Long
DOB: 5/12
Age: 32
Sex: Male
Education: Bachelor's degree
Occupation: Currently on disability but previously worked as nurse in dialysis clinic
Hours of work: N/A
Household members: Father age 69, mother age 66, both well
Ethnic background: African American
Religious affiliation: AME (African Methodist Episcopal)
Referring physician: Agnes Fremont, MD (family medicine/internal medicine)

Chief complaint:
"I was diagnosed with HIV 4 years ago when I was living in St. Louis. I just recently moved back home because I am not able to work right now. I have not been treated before, but I am pretty sure I will need to be. I feel exhausted all the time—I have a really sore mouth and throat. I have lost a lot of weight. I think I've just been denying that I may have AIDS. But a lot of people I know are doing OK on drugs, so I came to this new physician. The case manager at the Health Department set it up for me. Dr. Fremont thinks that I may have pneumonia as well, so she admitted me for a full workup."

Patient history:
Onset of disease: Seropositive for HIV-1 confirmed by ELISA and Western blot 4 years previously. Etiology of contraction not known but was employed in high-risk environment. Admits to intercourse with multiple partners but denies same-sex intercourse.
Type of Tx: None
PMH: Tonsillectomy age 6; appendectomy age 18; hernia repair age 26
Meds: Multivitamin, vitamin E (1,500 IU), vitamin C (500 mg twice daily), ginseng (500 mg daily), milk thistle (200 mg twice daily), echinacea (three capsules every day); St. John's wort (300 mg daily).
Smoker: No—quit 5 years ago
Family Hx: What CAD, HTN *Who?* Father

Physical exam:
General appearance: Thin African American male in no acute distress
Vitals: Temp 98.6°F, BP 120/84 mm Hg, HR 92 bpm/normal, RR 18 bpm
Heart: Regular rate and rhythm—normal heart sounds
HEENT:
Eyes: PERRLA
Ears: Unremarkable
Nose: Mucosa pink without drainage
Throat: Erythematous with white, patchy exudate
Genitalia: Rectal exam normal; stool: heme negative
Neurologic: Oriented × 3, no focal motor or sensory deficits, cranial nerves intact, DTR +2 in all groups
Extremities: Good pulses, no edema
Skin: Warm, dry, with flaky patches

Chest/lungs: Rhonchi in lower left lung
Abdomen: Nondistended, nontender, hyperactive bowel sounds

Nutrition Hx:
General: Patient describes appetite as OK, but not normal. "I have always been a picky eater. There are a lot of foods that I don't like. But in the last few days, it is the sores in my mouth and throat that have made the biggest difference. It hurts pretty badly, and I can hardly even drink. I have been reading about nutrition and HIV on the Internet—I've been trying to do some research. That's when I started taking more supplements. I thought if I wasn't eating like I should that I could at least take supplements. They are expensive, though, so I don't have them every day like I probably should. My highest weight ever was about 175 lbs, which was during college almost 10 years ago. But I have never been this thin as an adult."
Anthropometrics: MAC 10″; TSF 7 mm; body fat 12.5%, Ht 6′1″, Wt 151 lbs, UBW 160–165 lbs

Usual dietary intake (before mouth sores):

Breakfast/lunch: ("I usually don't get up before noon because I stay up really late.") cold cereal 1–2 c, ½ c whole milk

Supper: Meat—2 pork chops or other type of meat (except beef); mashed potatoes—1 c, rice or pasta, tea or soda

Snacks: Pizza, candy bar, or cookies with tea or soda. Drinks 1–2 beers or glass of wine several times a week.

24-hour recall: Sips of apple juice, pudding 1 c, rice and gravy 1 c, iced tea with sugar—sips throughout the day
Food allergies/intolerances/aversions (specify): Can only tolerate small amounts of milk at a time; does not like beef, coffee, or vegetables (except salad)
Previous nutrition therapy? No
Food purchase/preparation: Parent(s), self
Vit/min intake: Multivitamin 1 daily, vitamin E 1,500 IU daily, vitamin C 500 mg four times daily, ginseng 500 mg twice daily, milk thistle 200 mg twice daily, echinacea 3 capsules daily (88.5 mg per capsule); St. John's wort 300 mg daily

Dx:
AIDS—clinical category C2 with oral thrush; no clinical evidence of pneumonia; HAART regimen initiated with indinavir, stavudine, and didanosine.

Tx plan:
Admit: AIDS, oral candidasis, R/O pneumonia; CXR, WBC with diff, CD4, and viral load; begin $D_5$1/2 NS @ 100 cc/hr; fluconazole IV

UH UNIVERSITY HOSPITAL

NAME: Terry Long DOB: 5/12
AGE: 32 SEX: M
PHYSICIAN: A. Fremont, MD

CHEMISTRY

DAY: Admit
DATE: 10/17
TIME:
LOCATION:

	NORMAL	Admit 10/17	UNITS
Albumin	3.5-5	3.6	g/dL
Total protein	6-8	6.0	g/dL
Prealbumin	16-35	15 L	mg/dL
Transferrin	250-380 (women) 215-365 (men)	217	mg/dL
Sodium	136-145	142	mEq/L
Potassium	3.5-5.5	3.6	mEq/L
Chloride	95-105	101	mEq/L
PO_4	2.3-4.7	3.2	mg/dL
Magnesium	1.8-3	1.8	mg/dL
Osmolality	285-295	292	mmol/kg/H_2O
Total CO_2	23-30	27	mEq/L
Glucose	70-110	75	mg/dL
BUN	8-18	11	mg/dL
Creatinine	0.6-1.2	0.8	mg/dL
Uric acid	2.8-8.8 (women) 4.0-9.0 (men)	5.2	mg/dL
Calcium	9-11	9.1	mg/dL
Bilirubin	≤0.3	0.9 H	mg/dL
Ammonia (NH_3)	9-33	18	μmol/L
ALT	4-36	12	U/L
AST	0-35	17	U/L
Alk phos	30-120	102	U/L
CPK	30-135 (women) 55-170 (men)	110	U/L
LDH	208-378	710 H	U/L
CHOL	120-199	150	mg/dL
HDL-C	>55 (women) >45 (men)	42 L	mg/dL
VLDL	7-32		mg/dL
LDL	<130	114	mg/dL
LDL/HDL ratio	<3.22 (women) <3.55 (men)		
Apo A	101-199 (women) 94-178 (men)		mg/dL
Apo B	60-126 (women) 63-133 (men)		mg/dL
TG	35-135 (women) 40-160 (men)	78	mg/dL
T_4	4-12		mcg/dL
T_3	75-98		mcg/dL
HbA_{1C}	3.9-5.2		%

UH UNIVERSITY HOSPITAL

NAME: Terry Long
AGE: 32
PHYSICIAN: A. Fremont, MD

DOB: 5/12
SEX: M

HEMATOLOGY

DAY: Admit
DATE: 10/17
TIME:
LOCATION:

	NORMAL		UNITS
WBC	4.8–11.8	8.5	$\times 10^3/mm^3$
RBC	4.2–5.4 (women) 4.5–6.2 (men)	5.2	$\times 10^6/mm^3$
HGB	12–15 (women) 14–17 (men)	14.2	g/dL
HCT	37–47 (women) 40–54 (men)	40	%
MCV	80–96	96	μm^3
RETIC	0.8–2.8		%
MCH	26–32	34.2 H	pg
MCHC	31.5–36	35.5	g/dL
RDW	11.6–16.5	16.3	%
Plt Ct	140–440	220	$\times 10^3/mm^3$
Diff TYPE			
ESR	0–25 (women) 0–15 (men)	18 H	mm/hr
% GRANS	34.6–79.2	82 H	%
% LYM	19.6–52.7	3 L	%
SEGS	50–62	51	%
BANDS	3–6	4	%
LYMPHS	24–44	3 L	%
MONOS	4–8	10 H	%
EOS	0.5–4	3	%
Ferritin	20–120 (women) 20–300 (men)		mg/mL
ZPP	30–80		µmol/mol
Vitamin B_{12}	24.4–100		ng/dL
Folate	5–25		µg/dL
Viral Load	0	29,000 H	mm^3
T Cells	800–2,500	255 L	mm^3
T Helper (CD4+)	600–1,500	153 L	mm^3
T Suppressor (CD8+)	300–1,000	102 L	mm^3
PT	11–16	11.9	sec

Case Questions

I. Understanding the Disease and Pathophysiology

1. How is HIV transmitted? Based on Mr. Long's history and physical, what risk factors for contracting HIV might he have had?

2. The history and physical indicate that he is seropositive. What does that mean? The Western blot and ELISA confirmed that he was seropositive. Describe these tests.

3. Mr. Long says he found out he was HIV positive 4 years ago. Why has he only now become symptomatic?

4. What is thrush, and why might Mr. Long have this condition?

5. After this admission, Mr. Long was diagnosed with AIDS, category C2. What information from his medical record confirms this diagnosis?

II. Understanding the Nutrition Therapy

6. What are common nutritional complications of HIV and AIDS?

7. Are there specific recommendations for energy, protein, vitamin, and mineral requirements for someone with AIDS?

III. Nutrition Assessment

A. Evaluation of Weight/Body Composition

8. Evaluate the patient's anthropometric information.
   - **a.** Calculate percent UBW and BMI.

   - **b.** Compare the TSF to population standards. What does this comparison mean? Is this a viable comparison? Explain.

c. Using MAC and TSF, calculate upper arm muscle area. What can you infer from this calculation?

d. Mr. Long's body fat percentage is 12.5%. What does this mean? Compare to standards.

9. From the information gathered after assessing weight and body composition, list possible nutrition problems using the diagnostic term.

B. Calculation of Nutrient Requirements

10. Determine Mr. Long's energy and protein requirements.

C. Intake Domain

11. Evaluate Mr. Long's dietary information. What tools could you use to evaluate his dietary intake?

12. Does he seem to be consuming adequate amounts of food? Can you identify anything from his history that indicates he is having difficulty eating? Explain.

13. Mr. Long states that he consumes alcohol several times a week. Are there any contraindications for alcohol consumption for him?

14. From the information gathered within the intake domain, list possible nutrition problems using the diagnostic term.

D. Clinical Domain

15. Using this patient's laboratory values, identify those labs used to monitor his disease status. What do these specifically measure, and how would you interpret them for him? Explain how the virus affects these laboratory values.

16. What laboratory values can be used to evaluate nutritional status? Do any of Mr. Long's values indicate nutritional risk?

17. Mr. Long was started on three medications that he will be discharged on.

a. Identify these medications and the purpose of each.

b. Are there any specific drug–nutrient interactions to be concerned about? Explain.

c. Is there specific nutrition information you would want Mr. Long to know about taking these medications?

18. From the information gathered within the clinical domain, list possible nutrition problems using the diagnostic term.

E. Behavioral–Environmental Domain

19. Mr. Long is taking several vitamin and herbal supplements. Find out why someone with AIDS might take each of the supplements. What would you tell Mr. Long about these supplements? Do they pose any risk? Use the following table to organize your answers.

Supplement	Proposed Use in HIV/AIDS	Potential Risk
Vitamin C		
Vitamin E		
Milk thistle		
Ginseng		
Echinacea		
St. John's wort		
Multivitamin		

20. From the information gathered within the behavioral–environmental domain, list possible nutrition problems using the diagnostic term.

IV. Nutrition Diagnosis

21. Select two high-priority nutrition problems and complete the PES statement.

V. Nutrition Intervention

22. For each of the PES statements that you have written, establish an ideal goal (based on the signs and symptoms) and an appropriate intervention (based on the etiology).

23. Identify three interventions you would recommend for modifying Mr. Long's tolerance of food until his oral thrush has subsided.

24. Describe at least two areas of nutrition education that you would want to ensure that Mr. Long receives. Explain your rationale for these choices.

25. Patients with AIDS are at increased risk for infection. What nutritional practices would you teach Mr. Long to help him prevent illness related to food or water intake?

26. Why is exercise important as a component of the nutritional care plan? What general recommendations could you give to Mr. Long regarding physical activity?

VI. Nutrition Monitoring and Evaluation

27. Lipodystrophy syndrome has been associated with AIDS. Define this condition and describe the most common signs and symptoms.

28. How would the clinician monitor Mr. Long for the development of this disorder?

Bibliography

American Dietetic Association and the Dietitians of Canada. Nutrition intervention in the care of persons with human immunodeficiency virus infection—Position of the American Dietetic Association and Dietitians of Canada. *J Am Diet Assoc.* 2000;100:708–71.

American Dietetic Association. HIV/AIDS Nutrition Evidence Analysis Project. Available at: http://www.adaevidencelibrary.com/topic.cfm?cat=1404. Accessed September 5, 2007.

Batterham MJ, Morgan-Jones J, Greenop P, Garsia R, Gold J, Caterson I. Calculating energy requirements for men with HIV/AIDS in the era of highly active antiretroviral therapy. *Eur J Clin Nutr.* 2003;57:209–217.

Batterham MJ. Investigating heterogeneity in studies of resting energy expenditure in persons with HIV/AIDS: A meta-analysis. *Am J Clin Nutr.* 2005;81:702–713.

Carr A, Law M; HIV Lipodystrophy Case Definition Study Group. An objective lipodystrophy severity grading scale derived from the lipodystrophy case definition score. *J Acquir Immune Defic Syndr.* 2003;33(5):571–576.

Centers for Disease Control and Prevention. 1993 Revised Classification System for HIV Infection and Expanded Surveillance Case Definition for AIDS Among Adolescents and Adults. Available at: http://www.cdc.gov/mmwr/preview/mmwrhtml/00018871.htm. Accessed September 5, 2007.

Drug Facts and Comparisons 2004, 58th ed. St. Louis, MO: Drug Facts and Comparisons; 2003.

Engleson ES. HIV lipodystrophy diagnosis and management. Body composition and metabolic alterations: Diagnosis and management. *AIDS Reader.* 2003;13(4 Suppl):S10–S14.

Escott-Stump S. *Nutrition and Diagnosis Related Care,* 6th ed. Baltimore, MD: Lippincott Williams & Wilkins; 2007.

Fields-Gardner C. HIV and AIDS. In: Nelms M, Sucher K, Long S. *Nutrition Therapy and Pathophysiology.* Belmont, CA: Thomson/Brooks-Cole; 2007:805–842.

Hadigan C. Dietary habits and their association with metabolic abnormalities in human immunodeficiency virus-related lipodystrophy. *Clin Infect Dis.* 2003;37(Suppl 2):S101–S104.

Hayes C, Elliot E, Krales E, Downer G. Food and water safety for persons infected with human immunodeficiency virus. *Clin Infect Dis.* 2003;36:S106–S109

Jacobson DL, Bica I, Knox TA, Wanke C, Tchetgen E, Spiegelman D, Silva M, Gorbach S, Wilson IB. Difficulty swallowing and lack of receipt of highly active antiretroviral therapy predict acute weight loss in human immunodeficiency virus disease. *Clin Infect Dis.* 2003;37(10):1349–1356.

Kittler PG. Complementary and alternative medicine. In: Nelms M, Sucher K, Long S. *Nutrition Therapy and Pathophysiology.* Belmont, CA: Thomson/Brooks-Cole; 2007:65–97.

Mwamburi DM, Wilson IB, Jacobson DL, Spiegelman D, Gorbach SL, Knox TA, Wanke CA. Understanding the role of HIV load in determining weight change in the era of highly active antiretroviral therapy. *Clin Infect Dis.* 2005;40(1):167–173.

Nelson MH. HIV-protease inhibitors: Viral resistance, pharmacokinetic boosting, and medication adherence. *US Pharm.* 2006;1:HS-5-HS-12. Available at: http://www.uspharmacist.com/index.asp?show=article&page=8_1668.htm. Accessed September 5, 2007.

Pronsky ZM. *Food Medication Interactions,* 15th ed. Birchrunville, PA: Food-Medication Interactions; 2008.

Scevola D, Di Matteo A, Lanzarini P, Uberti F, Scevola S, Bernini V, Spoladore G, Faga A. Effect of exercise and strength training on cardiovascular status in HIV-infected patients receiving highly active antiretroviral therapy. *AIDS.* 2003;17(Suppl 1):S123–S129.

Suttmann U, Holtmannspotter M, Ockenga J, Gallati H, Deicher H, Selberg O. Tumor necrosis factor, interleukin-6, and epinephrine are associated with hypermetabolism in AIDS patients with acute opportunistic infections. *Ann Nutr Metab.* 2000;44(2):43–53.

Internet Resources

The Body. http://www.thebody.com/index.html

HIV Nutrition Resources. http://www.hivresources.com/

AIDS Education Global Information System. http://www.aegis.com/

HIV/AIDS Dietetic Practice Group of the American Dietetic Association. http://www.hivaidsdpg.org

U.S. Department of Agriculture: Nutrient Data Laboratory. http://www.ars.usda.gov/ba/bhnrc/ndl

Case 30

Metabolic Stress and Trauma

Deborah A. Cohen, MMSc, RD—
Southeast Missouri State University

Objectives:

After completing the case, the student will be able to:

1. Apply knowledge of the pathophysiology of trauma and metabolic stress in order to provide nutrition support for the critically ill patient.
2. Identify the basic components of indirect calorimetry.
3. State specific indications for the use of indirect calorimetry in critically ill patients.
4. Interpret the respiratory quotient (RQ).
5. Compare different predictive equations that are appropriate for use in the critically ill population and identify their indications.
6. Assess the benefits of utilizing enteral nutrition support in a patient on parenteral nutrition.
7. Understand the current research that interprets the role of polyunsaturated fats in the inflammatory process.
8. Determine and prioritize nutrition diagnoses and write a PES statement for a critically ill patient.

Juan Perez is a 29-year-old male admitted to the Trauma Intensive Care Unit with a gunshot wound to the upper abdomen. He experienced gastric, duodenal, and jejunal injuries, liver laceration, and a left pleural effusion.

Name: Juan Perez
DOB: 3/22 (age 29)
Physician: Deborah Kuhls, MD

ADMISSION DATABASE

BED #	DATE:	TIME:	TRIAGE STATUS (ER ONLY):
5	7/1	0730	☒ Red ☐ Yellow ☐ Green ☐ White

Initial Vital Signs

TEMP:	RESP:	SAO$_2$:	
39	22		
HT (in):	**WT (lb):**	**B/P:**	**PULSE:**
5′10″	225	115/65	90
LAST TETANUS		**LAST ATE**	**LAST DRANK**
unknown		unknown	unknown

PRIMARY PERSON TO CONTACT:
Name: N/A
Home #:
Work #:

ORIENTATION TO UNIT: ☐ Call light ☐ Television/telephone
☐ Bathroom ☐ Visiting ☐ Smoking ☐ Meals
☐ Patient rights/responsibilities

CHIEF COMPLAINT/HX OF PRESENT ILLNESS

unresponsive

PERSONAL ARTICLES: (Check if retained/describe)
☐ Contacts ☐ R ☐ L ☐ Dentures ☐ Upper ☐ Lower
☒ Jewelry: necklace
☐ Other:

ALLERGIES: Meds, Food, IVP Dye, Seafood: Type of Reaction

unknown

VALUABLES ENVELOPE:
☐ Valuables instructions

PREVIOUS HOSPITALIZATIONS/SURGERIES

unknown

INFORMATION OBTAINED FROM:
☐ Patient ☐ Previous record
☐ Family ☐ Responsible party

Signature ____________________

Home Medications (including OTC) **Codes: A=Sent home** **B=Sent to pharmacy** **C=Not brought in**

Medication	Dose	Frequency	Time of Last Dose	Code	Patient Understanding of Drug
unknown					

Do you take all medications as prescribed? ☐ Yes ☐ No If no, why?

PATIENT/FAMILY HISTORY

☐ Cold in past two weeks
☐ Hay fever
☐ Emphysema/lung problems
☐ TB disease/positive TB skin test
☐ Cancer
☐ Stroke/past paralysis
☐ Heart attack
☐ Angina/chest pain
☐ Heart problems
☐ High blood pressure
☐ Arthritis
☐ Claustrophobia
☐ Circulation problems
☐ Easy bleeding/bruising/anemia
☐ Sickle cell disease
☐ Liver disease/jaundice
☐ Thyroid disease
☐ Diabetes
☐ Kidney/urinary problems
☐ Gastric/abdominal pain/heartburn
☐ Hearing problems
☐ Glaucoma/eye problems
☐ Back pain
☐ Seizures
☐ Other

RISK SCREENING

Have you had a blood transfusion? ☐ Yes ☐ No
Do you smoke? ☐ Yes ☐ No
If yes, how many pack(s)?
Does anyone in your household smoke? ☐ Yes ☐ No
Do you drink alcohol? ☐ Yes ☐ No
If yes, how often? How much?
When was your last drink? ______/______/______
Do you take any recreational drugs? ☐ Yes ☐ No
If yes, type:______ Route:
Frequency:______ Date last used:______/______/______

FOR WOMEN Ages 12–52

Is there any chance you could be pregnant? ☐ Yes ☐ No
If yes, expected date (EDC):
Gravida/Para:

ALL WOMEN

Date of last Pap smear:
Do you perform regular breast self-exams? ☐ Yes ☐ No

ALL MEN

Do you perform regular testicular exams? ☐ Yes ☐ No

Additional comments:

x M. Barker, RN
Signature/Title

Client name: Juan Perez
DOB: 3/22
Age: 29
Sex: Male
Education: High school diploma
Occupation: Convenience store clerk
Hours of work: Varies; primarily the night shift, 11 PM to 7 AM
Household members: Lives with his brother, his brother's wife, and their two children ages 2 and 4
Ethnic background: Hispanic
Religious affiliation: Catholic
Referring physician: Deborah Kuhls, MD

Chief complaint:
The patient was brought into the emergency room by a friend after he had been shot in the abdomen. He was vomiting blood, and complained of severe back and "stomach" pain. He was able to respond to a few questions initially but stated the pain "was too bad for me to think." He denied being allergic to any medications or having any chronic medical problems.

Patient history:
Onset of disease: Brought into the ER by a friend at 2 AM yesterday vomiting blood, and with obvious bleeding wounds from abdominal area.
PMH: Unremarkable
Meds: None
Smoker: Yes
Family Hx: What CAD *Who?* Unknown

Physical exam:
General appearance: Mildly obese 29-year-old Hispanic male on mechanical ventilation
Vitals: Temp 102.6°F, BP 115/65 mm Hg, HR 135 bpm/normal, RR 20 bpm
Heart: Noncontributory
HEENT: NG tube in place for decompression
Rectal: Not done
Neurologic: Sedated
Extremities: 4+ bilateral pedal edema noted
Skin: Warm, moist
Chest/lungs: Lungs clear to auscultation and percussion
Peripheral vascular: Pulses full—no bruits
Abdomen: Abdominal distension, wound VAC in place, three tubes draining peritoneal fluid, hypoactive BS present in all regions. Liver percusses approx 8 cm at the midclavicular line, one fingerbreadth below the right costal margin.

Nutrition Hx:
General: Weight obtained from patient's brother who stated that patient usually weighs about 225 lbs, height 5′10″, and has not lost or gained a significant amount of weight recently. He denies that his

brother follows any special diet. Reports that his brother usually drinks "several beers" every night, more on the weekend.

Dx:
Abdominal GSW

Tx plan:
He was immediately taken to surgery where he underwent an exploratory damage-control laparotomy, gastric repair, control of liver hemorrhage, and resection of proximal jejunum, leaving his GI tract in discontinuity.

Hospital course:
After surgery, the patient was transferred to the Trauma Intensive Care Unit and maintained on mechanical ventilation. He returned to surgery on hospital day 2 to remove packs, and to reestablish bowel continuity. An abdominal vacuum-assisted closure (VAC) device was placed. Three Jackson-Pratt drains were left in place. On hospital day 3, the patient was taken back to surgery where an anastomotic leak was detected. A gastrojejunostomy tube was inserted through the patient's stomach, with the jejunal limb shortened in order to provide antegrade intraluminal drainage, as well as a retrograde jejunostomy tube for drainage. On hospital day 7, the patient was again taken to surgery for an abdominal washout, insertion of a distally placed J-tube for feeding, and a VAC change. The patient subsequently returned to the OR for multiple washouts and reapplication of a wound VAC. Nutrition consult was ordered by the trauma surgeon after this initial surgery on hospital day 1.

As per the clinical RD's recommendations, total parenteral nutrition (TPN) was initiated on hospital day 2 with dextrose 300 g and amino acids 170 g per day. Lipid emulsions were not recommended at this time. Although the patient was determined to have good nutritional status prior to his admission, he was now assessed to be at high nutritional risk due to the need for mechanical ventilation, large wounds, fluid and electrolyte losses, altered GI function, and the need for parenteral nutrition support. Energy needs were determined based on the patient's usual weight, rather than the current weight of 110 kg, due to the significant amount of generalized anasarca noted. The patient's medications included morphine, lorazepam, propofol @ 35 mL/hr, esomeprazole, meropenum, and vancomycin. A metabolic cart measurement was obtained on hospital day 4, which revealed the following: REE 3657 RQ 0.76. Blood glucose levels ranged from 107–185, and patient was placed on the insulin drip protocol. Dextrose was increased in the TPN to 350 g, and amino acids were increased to 180 g. On hospital day 10, the propofol was discontinued, and a second metabolic cart was obtained (REE 3765 RQ 0.70). At this point, IV lipids were added (250 mL three times per week). Blood glucose levels ranged from 110–145. Triglyceride levels were less than 400 mg/dL. Enteral nutrition support (Crucial with 1.5 calories per mL and 94 g of protein per liter) was initiated on hospital day 11 utilizing the jejunostomy tube at 10 mL/hr. On hospital day 12, the enteral nutrition formula was advanced to 15 mL/hr, and on hospital day 13, it was advanced to 20 mL/hr, at which point it was noted that enteral formula was draining from the anastomotic leak, and the enteral feeds were decreased to 15 mL/hr where they remained for the duration of his ICU stay.

UH UNIVERSITY HOSPITAL

NAME: Juan Perez DOB: 3/22
AGE: 29 SEX: M
PHYSICIAN: Deborah Kuhls, MD

*****CHEMISTRY*****

	NORMAL			UNITS
DAY:		4	10	
DATE:		7/5	7/11	
TIME:		0600	0545	
LOCATION:		TICU	TICU	
Albumin	3.5-5	1.4 L	1.9 L	g/dL
Total protein	6-8	5.2 L	5.1 L	g/dL
Prealbumin	16-35	3.0 L	5.0 L	mg/dL
Transferrin	250-380 (women) 215-365 (men)	190 L	160 L	mg/dL
Sodium	136-145	146 H	140	mEq/L
Potassium	3.5-5.5	4.0	3.7	mEq/L
Chloride	95-105	99	99	mEq/L
PO_4	2.3-4.7	2.2 L	2.4	mg/dL
Magnesium	1.8-3	1.9	1.5 L	mg/dL
Osmolality	285-295	317 H	305 H	mmol/kg/H_2O
Total CO_2	23-30	25	26	mEq/L
Glucose	70-110	164 H	140 H	mg/dL
BUN	8-18	23 H	25 H	mg/dL
Creatinine	0.6-1.2	1.4 H	1.6 H	mg/dL
Uric acid	2.8-8.8 (women) 4.0-9.0 (men)	8.9		mg/dL
Calcium	9-11	7.1		mg/dL
Bilirubin	≤0.3	.04		mg/dL
Ammonia (NH_3)	9-33	10		μmol/L
ALT	4-36	435 H		U/L
AST	0-35	190 H		U/L
Alk phos	30-120	540 H		U/L
CPK	30-135 (women) 55-170 (men)	177 H		U/L
C-reactive protein	<1.0	245 H	220 H	mg/dL
LDH	208-378	750 H		U/L
CHOL	120-199	180		mg/dL
HDL-C	>55 (women) >45 (men)	40 L		mg/dL
VLDL	7-32	110 H		mg/dL
LDL	<130	140 H		mg/dL
LDL/HDL ratio	<3.22 (women) <3.55 (men)			
Apo A	101-199 (women) 94-178 (men)			mg/dL
Apo B	60-126 (women) 63-133 (men)			mg/dL
TG	35-135 (women) 40-160 (men)	274 H	265 H	mg/dL
T_4	4-12			mcg/dL
T_3	75-98			mcg/dL
HbA_{1C}	3.9-5.2	7 H		%

UH UNIVERSITY HOSPITAL

NAME: Juan Perez DOB: 3/22
AGE: 29 SEX: M
PHYSICIAN: Deborah Kuhls, MD

HEMATOLOGY

DAY: 4
DATE: 7/5
TIME:
LOCATION:

	NORMAL		UNITS
WBC	4.8–11.8	15.2 H	$\times 10^3/mm^3$
RBC	4.2–5.4 (women) 4.5–6.2 (men)	3.2 L	$\times 10^6/mm^3$
HGB	12–15 (women) 14–17 (men)	14	g/dL
HCT	37–47 (women) 40–54 (men)	35 L	%
MCV	80–96	82	μm^3
RETIC	0.8–2.8	0.9	%
MCH	26–32	27	pg
MCHC	31.5–36	33	g/dL
RDW	11.6–16.5	12	%
Plt Ct	140–440	180	$\times 10^3/mm^3$
Diff TYPE			
ESR	0–25 (women) 0–15 (men)		mm/hr
% GRANS	34.6–79.2		%
% LYM	19.6–52.7		%
SEGS	50–62		%
BANDS	3–6		%
LYMPHS	24–44		%
MONOS	4–8		%
EOS	0.5–4		%
Ferritin	20–120 (women) 20–300 (men)	45	mg/mL
ZPP	30–80		μmol/mol
Vitamin B_{12}	24.4–100		ng/dL
Folate	5–25		μg/dL
Total T cells	812–2,318		mm^3
T-helper cells	589–1,505		mm^3
T-suppressor cells	325–997		mm^3
PT	11–16	9 L	sec

UH UNIVERSITY HOSPITAL

NAME: Juan Perez DOB: 3/22
AGE: 29 SEX: M
PHYSICIAN: Deborah Kuhls, MD

URINALYSIS

	NORMAL				UNITS
DAY:		4			
DATE:		7/5			
TIME:		0600			
LOCATION:		TICU			
Coll meth			Random specimen	First morning	
Color			Pale yellow	Pale yellow	
Appear			Cloudy	Clear	
Sp grv	1.003-1.030		1.045		
pH	5-7				
Prot	NEG		+1		mg/dL
Glu	NEG		+1		mg/dL
Ket	NEG	0			
Occ bld	NEG	0			
Ubil	NEG	0			
Nit	NEG	0			
Urobil	< 1.1	0			EU/dL
Leu bst	NEG	0			
Prot chk	NEG	0			
WBCs	0-5	0			/HPF
RBCs	0-5	0			/HPF
EPIs	0	0			/LPF
Bact	0	5			
Mucus	0	5			
Crys	0	0			
Casts	0	0			/LPF
Yeast	0	2			

Name: Juan Perez
Physician: Deborah Kuhls, MD

PATIENT CARE SUMMARY SHEET

Date: 7-5 Room: 5 Wt Yesterday: 107 kg Today: 109 kg

Temp °F	NIGHTS								DAYS								EVENINGS							
	00	01	02	03	04	05	06	07	08	09	10	11	12	13	14	15	16	17	18	19	20	21	22	23
105																								
104																								
103							x																	
102				x	x	x		x					x	x	x	x								
101		x	x						x	x	x	x					x				x			
100	x																	x	x	x		x	x	x
99																								
98																								
97																								
96																								
Pulse	95				90				96			95				85				90				
Respiration-*On Vent*																								
BP																								
Blood Glucose	175				166				150			150				160				145				
Appetite/Assist																								
INTAKE																								
Oral																								
IV TPN	75	75	75	75	75	75	75	75	75	75	75	75	75	75	75	75	75	75	75	75	75	75	75	75
TF Formula/Flush																								
Shift Total																								
OUTPUT																								
Cath	25			40				60			65			80		70		35			85			60
Void.																								
Emesis																								
BM																								
Drains (JP)	110	40	50	65	60	30	90	80	25	95	75	70	80	75	70	65	60	60	90	70	75	85	90	85
Shift Total																								
Gain	3640								3890								4050							
Loss	3750								3650								3775							
Signatures	D Hellerman, RN								K Sugae, RN								E Shewmake, RN							

Case Questions

I. Understanding the Disease and Pathophysiology

1. The patient has suffered a gunshot wound to the abdomen. This has resulted in an open abdomen. Define *open abdomen.*

2. The patient underwent gastric resection and repair, control of liver hemorrhage, and resection of proximal jejunum, leaving his GI tract in discontinuity. Describe the potential effects of surgery on this patient's ability to meet his nutritional needs.

3. Complications for this patient included anasarca. Define *anasarca* and describe how this condition may affect interpretation of his nutritional status.

4. The metabolic stress response to trauma has been described as a progression through three phases: the ebb phase, the flow phase, and finally the recovery or resolution. Define each of these and determine how they may correspond to this patient's hospital course.

5. Acute phase proteins are often used as a marker of the stress response. What is an acute phase protein? What is the role of C-reactive protein in the nutritional assessment of critically ill trauma patients?

II. Understanding the Nutrition Therapy

6. Metabolic stress and trauma significantly affect nutritional requirements. Describe the changes in nutrient metabolism that occur in metabolic stress. Specifically address energy requirements and changes in carbohydrate, protein, and lipid metabolism.

7. Are there specific nutrients that should be considered when designing nutrition support for a trauma patient? Explain the rationale and current recommendations regarding glutamine, arginine, and omega-3 fatty acids for this patient population.

8. Explain the decision-making process that would be applied in determining the route for nutrition support for the trauma patient.

III. Nutrition Assessment

A. Evaluation of Weight/Body Composition

9. Calculate and interpret the patient's BMI.

10. What factors make assessing his actual weight difficult on a daily basis?

B Calculation of Nutrient Requirements

11. Calculate energy and protein requirements for Mr. Perez. Use three different predictive equations for estimating his energy needs and explain your rationale for using each one.

12. What does indirect calorimetry measure?

13. What are the indications for obtaining a metabolic cart (indirect calorimetry) for this patient?

14. Compare the estimated energy needs calculated using the predictive equations with each other and with those obtained by indirect calorimetry measurements.

15. Interpret the RQ values. What do they indicate?

16. What factors contribute to the elevated energy expenditure in this patient?

C. Intake Domain

17. Mr. Perez was prescribed parenteral nutrition and was to receive 300 g of dextrose and 170 g of amino acids per day. Determine how many kilocalories and grams of protein are provided with this prescription. Read the patient care summary sheet. What was the total volume of PN provided that day?

18. Compare this nutrition support to his measured energy requirements obtained by the metabolic cart on day 4. Based on the metabolic cart results, what changes would you recommend be made to the TPN regimen, if any? What are the limitations that prevent the health care team from making significant changes to the nutrition support regimen?

19. The patient was also receiving propofol. What is this, and why should it be included in an assessment of his nutritional intake? How much energy did it provide?

20. On day 11, the patient was started on an enteral feeding. If his nutritional needs were met by parenteral nutrition, why was enteral feeding started?

21. This patient received the formula Crucial. What type of enteral formula is this? Why was this type of formula used? How many kcalories are being provided by the enteral nutrition support and what percent of his total nutritional intake does this represent?

22. From the information gathered within the intake domain, list possible nutrition problems using the appropriate diagnostic terms.

D. Clinical Domain

23. List abnormal biochemical values and describe why they might be abnormal.

Parameter	Normal Value	Patient's Value	Reason for Abnormality	Nutrition Implication

24. Assess the patient's urinalysis. Provide the most likely rationale for the presence of protein and glucose in the urine.

25. From the information gathered within the clinical domain, list possible nutrition problems using the diagnostic terms.

IV. Nutrition Diagnosis

26. Select two of the nutrition problems identified in questions 22 and 25, and complete the PES statement for each.

V. Nutrition Intervention

27. For each of the PES statements that you have written, establish an ideal goal (based on the signs and symptoms) and an appropriate intervention (based on the etiology).

VI. Nutrition Monitoring and Evaluation

28. What are the standard recommendations for monitoring the nutritional status of a patient receiving nutrition support?

29. Hyperglycemia was noted on the patient care monitoring sheet. List those values on day 4. Why is hyperglycemia of concern in the critically ill patient? How was this handled for this patient?

Bibliography

American Dietetic Association. *Critical Illness Evidence-Based Nutrition Practice Guideline.* Evidence Analysis Library. Available at: http://www.adaevidencelibrary.com/topic.cfm?cat=2799. Accessed October 24, 2007.

Cheatham ML, Safcsak K, Brzezinski SJ, Lube MW. Nitrogen balance, protein loss, and the open abdomen. *Crit Care Med.* 2007;35:127–131.

Dietitians in Nutrition Support: The relative effectiveness of practice change interventions in overcoming common barriers to change: A survey of 14 hospitals with experience implementing evidence-based guidelines. Available at: http://www.dnsdpg.org/news/research.php. Accessed May 8, 2008.

Frankenfield D, Roth-Yousey L, Compher C. Comparison of predictive equations for resting metabolic rate in healthy nonobese and obese adults: A systematic review of the literature. *J Am Diet Assoc.* 2005;105:775–789.

Ireton-Jones C, Jones JD. Improved equations for predicting energy expenditure in patients: The Ireton-Jones equations. *Nutr Clin Pract.* 2002;17:29–31.

McClave SA, Lowen CC, Kleber MJ, McConnell JW, Jung LY, Goldsmith LJ. Clinical use of the respiratory quotient obtained from indirect calorimetry. *JPEN: J Parenter Enter Nutr.* 2003;27(1):21–26.

McNelis J, Marini CP, Simms HH. Abdominal compartment syndrome: Clinical manifestations and predictive factors. *Curr Opin Crit Care.* 2003;9:133–136.

Mifflin MD, St. Jeor ST, Hill LA, Scott BJ, Daugherty SA, Koh YO. A new predictive equation for resting energy expenditure in healthy individuals. *Am J Clin Nutr.* 1990;51:241–247.

Nelms MN. Assessment of nutrition status and risk. In: Nelms M, Sucher K, Long S. *Nutrition Therapy and Pathophysiology.* Belmont, CA: Thomson/Brooks-Cole; 2007:101–133.

Nelms MN. Metabolic stress. In: Nelms M, Sucher K, Long S. *Nutrition Therapy and Pathophysiology.* Belmont, CA: Thomson/Brooks-Cole; 2007:785–804.

Pinilla JC, Hayes P, Laverty W, Arnold C, Laxdal V. The C-reactive protein to prealbumin ratio correlates with the severity of multiple organ dysfunction. *Surgery.* 1998;124(4):799–806.

Raguso CA, Dupertuis YM, Pichard C. The role of visceral proteins in the nutritional assessment of intensive care patients. *Curr Opin Clin Nutr.* 2003;6:211–216.

Saggi BH, Sugerman HJ, Ivatury RR, Bloomfield GL. Abdominal compartment syndrome. *J Trauma.* 1998;45:597–607.

Schecter WP, Ivatury RR, Rotondo MF, Hirshberg A. Open abdomen after trauma and abdominal sepsis: A strategy for management. *J Am Coll Surg.* 2006;203:390–396.

Skipper A, Nelms MN. Methods of nutrition support. In: Nelms M, Sucher K, Long S. *Nutrition Therapy and Pathophysiology.* Belmont, CA: Thomson/Brooks-Cole; 2007:149–179.

Wooley JA, Sax HC. Indirect calorimetry: Applications to practice. *Nutr Clin Pract.* 2003;18:434–439.

Internet Resources

American Dietetic Association: Evidence Analysis Library: Evidence-Based Nutrition Practice Guideline. http://www.adaevidencelibrary.com/topic.cfm?cat=2799

ASPEN: American Society for Parenteral and Enteral Nutrition. http://www.nutritioncare.org/

U.S. National Library of Medicine: MedlinePlus: Nutritional Support. http://www.nlm.nih.gov/medlineplus/nutritionalsupport.html

Unit Ten

NUTRITION THERAPY FOR HEMATOLOGY–ONCOLOGY

The layperson often uses *cancer* as a name for one disease. The term *cancer,* or *neoplasm,* actually describes any condition in which cells proliferate at a rapid rate and in an unrestrained manner. Each type of cancer is a different disease with different origins and responses to therapy. It is difficult to generalize about the role of nutrition in cancer treatment, because each diagnosis is truly an individual case. However, it is obvious to any clinician participating in the care of cancer patients that nutrition problems are common.

More than 80 percent of patients with cancer experience some degree of malnutrition. Nutrition problems may be some of the first symptoms the patient recognizes. Unexplained weight loss, changes in ability to taste, or a decrease in appetite are often present at diagnosis. The malignancy itself may affect not only energy requirements, but also the metabolism of nutrients.

As the patient begins therapy for a malignancy—surgery, radiation therapy, chemotherapy, immunotherapy, or bone marrow transplant—treatment side effects occur that can affect nutritional status. Can nutrition make a difference? Adequate nutrition helps prevent surgical complications, meet increased energy and protein requirements, and repair and rebuild tissues, which cancer therapies often damage. Furthermore, good nutrition allows increased tolerance of therapy and helps maintain the patient's quality of life. And finally, as with many medical conditions, cancer patients also face significant psychosocial issues.

All these factors must be considered when planning nutritional and medical care. The cases in this section allow you to plan nutritional care for some of the most common problems during cancer diagnosis and therapy. Esophageal cancer represents a new case in this edition of the book. Postoperatively, nutrition support and nutritional rehabilitation are crucial components of the care for this diagnosis. In addition, the cases in Unit Ten let you practice nutrition support and tackle psychosocial issues in complementary and alternative medicine.

Case 31

Lymphoma Treated with Chemotherapy

Objectives

After completing this case, the student will be able to:

1. Apply knowledge of the pathophysiology of cancer to identify and explain common metabolic and nutritional problems associated with malignancy.
2. Explain the complications of medical treatment for cancer and the potential nutritional consequences.
3. Apply the understanding of nutrition interventions during the treatment of and recovery from malignancy.
4. Analyze nutrition assessment data to evaluate nutritional status and identify specific nutrition problems.
5. Determine nutrition diagnoses and write appropriate PES statements.
6. Determine appropriate strategies for counseling cancer patients seeking complementary and alternative health care.

Denise Mitchell, a 21-year-old college student, is admitted for evaluation of viral illness in which she has experienced night sweats, fevers, and weight loss. A chest X-ray indicates a possible mass. After chest CT, MRI, and bone marrow and lymph node biopsy, she is diagnosed with Stage II diffuse large B-cell lymphoma with mediastinal disease and positive lymph node involvement.

ADMISSION DATABASE

Name: Denise Mitchell
DOB: 2/18 (age 21)
Physician: S. Miller, MD

BED #	DATE:	TIME:	TRIAGE STATUS (ER ONLY):
1	3/8	0300	☐ Red ☐ Yellow ☐ Green ☐ White

Initial Vital Signs

TEMP:	RESP:	SAO$_2$:	
100.5	18		
HT:	**WT (lb):**	**B/P:**	**PULSE:**
5′6″	120 (UBW 130)	95/70	85
LAST TETANUS		**LAST ATE**	**LAST DRANK**
2 years ago		this AM	this AM

PRIMARY PERSON TO CONTACT:
Name: Mel and Francis Mitchell
Home #: 212-555-1322
Work #: (same)

ORIENTATION TO UNIT: ☒ Call light ☒ Television/telephone ☒ Bathroom ☒ Visiting ☒ Smoking ☒ Meals ☒ Patient rights/responsibilities

CHIEF COMPLAINT/HX OF PRESENT ILLNESS

"I don't seem to have ever gotten over the flu that I had several weeks ago. I still have a fever sometimes, and the cough won't go away."

ALLERGIES: Meds, Food, IVP Dye, Seafood: Type of Reaction

NKA

PREVIOUS HOSPITALIZATIONS/SURGERIES

tonsillectomy-age 5

PERSONAL ARTICLES: (Check if retained/describe)
☒ Contacts ☒ R ☒ L ☐ Dentures ☐ Upper ☐ Lower
☐ Jewelry:
☐ Other:

VALUABLES ENVELOPE:
☐ Valuables instructions

INFORMATION OBTAINED FROM:
☒ Patient ☐ Previous record
☒ Family ☐ Responsible party

Signature *Denise Mitchell*

Home Medications (including OTC) **Codes: A=Sent home** **B=Sent to pharmacy** **C=Not brought in**

Medication	Dose	Frequency	Time of Last Dose	Code	Patient Understanding of Drug
Dimetapp	2 tsp	occ	9 PM yesterday	C	yes
Tylenol	400 mg	occ	9 PM yesterday	C	yes

Do you take all medications as prescribed? ☒ Yes ☐ No If no, why?

PATIENT/FAMILY HISTORY

☒ Cold in past two weeks Patient
☐ Hay fever
☐ Emphysema/lung problems
☐ TB disease/positive TB skin test
☒ Cancer Maternal grandmother
☐ Stroke/past paralysis
☒ Heart attack Paternal grandfather
☒ Angina/chest pain Paternal grandfather
☒ Heart problems Paternal grandfather

☐ High blood pressure
☐ Arthritis
☐ Claustrophobia
☐ Circulation problems
☐ Easy bleeding/bruising/anemia
☐ Sickle cell disease
☐ Liver disease/jaundice
☐ Thyroid disease
☐ Diabetes

☐ Kidney/urinary problems
☐ Gastric/abdominal pain/heartburn
☐ Hearing problems
☐ Glaucoma/eye problems
☐ Back pain
☐ Seizures
☐ Other

RISK SCREENING

Have you had a blood transfusion? ☐ Yes ☒ No
Do you smoke? ☐ Yes ☒ No
If yes, how many pack(s)?
Does anyone in your household smoke? ☐ Yes ☒ No
Do you drink alcohol? ☐ Yes ☒ No
If yes, how often? How much?
When was your last drink? ______/______/______
Do you take any recreational drugs? ☐ Yes ☒ No
If yes, type:______ Route:
Frequency:______ Date last used:______/______/______

FOR WOMEN Ages 12–52

Is there any chance you could be pregnant? ☐ Yes ☒ No
If yes, expected date (EDC):
Gravida/Para:

ALL WOMEN

Date of last Pap smear: 08/20
Do you perform regular breast self-exams? ☒ Yes ☐ No

ALL MEN

Do you perform regular testicular exams? ☐ Yes ☐ No

Additional comments:

x *C. Peck, RN*
Signature/Title

Client name: Denise Mitchell
DOB: 2/18
Age: 21
Sex: Female
Education: College student
Occupation: Student
Hours of work: N/A
Household members: Mother age 45; father age 50; brothers ages 9, 12, and 16
Ethnic background: Caucasian
Religious affiliation: Methodist
Referring physician: Simon Miller, MD (hematology/oncology)

Chief complaint:
"I don't seem to have ever gotten over the flu that I had. I still have a fever sometimes, and the cough won't go away."

Patient history:
Onset of disease: Ms. Mitchell is a 21-year-old female who currently is a sophomore at Midwest University. She has had an uneventful medical history with no significant illness until the past 2–3 months. Patient describes having the "flu" and feeling run down ever since. She has continued to have fevers, especially at night; she describes having to change her nightgown and bedclothes due to excessive sweating. She now presents for admission on referral from her family physician.
Type of Tx: None at present
PMH: Tonsillectomy age 5
Meds: OTC cough medicine and Tylenol
Smoker: No
Family Hx: Noncontributory

Physical exam:
General appearance: Patient is a thin, pale young woman who appears tired.
Vitals: Temp 100.5°F, BP 95/70 mm Hg, HR 85 bpm, RR 18 bpm
Heart: Regular rate and rhythm, no gallops or rubs, point of maximal impulse at the fifth intercostal space in the midclavicular line
HEENT:
Head: Normocephalic
Eyes: Extraocular movements intact; wears glasses for myopia; fundi grossly normal bilaterally
Ears: Tympanic membranes normal
Nose: Dry mucous membranes without lesions
Throat: Slightly dry mucous membranes without exudates or lesions; abnormal lymph nodes
Genitalia: Normal without lesions
Neurologic: Alert and oriented; cranial nerves II–XII grossly intact; strength 5/5 throughout; sensation to light touch intact; normal gait; and normal reflexes
Extremities: Normal muscular tone with normal ROM, nontender
Skin: Warm and dry without lesions
Chest/lungs: Respirations are shallow; dullness present to percussion

Peripheral vascular: Pulse +2 bilaterally, warm and nontender
Abdomen: Normal active bowel sounds, soft and nontender, without masses or organomegaly

Nutrition Hx:
General: Appetite decreased. No nausea, vomiting, constipation, or diarrhea.

Usual dietary intake:
AM: Cold cereal, toast or doughnut, skim milk, juice
Lunch: (In college cafeteria) sandwich or salad, frozen yogurt, chips or pretzels, soda
PM: Meat (eats only chicken and fish), 1–2 vegetables including a salad, iced tea, or skim milk
Snack: Popcorn, occasionally pizza, soda, juice, iced tea

24-hour recall:
AM: 1 slice dry toast, plain hot tea
Lunch: ½ c ice cream, ¼ c fruit cocktail, few bites of other foods on tray
Dinner: Few bites of chicken (1 oz), 2 tbsp mashed potatoes, ½ c Jell-O, plain hot tea

Food allergies/intolerances/aversions: NKA
Previous nutrition therapy? No
Food purchase/preparation: Self, parents, college cafeteria
Vit/min intake: None

Dx:
Stage II diffuse large B-cell lymphoma with mediastinal disease and positive lymph nodes. Bone marrow and other organs show no indication of disease.

Tx plan:
A chemotherapy regimen of cyclophosphamide, doxorubicin, vincristine, and prednisone (CHOP) is prescribed. Prednisone will be administered orally on the first 5 days of each 21-day cycle, and the other chemotherapeutic medications will be given intravenously on the first day of the cycle. Radiotherapy is planned to start 3 weeks after the third cycle of CHOP.

Hospital course:
Chest X-ray indicated possible mass. After chest CT, MRI, bone marrow biopsy, and biopsy of suspect lymph nodes, patient's course of treatment of chemotherapy and radiation therapy was determined. She was discharged for outpatient therapy on hospital day 5.

UH UNIVERSITY HOSPITAL

NAME: Denise Mitchell DOB: 2/18
AGE: 21 SEX: F
PHYSICIAN: S. Miller, MD

CHEMISTRY

DAY:		Admit	
DATE:		3/18	
TIME:			
LOCATION:			
	NORMAL		UNITS
Albumin	3.5-5	3.3 L	g/dL
Total protein	6-8	5.5 L	g/dL
Prealbumin	16-35		mg/dL
Transferrin	250-380 (women) 215-365 (men)		mg/dL
Sodium	136-145	141	mEq/L
Potassium	3.5-5.5	3.8	mEq/L
Chloride	95-105	100	mEq/L
PO_4	2.3-4.7	3.9	mg/dL
Magnesium	1.8-3	2.1	mg/dL
Osmolality	285-295	292	mmol/kg/H_2O
Total CO_2	23-30	27	mEq/L
Glucose	70-110	105	mg/dL
BUN	8-18	14	mg/dL
Creatinine	0.6-1.2	0.6	mg/dL
Uric acid	2.8-8.8 (women) 4.0-9.0 (men)	2.9	mg/dL
Calcium	9-11	9.2	mg/dL
Bilirubin	≤ 0.3	0.8 H	mg/dL
Ammonia (NH_3)	9-33	11	μmol/L
ALT	4-36	10	U/L
AST	0-35	21	U/L
Alk phos	30-120	101	U/L
CPK	30-135 (women) 55-170 (men)	122	U/L
LDH	208-378	245	U/L
CHOL	120-199	171	mg/dL
HDL-C	> 55 (women) > 45 (men)	57	mg/dL
VLDL	7-32		mg/dL
LDL	< 130	101	mg/dL
LDL/HDL ratio	< 3.22 (women) < 3.55 (men)	1.77	
Apo A	101-199 (women) 94-178 (men)		mg/dL
Apo B	60-126 (women) 63-133 (men)		mg/dL
TG	35-135 (women) 40-160 (men)	82	mg/dL
T_4	4-12	8.3	mcg/dL
T_3	75-98	81	mcg/dL
HbA_{1C}	3.9-5.2	4.3	%

UH UNIVERSITY HOSPITAL

NAME: Denise Mitchell DOB: 2/18
AGE: 21 SEX: F
PHYSICIAN: S. Miller MD

HEMATOLOGY

DAY: Admit
DATE: 3/18
TIME:
LOCATION:

	NORMAL		UNITS
WBC	4.8–11.8	12.0 H	$\times 10^3/mm^3$
RBC	4.2–5.4 (women) 4.5–6.2 (men)	4.2	$\times 10^6/mm^3$
HGB	12–15 (women) 14–17 (men)	11 L	g/dL
HCT	37–47 (women) 40–54 (men)	31 L	%
MCV	80–96	70 L	μm^3
RETIC	0.8–2.8	2.9 H	%
MCH	26–32	28	pg
MCHC	31.5–36	27 L	g/dL
RDW	11.6–16.5	11.9	%
Plt Ct	140–440	210	$\times 10^3/mm^3$
Diff TYPE			
ESR	0–25 (women) 0–15 (men)	19	mm/hr
% GRANS	34.6–79.2		%
% LYM	19.6–52.7	23	%
SEGS	50–62		%
BANDS	3–6	3	%
LYMPHS	24–44	38	%
MONOS	4–8	4	%
EOS	0.5–4	3	%
Ferritin	20–120 (women) 20–300 (men)	19 L	mg/mL
ZPP	30–80		μmol/mol
Vitamin B_{12}	24.4–100	54	ng/dL
Folate	5–25	18	μg/dL
Total T cells	812–2,318	1,500	mm^3
T-helper cells	589–1,505		mm^3
T-suppressor cells	325–997		mm^3
PT	11–16	13.1	sec

Case Questions

I. Understanding the Disease and Pathophysiology

1. What type of cancer is lymphoma?

2. Which symptoms found in Ms. Mitchell's history and physical are consistent with the classic signs of lymphoma?

3. Ms. Mitchell's diagnosis stated that she had Stage II lymphoma. What does Stage II mean, and how does her physical examination support this?

4. Generally, patients with cancer are treated with surgery, radiation therapy, chemotherapy, biological therapy, bone marrow transplant, or a combination of therapies. Ms. Mitchell's medical plan indicates that she will have both chemotherapy and radiation therapy. Describe how each of these therapy modalities work to treat malignant cells.

5. Radiation and chemotherapy may also affect healthy tissues.

   a. What other cells in the body may be affected by either or both of these treatments?

   b. What symptoms may the patient experience as a result of the destruction of these cells?

II. Understanding the Nutrition Therapy

6. Describe the major factors that may impact the nutritional status of the cancer patient.

7. You have read that most cancer patients require additional energy and protein. Explain the rationale for this. Is this true for every cancer patient?

8. In question 5, you listed the specific symptoms that a patient may experience from chemotherapy and radiation therapy. For each, describe the nutrition therapy recommendations that would be appropriate to assist in treatment of that symptom.

III. Nutrition Assessment

A. Evaluation of Weight/Body Composition

9. Calculate this patient's body mass index and the percent usual body weight. How would their interpretation differ? Which is the most appropriate to use in determining nutritional risk for this patient?

10. What factors may have contributed to these weight changes?

11. Using the information gathered during your assessment of weight and body composition, list possible nutrition problems using the diagnostic term.

B. Calculation of Nutrient Requirements

12. Calculate the patient's protein requirements. Explain how you determined the appropriate level of protein.

13. Calculate energy requirements for Denise. Identify the formula/calculation method you used and explain the rationale for using it. Which weight (UBW or current body weight) should you use to accurately calculate the patient's energy needs?

C. Intake Domain

14. How would you assess the dietary information gathered for usual nutritional intake?

15. What additional information would you ask the patient to provide regarding her usual intake?

16. Using one of the methods you have identified, determine whether this patient's usual intake is adequate to meet her needs. Explain.

17. What method would you use to assess her 24-hour recall? Is it adequate to meet her needs? Explain.

18. What common side effects of her illness may affect her dietary intake and subsequently her nutritional status?

19. What physical symptom(s) is this patient experiencing that might affect her dietary intake?

20. From the information gathered within the intake domain, list possible nutrition problems using the diagnostic term.

D. Clinical Domain

21. Which labs can be used to assess protein status?

a. Which labs will reflect acute changes in protein status versus chronic changes? Why?

b. Which are available for this patient? Considering her diagnosis, which labs would *not* be appropriate to use to evaluate protein status?

c. Determine the nutritional risk associated with this patient's laboratory value. Would you request any additional nutrition assessment labs?

22. Identify each of the drugs that the patient is prescribed, and note the possible nutritional side effects of each. In general, what might you tell this patient to expect from receiving her chemotherapy?

Drug	Possible Nutritional Side Effect(s)
Cyclophosphamide	
Doxorubicin	
Vincristine	
Prednisone	

23. From the information gathered within the clinical domain, list possible nutrition problems using the diagnostic term.

E. Behavioral–Environmental Domain

24. During a follow-up visit, Denise's mother asks about an "anti-cancer" diet that Denise's aunt has suggested. This diet recommends a cleansing protocol with frequent coffee enemas with a diet that focuses primarily on a liquid mixture made from fruits, vegetables, and raw calf's liver. Mrs. Mitchell is concerned that Denise cannot even tolerate drinking the mixture and refuses to even consider enemas. How would you advise this patient and her parents regarding adherence to this "anti-cancer" diet? What steps would you suggest for them as they research and make appropriate decisions for care? Why may cancer patients be especially vulnerable to nutrition and medical quackery?

IV. Nutrition Diagnosis

25. Select two high-priority nutrition problems and complete the PES statement.

V. Nutrition Intervention

26. For each of the PES statements that you have written, establish an ideal goal (based on the signs and symptoms) and an appropriate intervention (based on the etiology).

VI. Nutrition Monitoring and Evaluation

27. How would you follow up or evaluate the interventions you have determined?

28. What types of nutrition education would be important to provide for Denise? When would it be appropriate to provide this education? What factors might interfere with the patient's reception of nutrition education?

29. What is a low-microbial or low-bacterial diet? Why may Denise need to follow food safety guidelines during immunosuppression?

30. Recently glutamine has been a component of several clinical trials to reduce gastrointestinal complications of both chemo and radiation therapy. What is glutamine? What is the rationale for its use?

Bibliography

Baileys K, Nahikian-Nelms M. Lymphoma. In: Kogut VJ, Luthringer SL. *Nutritional Issues in Cancer Care.* Pittsburgh, PA: Oncology Nursing Society; 2005:201–218.

Bosaeus I, Daneryd P, Lundholm K. Dietary intake, resting energy expenditure, weight loss, and survival in cancer patients. *J Nutr.* 2002;132(Suppl 11):3465S–3466S.

Clarkson JE, Worthington HV, Eden OB. Interventions for preventing oral mucositis for patients with cancer receiving treatment. *Cochrane Database Syst Rev.* 2003;(3):CD000978.

Cohen DA Neoplastic disease. In: Nelms M, Sucher K, Long S. *Nutrition Therapy and Pathophysiology.* Belmont, CA: Thomson/Brooks-Cole; 2007:751–783.

Daniele B, Perrone F, Gallo C, Pignata S, De Martino S, De Vivo R, Barletta E, Tambaro R, Abbiati R, D'Agostino L. Oral glutamine in the prevention of fluorouracil induced intestinal toxicity: A double blind, placebo controlled, randomised trial. *Gut.* 2001;48(1):28–33.

de Luis DA, Izaola O, Aller R, Cuellar L, Terroba MC. A randomized clinical trial with oral immunonutrition (omega 3-enhanced formula vs. arginine-enhanced formula) in ambulatory head and neck cancer patients. *Ann Nutr Metab.* 2005;49(2):95–99.

Garcia-Peris P, Lozano MA, Velasco C, De L Cuerda, C, Iriondo T, Breton I, Camblor M, Navarro C. Prospective study of resting energy expenditure changes in head and neck cancer patients treated with chemoradiotherapy measured by indirect calorimetry. *Nutrition.* 2005;21:1107–1112.

Gill CA, Murphy-Ende K. Immunonutrition. In: Kogut VJ, Luthringer SL. *Nutritional Issues in Cancer Care.* Pittsburgh, PA: Oncology Nursing Society; 2005:291–318.

Kwong K. Prevention and treatment of oropharyngeal mucositis following cancer therapy: Are there new approaches? *Cancer Nurs.* 2004;27:183–205.

Nelms MN. Assessment of nutrition status and risk. In: Nelms M, Sucher K, Long S. *Nutrition Therapy and Pathophysiology.* Belmont, CA: Thomson/Brooks-Cole; 2007:101–135.

Porter SR, Scully C, Hegarty AM. An update of the etiology and management of xerostomia. *Oral Surg Oral Med Oral Pathol Oral Radiol Endod.* 2004;97:28–46.

Schnell FM. Chemotherapy-induced nausea and vomiting: The importance of acute antiemetic control. *Oncologist.* 2003;8(2):187–198.

Internet Resources

American Cancer Society: Nutrition for the Person with Cancer. http://www.cancer.org/docroot/MBC/MBC_6.asp

MD Anderson Cancer Center: Nutrition & Cancer. http://www.mdanderson.org/topics/food/

National Cancer Institute: Nutrition in Cancer Care (PDQ). http://www.cancer.gov/cancerinfo/pdq/supportivecare/nutrition

Quackwatch. http://www.quackwatch.com

Case 32

Esophageal Cancer Treated with Surgery and Radiation[1]

Objectives

After completing this case, the student will be able to:

1. Apply knowledge of the pathophysiology of cancer to identify and explain common metabolic and nutritional problems associated with malignancy.
2. Explain the complications of medical treatment for cancer and the potential nutritional consequences.
3. Apply the understanding of nutrition support in the treatment of and recovery from malignancy.
4. Analyze nutrition assessment data to evaluate nutritional status and identify specific nutrition problems.
5. Determine nutrition diagnoses and write appropriate PES statements.
6. Evaluate adequacy of an enteral feeding regimen providing for the nutritional needs of the cancer patient.

Nick Seyer is a 58-year-old gentleman who, after suffering from recurrent heartburn for over a year, seeks medical attention. He presents to his physician with difficulty swallowing and a significant unexplained weight loss.

[1] Adapted with permission from: Whitman M. Esophageal Cancer. Available from: Virtual Health Care Team School of Health Professions University of Missouri–Columbia. http://www.vhct.org/index.htm.

Name: Nick Seyer
DOB: 3/4 (age 58)
Physician: H. Brown, MD

ADMISSION DATABASE

BED #	DATE:	TIME:	TRIAGE STATUS (ER ONLY):
1	9/5	0930	☐ Red ☐ Yellow ☐ Green ☐ White

Initial Vital Signs

TEMP:	RESP:	SAO$_2$:	
98.3	14		
HT (in): 6′3″	WT (lb): 198 (UBW 230)	B/P: 132/92	PULSE: 88
LAST TETANUS: > 5 years ago		LAST ATE: last night	LAST DRANK: this AM

PRIMARY PERSON TO CONTACT:
Name: Betty Seyer
Home #: 907-555-2895
Work #: 907-555-9765

ORIENTATION TO UNIT: ☒ Call light ☒ Television/telephone ☒ Bathroom ☒ Visiting ☒ Smoking ☒ Meals ☒ Patient rights/responsibilities

CHIEF COMPLAINT/HX OF PRESENT ILLNESS

Heartburn for "a long time" and problems swallowing during the past 4 or 5 months.

PERSONAL ARTICLES: (Check if retained/describe)
☐ Contacts ☐ R ☐ L ☐ Dentures ☐ Upper ☐ Lower
☒ Jewelry: wedding band
☐ Other:

ALLERGIES: Meds, Food, IVP Dye, Seafood: Type of Reaction

NKA

VALUABLES ENVELOPE: yes
☒ Valuables instructions

PREVIOUS HOSPITALIZATIONS/SURGERIES

Leg fracture at age 14
Stitches due to lacerations from car accident

INFORMATION OBTAINED FROM:
☒ Patient ☐ Previous record
☒ Family ☐ Responsible party

Signature *Nick Seyer*

Home Medications (including OTC) **Codes: A=Sent home** **B=Sent to pharmacy** **C=Not brought in**

Medication	Dose	Frequency	Time of Last Dose	Code	Patient Understanding of Drug
TUMS	2-3 tablets	2-3 times daily	this morning	C	yes
Alka-Seltzer	2 tablets	1-2 times daily	last night	C	yes

Do you take all medications as prescribed? ☒ Yes ☐ No If no, why?

PATIENT/FAMILY HISTORY

☐ Cold in past two weeks
☐ Hay fever
☐ Emphysema/lung problems
☐ TB disease/positive TB skin test
☒ Cancer Mother (liver cancer, age 59)
☐ Stroke/past paralysis
☐ Heart attack
☐ Angina/chest pain
☐ Heart problems
☐ High blood pressure
☐ Arthritis
☐ Claustrophobia
☐ Circulation problems
☐ Easy bleeding/bruising/anemia
☐ Sickle cell disease
☐ Liver disease/jaundice
☐ Thyroid disease
☐ Diabetes
☐ Kidney/urinary problems
☐ Gastric/abdominal pain/heartburn
☐ Hearing problems
☐ Glaucoma/eye problems
☐ Back pain
☐ Seizures
☐ Other

RISK SCREENING

Have you had a blood transfusion? ☐ Yes ☒ No
Do you smoke? ☒ Yes ☐ No
If yes, how many pack(s)? 2/day
Does anyone in your household smoke? ☒ Yes ☐ No
Do you drink alcohol? ☒ Yes ☐ No
If yes, how often? Beer daily How much? 1-2
When was your last drink? ______/______/______
Do you take any recreational drugs? ☐ Yes ☒ No
If yes, type:______ Route:
Frequency:______ Date last used:______/______/______

FOR WOMEN Ages 12–52

Is there any chance you could be pregnant? ☐ Yes ☐ No
If yes, expected date (EDC):
Gravida/Para:

ALL WOMEN

Date of last Pap smear:
Do you perform regular breast self-exams? ☐ Yes ☐ No

ALL MEN

Do you perform regular testicular exams? ☐ Yes ☒ No

Additional comments:

x *Susan Elizabeth Bailey, RN*
Signature/Title

Client name: Nick Seyer
DOB: 3/4
Age: 58
Sex: Male
Education: Some college
Occupation: Contractor
Hours of work: Variable but usually 5–6 days per week—starts as early as 6:30 AM and works often until after 6 PM
Household members: Wife age 52; son age 18; two other sons are away at college, ages 19 and 22
Ethnic background: Caucasian
Religious affiliation: Catholic
Referring physician: H. Brown, MD

Chief complaint:
Heartburn for "a long time" and difficulty swallowing during the past 4 or 5 months. Occasionally food seems to "hang up" in his throat. He points to the upper portion of his neck, directly beneath his chin.

Patient history:
Patient describes significant heartburn for the previous year. He has been taking TUMS, Alka-Seltzer, and Pepcid consistently for the past year. He has noted weight loss of over 30 lbs in last several months. He states that he just has not been able to eat because of the pain and heartburn. Now, difficulty swallowing foods—especially anything with texture—brought him to his physician. Patient also describes a recurrent cough at night.
Onset of disease: Dysphagia × 3–4 months; odynophagia × 5–6 months.
Type of Tx: None at present
Meds: TUMS, Alka-Seltzer, and Pepcid
Smoker: Yes
Family Hx: What? Liver cancer *Who?* Mother—died age 58.

Physical exam:
General appearance: Distressed, thin, pale white male
Vitals: Temp 98.3°F, BP 132/92 mm Hg, HR 88 bpm, RR 13 bpm
Heart: Unremarkable
HEENT:
  Eyes: Sunken; sclera clear without evidence of tears
  Ears: Clear
  Nose: Dry mucous membranes
  Throat: Dry mucous membranes, no inflammation
Genitalia: Unremarkable
Neurologic: Alert, oriented × 3
Extremities: Joints appear prominent with evidence of some muscle wasting. No edema.
Skin: Warm, dry
Chest/lungs: Clear to auscultation and percussion
Abdomen: Epigastric tenderness on palpation

Nutrition Hx:

General: Prior to admission has noted decreased appetite, feeling full all the time, and regurgitation of some foods. He notes pain upon swallowing as well as pretty constant heartburn.

Usual dietary intake:

AM:	Used to eat eggs, bacon, toast every morning but has not eaten this for at least the past month. Most recently has had just coffee and cereal.
Lunch:	Previously, ate cold lunch packed for the work site. Included sandwich, cold meat or other leftovers from previous dinner, fruit, cookies, and tea.
Dinner:	All meats, pasta or rice, 2–3 vegetables, 1–2 beers
Snacks:	Ice cream, popcorn, or homemade dessert

24-hour recall:

AM:	1 packet of instant oatmeal; sips of coffee
Lunch:	6 oz tomato soup with 2–4 crackers
Dinner:	Macaroni and cheese—homemade ½ c
Bedtime:	1 scoop of chocolate ice cream

Food allergies/intolerances/aversions: NKA
Previous nutrition therapy? No
Food purchase/preparation: Wife
Vit/min intake: None

Dx:

After undergoing chest X-ray, endoscopy with brushings and biopsy, and CT scan, Mr. Seyer was diagnosed with Stage IIB (T1, N1, M0) adenocarcinoma of the esophagus.

Tx plan:

A surgery consult was requested to evaluate surgical resection. He also was evaluated for pre- and postoperative external beam radiation therapy.

Hospital course:

Mr. Seyer is now POD #4 s/p transhiatal esophagectomy. During surgery, a jejunal feeding tube was placed. He is prescribed Isosource HN 1.5 kcal at 75 mL/hr × 24 hrs.

UH *UNIVERSITY HOSPITAL*

NAME: Nick Seyer DOB: 3/4
AGE: 58 SEX: M
PHYSICIAN: H. Brown, MD

CHEMISTRY

	NORMAL			UNITS
DAY:		Admit		
DATE:		9/5	9/11	
TIME:		1200	0600	
LOCATION:				
Albumin	3.5-5	3.1 L	3.0 L	g/dL
Total protein	6-8	5.7 L	5.7 L	g/dL
Prealbumin	16-35	15 L	12 L	mg/dL
Transferrin	250-380 (women) 215-365 (men)	285	175 L	mg/dL
Sodium	136-145	137	136	mEq/L
Potassium	3.5-5.5	3.8	3.6	mEq/L
Chloride	95-105	101	99	mEq/L
PO_4	2.3-4.7	3.1	2.9	mg/dL
Magnesium	1.8-3	1.8	1.8	mg/dL
Osmolality	285-295			mmol/kg/H_2O
Total CO_2	23-30	26	25	mEq/L
Glucose	70-110	71	108	mg/dL
BUN	8-18	9	10	mg/dL
Creatinine	0.6-1.2	0.7	0.9	mg/dL
Uric acid	2.8-8.8 (women) 4.0-9.0 (men)	6.2		mg/dL
Calcium	9-11	9.1	9.4	mg/dL
Bilirubin	≤0.3	0.2	0.3	mg/dL
Ammonia (NH_3)	9-33	11	21	μmol/L
ALT	4-36	21	33	U/L
AST	0-35	32	27	U/L
Alk phos	30-120	101	99	U/L
CPK	30-135 (women) 55-170 (men)	172 H	145	U/L
LDH	208-378	350	342	U/L
CHOL	120-199	180	170	mg/dL
HDL-C	>55 (women) >45 (men)	47		mg/dL
VLDL	7-32			mg/dL
LDL	<130	129		mg/dL
LDL/HDL ratio	<3.22 (women) <3.55 (men)	2.7		
Apo A	101-199 (women) 94-178 (men)			mg/dL
Apo B	60-126 (women) 63-133 (men)			mg/dL
TG	35-135 (women) 40-160 (men)	158		mg/dL
T_4	4-12			mcg/dL
T_3	75-98			mcg/dL
HbA_{1C}	3.9-5.2			%

UH UNIVERSITY HOSPITAL

NAME: Nick Seyer DOB: 3/4
AGE: 58 SEX: M
PHYSICIAN: H. Brown, MD

HEMATOLOGY

DAY:		Admit	9/11	
DATE:		9/5	0600	
TIME:				
LOCATION:				
	NORMAL			UNITS
WBC	4.8-11.8	5.2	6.9	$\times 10^3/mm^3$
RBC	4.2-5.4 (women) 4.5-6.2 (men)	4.2 L	4.3 L	$\times 10^6/mm^3$
HGB	12-15 (women) 14-17 (men)	13.5 L	13.9 L	g/dL
HCT	37-47 (women) 40-54 (men)	38 L	38 L	%
MCV	80-96	90	86	μm^3
RETIC	0.8-2.8	0.9	1.0	%
MCH	26-32	32.4	32.3	pg
MCHC	31.5-36	35.5	36.5	g/dL
RDW	11.6-16.5	11.9	12.1	%
Plt Ct	140-440	250	232	$\times 10^3/mm^3$
Diff TYPE				
ESR	0-25 (women) 0-15 (men)	17 H	15	mm/hr
% GRANS	34.6-79.2	75	65	%
% LYM	19.6-52.7	25	35	%
SEGS	50-62	55	60	%
BANDS	3-6	4	3	%
LYMPHS	24-44	28	32	%
MONOS	4-8	4	5	%
EOS	0.5-4	0.5	0.6	%
Ferritin	20-120 (women) 20-300 (men)	220	208	mg/mL
ZPP	30-80			μmol/mol
Vitamin B_{12}	24.4-100			ng/dL
Folate	5-25			μg/dL
Total T cells	812-2,318			mm^3
T-helper cells	589-1,505			mm^3
T-suppressor cells	325-997			mm^3
PT	11-16	12	12.8	sec

Name: Nick Seyer
Physician: H. Brown, MD

PATIENT CARE SUMMARY SHEET

Date: 9/11 Room: 832 Wt Yesterday: 194 Today: 194.3

Temp °F	NIGHTS								DAYS								EVENINGS							
	00	01	02	03	04	05	06	07	08	09	10	11	12	13	14	15	16	17	18	19	20	21	22	23
105																								
104																								
103																								
102																								
101																								
100	X																							
99																								
98									X								X							
97																								
96																								
Pulse	68								72								78							
Respiration	19								20								19							
BP	118/72								125/80								110/77							
Blood Glucose	92								103								82							
Appetite/Assist	NPO								NPO								NPO							
INTAKE																								
Oral																								
IV	100	100	100	100	100	100	100	100	100	100	100	100	100	100	100	100	100	100	100	100	100	100	100	100
TF Formula/Flush	75/50	75	75	75	75	75	75	75	35/50	50	75	75	75	75	75	75	75	75	75	75	75	75	75	75/50
Shift Total	1450								1385								1450							
OUTPUT																								
Void																								
Cath.								1100								1700								900
Emesis																								
BM																								300
Drains																								
Shift Total	1100																							
Gain	+350																+250							
Loss									-315															
Signatures	F. Moore RN								M Seymour, RN								Mike Phillips, RN							

Case Questions

I. Understanding the Disease and Pathophysiology

1. Mr. Seyer has been diagnosed with adenocarcinoma of the esophagus. What does the term *adenocarcinoma* mean?

2. What are the two most common types of esophageal cancer? What are the risk factors for development of this malignancy? Does Mr. Seyer's medical record indicate that he has any of these risk factors?

3. Mr. Seyer's cancer was described as: Stage IIB (T1, N1, M0). Explain this terminology used to describe staging for malignancies.

4. Cancer is generally treated with a combination of therapies. This can include surgical resection, radiation therapy, chemotherapy, and immunotherapy. The type of malignancy and the staging of the disease will, in part, determine the type of therapies that are prescribed. Define and describe each of these therapies. Briefly describe the mechanism for each. In general, how do they act to treat a malignancy?

5. Mr. Seyer had a transhiatal esophagectomy. Describe this surgical procedure. How may this procedure affect his digestion and absorption?

II. Understanding the Nutrition Therapy

6. Many cancer patients experience changes in nutritional status. Briefly describe the potential affect of cancer on nutritional status.

7. Both surgery and radiation affect nutritional status. Describe the potential nutritional and metabolic effects of these treatments.

III. Nutrition Assessment

A. Evaluation of Weight/Body Composition

8. Calculate and evaluate Mr. Seyer's percent UBW and BMI.

9. Summarize your findings regarding his weight status. Classify the severity of his weight loss. What factors may have contributed to his weight loss? Explain.

10. What does research tell us about the relationship of significant weight loss and prognosis in cancer patients?

11. What other assessment measures would you recommend be conducted to complete his nutrition assessment?

B. Calculation of Nutrient Requirements

12. Estimate Mr. Seyer's energy and protein requirements based on his current weight. Identify the factors you used in determining which equations to use for these calculations.

13. Estimate Mr. Seyer's fluid requirements based on his current weight.

C. Intake Domain

14. What factors can you identify from Mr. Seyer's history and physical that may indicate any problems with eating an oral diet prior to admission?

15. How are these factors consistent with his diagnosis?

16. Mr. Seyer is currently receiving enteral nutrition therapy. He is prescribed Isosource HN at 75 mL/hr.

    a. Calculate the amount of energy and protein that will be provided at this rate.

    b. Next, by assessing the information on the patient care summary sheet, determine the actual amount of enteral nutrition that he received on September 11.

    c. Compare this to his estimated nutrient requirements. Identify any nutrition problems.

d. Compare fluids required to fluids received. Is he meeting his fluid requirements? How did you determine this? Why would you evaluate his output when assessing his fluid intake?

17. What type of formula is Isosource HN? One of the residents taking care of Mr. Seyer asks about a formula with a higher concentration of omega-3 fatty acids, antioxidants, arginine, and glutamine that could promote healing after surgery. What does the evidence indicate regarding nutritional needs for cancer patients and, in particular, nutrients to promote postoperative wound healing? What formulas may meet this profile? List them and discuss why you chose them.

18. From the information gathered within the intake domain, list possible nutrition problems using the diagnostic term.

D. Clinical Domain

19. After reviewing the patient's admission history and physical, discuss any factors noted there that are consistent with decreased oral intake.

20. After reviewing the patient's admission history and physical, are there any clinical signs of malnutrition?

21. Review the patient's chemistries upon admission. Identify any that are abnormal. Using the following table, describe their clinical significance for this patient.

Chemistry/Date	Normal Value	Mr. Seyer's Value	Reason for Abnormality	Nutritional Implications

22. From the information gathered within the clinical domain, list possible nutrition problems using the diagnostic term.

E. Behavioral–Environmental Domain

23. Mr. Seyer has been diagnosed with a life-threatening illness. What is the definition of a terminal illness?

24. The literature describes how a patient and his family may experience varying levels of emotional response to a terminal illness. These may include anger, denial, depression, and acceptance. How may this affect the patient's nutritional intake? How would you handle these components in your nutritional care? What questions might you have for Mr. Seyer or his family? List three.

IV. Nutrition Diagnosis

25. Select two high-priority nutrition problems and complete the PES statement.

V. Nutrition Intervention

26. For each of the PES statements that you have written, establish an ideal goal (based on the signs and symptoms) and an appropriate intervention (based on the etiology) at this point of Mr. Seyer's hospital course.

27. Does his current nutrition support meet his estimated nutritional needs? If not, determine the recommended changes. Discuss any areas of deficiency and ideas for implementing a new plan.

28. How may these interventions (from question 27) change as he progresses postoperatively? Discuss how Mr. Seyer may transition from enteral feeding to an oral diet.

VI. Nutrition Monitoring and Evaluation

29. List at least four factors that you should monitor for Mr. Seyer while he is receiving enteral nutrition therapy. For example, you might indicate that you should "monitor weight weekly."

30. Mr. Seyer will receive radiation therapy as an outpatient. In question 7, you identified potential nutritional complications with radiation therapy. Choose one of the nutritional complications and describe the nutrition intervention that would be appropriate.

31. Identify the major assessment indices you would use to monitor his nutritional status once he begins radiation therapy.

Bibliography

Aiko S, Yoshizumi Y, Tsuwano S, Shimanouchi M, Sugiura Y, Maehara T. The effects of immediate enteral feeding with a formula containing high levels of omega-3 fatty acids in patients after surgery for esophageal cancer. *J Parenter Enteral Nutr.* 2005;29(3):141–147.

American College of Physicians. Home Care Guide for Advanced Cancer. Available at: http://www.acponline.org/patients_families/end_of_life_issues/cancer/. Accessed April 29, 2008.

Andreassen S, Randers I, Näslund E, Stockeld D, Mattiasson AC. Patients' experiences of living with oesophageal cancer. *J Clin Nurs.* 2006;15:685–695.

Churma SA, Horrell CJ. Esophageal and gastric cancers. In: Kogut VJ, Luthringer SL. *Nutritional Issues in Cancer Care.* Pittsburgh, PA: Oncology Nursing Society; 2005.

Cohen DA. Neoplastic Disease. In: Nelms M, Sucher K, Long S. *Nutrition Therapy and Pathophysiology.* Belmont, CA: Thomson/Brooks-Cole; 2007:751–779.

Gabor S, Renner H, Matzi V, Ratzenhofer B, Lindenmann J, Sankin O, Pinter H, Maier A, Smolle J, Smolle-Jüttner FM. Early enteral feeding compared with parenteral nutrition after oesophageal or oesophagogastric resection and reconstruction. *Br J Nutr.* 2005;93:509–513.

Gurski RR, Schirmer CC, Rosa AR, Brentano L. Nutritional assessment in patients with squamous cell carcinoma of the esophagus. *Hepatogastroenterology.* 2003;50:1943–1947.

Jenkinson AD, Lim J, Agrawal N, Menzies D. Laparoscopic feeding jejunostomy in esophagogastric cancer. *Surg Endosc.* 2007;21:299–302.

Korn WM. Prevention and management of early esophageal cancer. *Curr Treat Options Oncol.* 2004;5:405–416.

National Cancer Institute. Loss, Grief, and Bereavement. Available at: http://www.cancer.gov/cancertopics/pdq/supportivecare/bereavement/Patient. Accessed April 30, 2008.

Odelli C, Burgess D, Bateman L, Hughes A, Ackland S, Gillies J, Collins CE. Nutrition support improves patient outcomes, treatment tolerance and admission characteristics in oesophageal cancer. *Clin Oncol (R Coll Radiol).* 2005;17:639–645.

Pavia R, Barresi P, Piermanni V, Mondello B, Urgesi R. Role of artificial nutrition in patients undergoing surgery for esophageal cancer. *Rays.* 2006;31(1):25–29.

Pera M, Manterola C, Vidal O, Grande L. Epidemiology of esophageal adenocarcinoma. *Surg Oncol.* 2005;92:151–159

University of Michigan. Transhiatal esophagectomy (THE). Available at: http://surgery.med.umich.edu/thoracic/clinical/what_we_do/esophagectomy_faq.shtml. Accessed September 9, 2007.

Internet Resources

National Cancer Institute: Types of Cancer. http://www.nci.nih.gov/cancerinfo

National Cancer Institute: Cancer Facts. http://www.cancerfacts.com

University of Virginia Health System: All about Cancer. http://www.healthsystem.virginia.edu/uvahealth/hub_cancer/index.cfm

Appendix A
COMMON MEDICAL ABBREVIATIONS

AAL	anterior axillary line
ab lib	at pleasure; as desired (ab libitum)
ac	before meals
ACTH	adrenocorticotropic hormone
AD	Alzheimer's disease
ad lib	as desired (ad libitum)
ADA	American Dietetic Association, American Diabetes Association
ADL	activities of daily living
AGA	antigliadin antibody
AIDS	acquired immunodeficiency syndrome
ALP (Alk phos)	alkaline phosphatase
ALS	amyotrophic lateral sclerosis
ALT	alanine aminotransferase
amp	ampule
ANC	absolute neutrophil count
ANCA	antineutrophil cytoplasmic antibodies
AP	anterior posterior
ARDS	adult respiratory distress syndrome
ARF	acute renal failure, acute respiratory failure
ASA	acetylsalicylic acid, aspirin
ASCA	antisaccharomyces antibodies
ASHD	arteriosclerotic heart disease
AV	arteriovenous
BANDS	neutrophils
BCAA	branched-chain amino acids
BE	barium enema
BEE	basal energy expenditure
BG	blood glucose
bid	twice a day (bis in die)
bili	bilirubin
BM	bowel movement
BMI	body mass index
BMR	basal metabolic rate
BMT	bone marrow transplant
BP (B/P)	blood pressure
BPD	bronchopulmonary dysplasia
BPH	benign prostate hypertrophy
bpm	beats per minute, breaths per minute
BS	bowel sounds, breath sounds, or blood sugar
BSA	body surface area
BUN	blood urea nitrogen
c	cup
c	with
C	centigrade
C.C.E.	clubbing, cyanosis, or edema
c/o	complains of
CA	cancer; carcinoma
CABG	coronary artery bypass graft
CAD	coronary artery disease
CAPD	continuous ambulatory peritoneal dialysis
cath	catheter, catheterize
CAVH	continuous arteriovenous hemofiltration
CBC	complete blood count
cc	cubic centimeter
CCK	cholecystokinin
CCU	coronary care unit
CDAI	Crohn's disease activity index
CDC	Centers for Disease Control and Prevention
CHD	coronary heart disease
CHF	congestive heart failure
CHI	closed head injury
CHO	carbohydrate
CHOL	cholesterol
CKD	chronic kidney disease
cm	centimeter
CNS	central nervous system
COPD	chronic obstructive pulmonary disease
CPK	creatinine phosphokinase
Cr	creatinine
CR	complete remission
CSF	cerebrospinal fluid
CT	computed tomography
CVA	cerebrovascular accident

Note: Abbreviations can vary from institution to institution. Although the student will find many of the accepted variations listed in this appendix, other references may be needed to supplement this list.

CVD	cardiovascular disease
CVP	central venous pressure
CXR	chest X-ray
d/c	discharge
D/C	discontinue
D_5NS	dextrose, 5% in normal saline
D5W	dextrose, 5% in water
DASH	Dietary Approaches to Stop Hypertension
DBW	desirable body weight
DCCT	Diabetes Control and Complications Trial
DKA	diabetic ketoacidosis
dL	deciliter
DM	diabetes mellitus
DRI	Dietary Reference Intake
DTR	deep tendon reflex
DTs	delirium tremens
DVT	deep vein thrombosis
Dx	diagnosis
e.g.	for example
ECF	extracellular fluid
ECG/EKG	electrocardiogram
EEG	electroencephalogram
EGD	esophagogastroduodenoscopy
ELISA	enzyme-linked immunosorbent assay
EMA	antiendomysial antibody
EMG	electromyography
EOMI	extra-ocular muscles intact
ER	emergency room
ERT	estrogen replacement therapy
ESR	erythrocyte sedimentation rate
F	Fahrenheit
FACSM	Fellow, American College of Sports Medicine
FBG	fasting blood glucose
FBS	fasting blood sugar
FDA	Food and Drug Administration
FEF	forced mid-expiratory flow
FEV	forced mid-expiratory volume
FFA	free fatty acid
FH	family history
FTT	failure to thrive
FUO	fever of unknown origin
FVC	forced vital capacity
FX	fracture
g	gram
g/dL	grams per deciliter
GB	gallbladder
GERD	gastroesophageal reflux disease
GFR	glomerular filtration rate
GI	gastrointestinal
GM-CSF	granulocyte/macrophage colony stimulating factor
GTF	glucose tolerance factor
GTT	glucose tolerance test
GVHD	graft versus host disease
h	hour
H & P (HPI)	history and physical
HAV	hepatitis A virus
HbA_{1c}	glycosylated hemoglobin
HBV	hepatitis B virus
HC	head circumference
Hct	hematocrit
HCV	hepatitis C virus
HDL	high-density lipoprotein
HEENT	head, eyes, ears, nose, throat
Hg	mercury
Hgb	hemoglobin
HHNK	hyperosmolar hyperglycemic nonketotic (syndrome)
HIV	human immunodeficiency virus
HLA	human leukocyte antigen
HOB	head of bed
HR	heart rate
HS or h.s.	hours of sleep
HTN	hypertension
Hx	history
I & O (I/O)	intake and output
i.e.	that is
IBD	inflammatory bowel disease
IBS	irritable bowel syndrome
IBW	ideal body weight
ICF	intracranial fluid
ICP	intracranial pressure
ICS	intercostal space
ICU	intensive care unit
IGT	impaired glucose tolerance
IM	intramuscularly
inc	incontinent
IU	international unit
IV	intravenous
J	joule
K	potassium
kcal	kilocalorie
KCl	potassium chloride
kg	kilogram
KS	Kaposi's sarcoma
KUB	kidney, ureter, bladder
L	liter
lb	pounds
LBM	lean body mass
LCT	long-chain triglyceride
LDH	lactic dehydrogenase
LES	lower esophageal sphincter
LFT	liver function test
LIGS	low intermittent gastric suction
LLD	left lateral decubitus position
LLQ	lower left quadrant
LMP	last menstrual period
LOC	level of consciousness
LP	lumbar puncture
LUQ	lower upper quadrant
lytes	electrolytes
MAC	midarm circumference
MAMC	midarm muscle circumference
MAOI	monoamine oxidase inhibitor

MCHC mean corpuscular hemoglobin concentration
MCL midclavicular line
MCT medium-chain triglyceride
MCV mean corpuscular volume
mEq milliequivalent
mg milligram
Mg magnesium
MI myocardial infarction
mm millimeter
mmHg millimeters of mercury
MNT medical nutrition therapy
MODY maturity onset diabetes of the young
MOM Milk of Magnesia
mOsm milliosmol
MR mitral regurgitation
MRI magnetic resonance imaging
MS multiple sclerosis, morphine sulfate
MVA motor vehicle accident
MVI multiple vitamin infusion
N nitrogen
N/V nausea and vomiting
NG nasogastric
NH_3 ammonia
NICU neurointensive care unit, neonatal intensive care unit
NKA no known allergies
NKDA no known drug allergies
NPH neutral protamine Hagedorn insulin
NPO nothing by mouth
NSAID nonsteroidal antiinflammatory drug
NTG nitroglycerin
O_2 oxygen
OA osteoarthritis
OC oral contraceptive
OHA oral hypoglycemic agent
OR operating room
ORIF open reduction internal fixation
OT occupational therapist
OTC over the counter
$paCO_2$ partial pressure of dissolved carbon dioxide in arterial blood
paO_2 partial pressure of dissolved oxygen in arterial blood
pc after meals
PCM protein-calorie malnutrition
PD Parkinson's disease
PE pulmonary embolus
PED percutaneous endoscopic duodenostomy
PEEP positive end expiratory pressure
PEG percutaneous endoscopic gastrostomy
PEM protein-energy malnutrition
PERRLA pupils equal, round, and reactive to light and accommodation
pH hydrogen ion concentration
PKU phenylketonuria
PMI point of maximum impulse
PMN polymorphonuclear
PN parenteral nutrition
PO by mouth (per os)
PPD packs per day
PPN peripheral parenteral nutrition
prn may be repeated as necessary (pro re nata)
pt patient
PT patient, physical therapy, prothrombin time
PTA prior to admission
PTT prothromboplastin time
PUD peptic ulcer disease
PVC premature ventricular contraction
PVD peripheral vascular disease
q every (quaque)
qd every day (quaque die)
qh every hour (quaque hora)
qid four times daily (quater in die)
qns quantity not sufficient (quantum non sufficiat)
qod every other day
R/O rule out
RA rheumatoid arthritis
RBC red blood cell
RBW reference body weight
RD registered dietitian
RDA Recommended Dietary Allowance
RDS respiratory distress syndrome
REE resting energy expenditure
RLL right lower lobe
RLQ right lower quadrant
ROM range of motion
ROS review of systems
RQ respiratory quotient
RR respiratory rate
RUL right upper lobe
RUQ right upper quadrant
Rx take, prescribe, or treat
s̄ without
S/P status post
SBGM self blood glucose monitoring
SBO small bowel obstruction
SBS short bowel syndrome
SGOT serum glutamic oxaloacetic transaminase
SGPT serum glutamic pyruvic transaminase
SOB shortness of breath
SQ subcutaneous
ss half
stat immediately
susp suspension
T temperature
T & A tonsillectomy and adenoidectomy
T, tbsp tablespoon
t, tsp teaspoon
T_3 triiodothyronine
T_4 thyroxine
TB tuberculosis
TEE total energy expenditure

TF	tube feeding
TG	triglyceride
TIA	transient ischemic attack
TIBC	total iron binding capacity
tid	three times daily (ter in die)
TKO	to keep open
TLC	total lymphocyte count
TNM	tumor, node, metastasis
TPN	total parenteral nutrition
TSF	triceps skinfold
TSH	thyroid stimulating hormone
TURP	transurethral resection of the prostate
U	unit
UA	urinalysis
UBW	usual body weight
UL	Tolerable Upper Intake Level
URI	upper respiratory intake
UTI	urinary tract infection
UUN	urine urea nitrogen
VLCD	very-low-calorie diet
VOD	venous occlusive disease
VS	vital signs
w.a.	while awake
WBC	white blood cell
WNL	within normal limits
wt	weight
WW	whole wheat
yo	year old

Appendix B

NORMAL VALUES FOR PHYSICAL EXAMINATION

Vital Signs

Temperature
Rectal: C = 37.6°/F = 99.6°
Oral: C = 37°/F = 98.6° (± 1°)
Axilla: C = 37.4°/F = 97.6°

Blood Pressure: average 120/80 mmHg

Heart Rate (beats per minute)

Age	At Rest Awake	At Rest Asleep	Exercise or Fever
Newborn	100–180	80–160	≤ 220
1 week–3 months	100–220	80–200	≤ 220
3 months–2 years	80–150	70–120	≤ 200
2–10 years	70–110	60–90	≤ 200
11 years-adult	55–90	50–90	≤ 200

Respiratory Rate (breaths per minute)

Age	Respirations
Newborn	35
1–11 months	30
1–2 years	25
3–4 years	23
5–6 years	21
7–8 years	20
9–10 years	19
11–12 years	19
13–14 years	18
15–16 years	17
17–18 years	16–18
Adult	12–20

Cardiac Exam: carotid pulses equal in rate, rhythm, and strength; normal heart sounds; no murmurs present

HEENT Exam (head, eyes, ears, nose, throat)

Mouth: pink, moist, symmetrical; mucosa pink, soft, moist, smooth
Gums: pink, smooth, moist; may have patchy pigmentation
Teeth: smooth, white, shiny
Tongue: medium red or pink, smooth with free mobility, top surface slightly rough
Eyes: pupils equal, round, reactive to light and accommodation
Ears: tympanic membrane taut, translucent, pearly gray; auricle smooth without lesions; meatus not swollen or occluded; cerumen dry (tan/light yellow) or moist (dark yellow/brown)
Nose: external nose symmetrical, nontender without discharge; mucosa pink; septum at the midline
Pharynx: mucosa pink and smooth
Neck: thyroid gland, lymph nodes not easily palpable or enlarged

Lungs: chest contour symmetrical; spine straight without lateral deviation; no bulging or active movement within the intercostal spaces during breathing; respirations clear to auscultation and percussion

Peripheral Vascular: normal pulse graded at 3+, which indicates that pulse is easy to palpate and not easily obliterated; pulses equal bilaterally and symmetrically

Neurological: normal orientation to people, place, time, with appropriate response and concentration

Skin: warm and dry to touch; should lift easily and return back to original position, indicating normal turgor and elasticity

Abdomen: umbilicus flat or concave, positioned midway between xyphoid process and symphysis pubis; bowel motility notes normal air and fluid movement every 5–15 seconds; graded as normal, audible, absent, hyperactive, or hypoactive

Appendix C

ROUTINE LABORATORY TESTS WITH NUTRITIONAL IMPLICATIONS[1]

This table presents a partial listing of some uses of commonly performed lab tests that have implications for nutritional problems.

Laboratory Test	Acceptable Range	Description
Hematology		
Red blood cell (RBC) count	$4.2–5.4 \times 10^6/mm^3$ (women) $4.5–6.2 \times 10^6/mm^3$ (men)	Number of RBC; aids anemia diagnosis.
Hemoglobin (Hgb)	12–15 g/dL (women) 14–17 g/dL (men)	Hemoglobin content of RBC; aids anemia diagnosis.
Hematocrit (Hct)	37%–47% (women) 40%–54% (men)	Percentage RBC in total blood volume; aids anemia diagnosis.
Mean corpuscular volume (MCV)	80–96 μm^3	RBC size, helps to distinguish between microcytic and macrocytic anemias.
Mean corpuscular hemoglobin concentration (MCHC)	31.5–36 pg	Hb concentration within RBCs, helps to distinguish iron-deficiency anemia.
White blood cell (WBC) count	$4.8–11.8 \times 10^3/mm^3$	Number of WBC; general assessment of immunity.
Blood Chemistry		
Serum Proteins		
Total protein	6–8 g/dL	Protein levels are not specific to disease or highly sensitive; they can reflect poor protein intake, illness or infections, changes in hydration or metabolism, pregnancy, or medications.
Albumin	3.5–5.0 g/dL	May reflect illness or PEM; slow to respond to improvement or worsening of disease.
Transferrin	250–380 mg/dL (women) 215–365 mg/dL (men)	May reflect illness, PEM, or iron deficiency; slightly more sensitive to changes than albumin.
Prealbumin (transthyretin)	16–35 mg/dL	May reflect illness or PEM; more responsive to health status changes than albumin or transferrin.
C-reactive protein	< 1.0 mg/dL	Indicator of inflammation or disease.
Serum Enzymes		
Creatine phosphokinase (CK, CPK)	30–135 U/L (women) 55–170 U/L (men)	Different forms of CK are found in muscle, brain, and heart. High levels in blood may indicate heart attack, brain tissue damage, or skeletal muscle injury.

[1] Adapted from *Nutrition Therapy & Pathophysiology,* 1st ed. (ISBN 053462154), Appendix B3, which was adapted from *Understanding Normal & Clinical Nutrition,* 7th ed. (ISBN 0534622089), Table 17-8, p. 594.

Lactate dehydrogenase (LDH)	208–378 U/L	LDH is found in many tissues. Specific types may be elevated after heart attack, lung damage, or liver disease.
Alkaline phosphatase	30–120 U/L	Found in many tissues; often measured to evaluate liver function.
Aspartate aminotransferase (AST, formerly SGOT)	0–35 U/L	Usually monitored to assess liver damage; elevated in most liver diseases. Levels are somewhat increased after muscle injury.
Alanine aminotransferase (ALT, formerly SGPT)	4–36 U/L	Usually monitored to assess liver damage; elevated in most liver diseases. Levels are somewhat increased after muscle injury.
Serum Electrolytes		
Sodium	136–145 mEq/L	Helps to evaluate hydration status or neuromuscular, kidney, and adrenal functions.
Potassium	3.5–5.5 mEq/L	Helps to evaluate acid-base balance and kidney function; can detect potassium imbalances.
Chloride	95–105 mEq/L	Helps to evaluate hydration status and detect acid-base and electrolyte imbalances.
Other		
Glucose	70–110 mg/dL	Detects risk of glucose intolerance, diabetes mellitus, and hypoglycemia; helps to monitor diabetes treatment.
Glycosylated hemoglobin (HbA_{1c})	3.9–5.2%	Used to monitor long-term blood glucose control (approximately 1 to 3 months prior).
Blood urea nitrogen (BUN)	8–18 mg/dL	Primarily used to monitor kidney function; value is altered by liver failure, dehydration, or shock.
Uric acid	2.8–8.8 mg/dL (women) 4.0–9.0 mg/dL (men)	Used for detecting gout or changes in kidney function; levels affected by age and diet; varies among different ethnic groups.
Creatinine (serum or plasma)	0.6–1.2 mg/dL	Used to monitor renal function.

Note: dL = deciliter; fL = femtoliter; mEq = milliequivalents; μL = microliter; ng = nanogram; U/L = units per liter.

Appendix D

EXCHANGE LISTS FOR DIABETES MEAL PLAN FORM AND FOOD LISTS NUTRIENT INFORMATION [1]

Meal Plan for: ________________ Date: ________________

RD: ________________ Phone: ________________

Carbohydrate ________ (grams) ________ (% of calories)

Carbohydrate choices ________ (servings)

Protein ________ (% of calories) Fat ________ (% of calories) Calories ________

	Starches	Fruits	Milk	Nonstarchy Vegetables	Meat and Meat Substitutes	Fats	Menu Ideas
Breakfast Time: ______							
Snack Time: ______							
Lunch Time: ______							
Snack Time: ______							
Dinner Time: ______							
Snack Time: ______							

[1] From *Choose Your Foods: Exchange Lists for Diabetes*, published by the American Diabetes Association and the American Dietetic Association, 2007. Copyright © 2008 by the American Diabetes Association and the American Dietetic Association. Used by permission.

The Food Lists

The following chart shows the amount of nutrients in one serving from each list.

Food List	Carbohydrate (grams)	Protein (grams)	Fat (grams)	Calories
Carbohydrates				
Starch: breads, cereals and grains, starchy vegetables, crackers, snacks, and beans, peas, and lentils	15	0–3	0–1	80
Fruits	15	—	—	60
Milk				
Fat-free, low-fat, 1%	12	8	0–3	100
Reduced-fat, 2%	12	8	5	120
Whole	12	8	8	160
Sweets, Desserts, and Other Carbohydrates	15	varies	varies	varies
Nonstarchy Vegetables	5	2	—	25
Meat and Meat Substitutes				
Lean	—	7	0–3	45
Medium-fat	—	7	4–7	75
High-fat	—	7	8+	100
Plant-based proteins	varies	7	varies	varies
Fats	—	—	5	45
Alcohol	varies	—	—	100

INDEX